CELLULAR AND MOLECULAR MECHANISMS IN ALZHEIMER'S DISEASE THERAPEUTICS

Volume 1

Cellular Mechanisms

Cellular and Molecular Mechanisms in Alzheimer's Disease Therapeutics, 2-volume set
ISBN: 978-1-77964-176-2 (hbk)

Cellular and Molecular Mechanisms in Alzheimer's Disease Therapeutics, Volume 1: Cellular Mechanisms
ISBN: 978-1-77964-178-6 (hbk)
ISBN: 978-1-77964-179-3 (ebk)

Cellular and Molecular Mechanisms in Alzheimer's Disease Therapeutics, Volume 2: Molecular Mechanisms
ISBN: 978-1-77964-180-9 (hbk)
ISBN: 978-1-77964-181-6 (ebk)

CELLULAR AND MOLECULAR MECHANISMS IN ALZHEIMER'S DISEASE THERAPEUTICS

Volume 1

Cellular Mechanisms

Prashant Tiwari, PhD
Sunil Kumar Kadiri, PhD
Kunjbihari Sulakhiya, PhD
Editors

AAP | APPLE ACADEMIC PRESS

First edition published 2027

Apple Academic Press Inc.
1265 Goldenrod Circle, NE,
Palm Bay, FL 32905 USA

760 Laurentian Drive, Unit 19,
Burlington, ON L7N 0A4, Canada

CRC Press
2385 NW Executive Center Drive,
Suite 320, Boca Raton FL 33431

4 Park Square, Milton Park,
Abingdon, Oxon, OX14 4RN UK

Library and Archives Canada Cataloguing in Publication

Title: Cellular and Molecular Mechanisms in Alzheimer's Disease Therapeutics / Prashant Tiwari, PhD, Sunil Kumar Kadiri, PhD, Kunjbihari Sulakhiya, PhD, editors.
Names: Tiwari, Prashant, editor. | Kadiri, Sunil Kumar, 1981- editor. | Sulakhiya, Kunjbihari, editor.
Description: First edition. | Includes bibliographical references and indexes. | Contents: Volume 1: Cellular mechanisms.
Identifiers: Canadiana (print) 20260143553 | Canadiana (ebook) 2026014360X | ISBN 9781779641762 (set) | ISBN 9781779641786 (v. 1 ; hardcover) | ISBN 9781779641793 (v. 1 ; ebook)
Subjects: LCSH: Alzheimer's disease—Cytopathology. | LCSH: Alzheimer's disease—Molecular aspects. | LCSH: Alzheimer's disease—Pathophysiology.
Classification: LCC RC523 .C45 2026 | DDC 616.8/3107—dc23

Library of Congress Cataloging-in-Publication Data

CIP data on file with US Library of Congress

ISBN: 978-1-77964-178-6 (hbk)
ISBN: 978-1-77964-179-3 (ebk)
DOI: 10.1201/9781779641793

About the Editors

Prashant Tiwari, PhD
Associate Professor, Department of Pharmacology,
College of Pharmaceutical Sciences, Dayananda Sagar University,
Bengaluru, India

Prashant Tiwari, PhD, is an Associate Professor in the Department of Pharmacology at the College of Pharmaceutical Sciences, Dayananda Sagar University, Bengaluru, Karnataka, India. He holds a PhD from Siksha O Anusandhan University, Bhubaneswar, and was a fellow of the Indian Council of Medical Research (ICMR), New Delhi. With over 13 years of experience in teaching and research, Dr. Tiwari's expertise includes drug screening, method validation for brain-targeted drug delivery, drug-induced dyskinesia revalidation, and the neuroprotective effects of benzimidazole derivatives. His research has explored the dopaminergic pathway, prolactin inhibition, and neuroprotection, with notable studies on nanoemulsions containing *Lepidium sativum* oil for Parkinson's disease and *Muntingia calabura* leaf extract for Alzheimer's disease. For his outstanding work, Dr. Tiwari received the Dr. P. D. Patil National Award in 2022 for Best Thesis in Pharmaceutical Sciences.

In addition to his research, Dr. Tiwari has authored five books, edited five books, and co-authored six book chapters. He has filed seven patents, two of which have been granted. He organized a national e-seminar on targeted drug delivery against neurodegenerative diseases, sponsored by the Department of Biotechnology (DBT), Government of India. Dr. Tiwari has been honored with various awards, including the *Indian Council of Medical Research (ICMR)* Senior Research Fellowship, Young Scientist Award, and Best Teacher Award. He supervises five PhD students and has guided nine MPharm students. He is also an active member of professional bodies and serves as a peer reviewer for international journals. His research interests span neurodegenerative disorders, drug interactions, pharmacokinetics/pharmacodynamics, and metabolic disorders.

Sunil Kumar Kadiri, PhD
Professor, College of Pharmaceutical Sciences,
Dayananda Sagar University, Bengaluru, Karnataka, India

Sunil Kumar Kadiri, PhD, is a Professor at the College of Pharmaceutical Sciences, Dayananda Sagar University, Bengaluru, Karnataka, India, with 16 years of teaching experience. His research interests encompass neuropharmacology, metabolic disorders, organ-protective activities, and drug interactions. He holds a PhD in Pharmaceutical Sciences from JNTU, Hyderabad and has authored 65 research and review articles. In addition to publishing five books and four book chapters, Dr. Kadiri holds six patents. An active member of Association of Pharmaceutical Teachers of India (APTI), his recent research includes drug interaction studies involving Alprazolam and Pomegranate, Carbamazepine and Losartan, Naproxen and St. John's Wort, as well as investigations into the antiparkinsonian effects of Zinda Tilismath.

Kunjbihari Sulakhiya, PhD
Assistant Professor, Department of Pharmacy, Indira Gandhi National Tribal University, Amarkantak, Madhya Pradesh, India

Kunjbihari Sulakhiya, PhD, is working as an Assistant Professor in the Department of Pharmacy, Indira Gandhi National Tribal University, Amarkantak, Madhya Pradesh, India. He obtained his MS (Pharm) and PhD in Pharmacology and Toxicology from the National Institute of Pharmaceutical Education and Research (NIPER), Guwahati, Assam, India. He has published 22 research papers, seven review articles, and 11 book chapters in reputed peer-reviewed national and international journals and books. Dr. Sulakhiya has received 13 awards/fellowships including six travel awards to present his research papers at international conferences in the United States, Canada, Japan, France, and Macau. He is a member of various professional societies, including the Indian Academy of Neurosciences (IAN), the International Society for Neurochemistry (ISN), Asia Pacific Society for Neurochemistry (APSN), APTI, and so forth. Dr. Sulakhiya is serving as an editorial member and reviewer for various reputed journals. His area of research is pre-clinical research in the field of neuropharmacology. He has completed three research projects funded by DBT, UGC, and ISN.

Contents

Contributors

Naureen Afrose
Bengal College of Pharmaceutical Sciences and Research, Basu Sarani, Bidhannagar, Durgapur, West Bengal, India

Asad Ahmad
Department of Pharmacy, Faculty of Pharmacy, Integral University, Lucknow, Uttar Pradesh, India

Mohammad Ahmad
Department of Pharmacy, Faculty of Pharmacy, Integral University, Lucknow, Uttar Pradesh, India

Juber Akhtar
Department of Pharmacy, Faculty of Pharmacy, Integral University, Lucknow, Uttar Pradesh, India

Komal Bhati
Metro college of Health Sciences and Research, Knowledge Park III, Greater Noida, Uttar Pradesh, India

Pratibha Bhowmick
Bengal College of Pharmaceutical Sciences and Research, BasuSarani, Bidhannagar, Durgapur, West Bengal, India

Mithun Bhowmick
Bengal College of Pharmaceutical Sciences and Research, BasuSarani, Bidhannagar, Durgapur, West Bengal, India

Snigdha Bhardwaj
Department of Pharmaceutics, KIET School of Pharmacy, Ghaziabad, Delhi-NCR, Uttar Pradesh, India

Avijit Bej
Mahadeva Lal Schroff College of Pharmacy, Aurangabad, Bihar, India

Badruddeen
Department of Pharmacy, Faculty of Pharmacy, Integral University, Lucknow, Uttar Pradesh, India

Rideb Chakraborty
Bengal College of Pharmaceutical Sciences and Research, BasuSarani, Bidhannagar, Durgapur, West Bengal, India

Phool Chandra
Teerthanker Mahaveer College of Pharmacy, Teerthanker Mahaveer University, Moradabad, Uttar Pradesh, India

Anwesha Das
Department of Pharmacy, Sanaka Educational Trust Group of Institutions (SETGOI), Malandighi, Durgapur, West Bengal, India

Sinchan Das
Department of Pharmacy, Guru Ghasidas University, Bilaspur, Chhattisgarh, India

Praveen Kumar Gaur
Metro college of Health Sciences and Research, Greater Noida, Uttar Pradesh, India

Radha Goel
Lloyd Institute of Management & Technology, Greater Noida, Uttar Pradesh, India

Pradeep Golani
Department of Pharmacy, Gyan Ganga Institute of Technology & Sciences, Jabalpur, Madhya Pradesh, India

Ahana Hazra
Bengal College of Pharmaceutical Sciences and Research, BasuSarani, Bidhannagar, Durgapur, West Bengal, India

Chowdhury Mobaswar Hossain
Department of Pharmaceutical Technology, Maulana Abul Kalam Azad University of Technology, Haringata, West Bengal, India

Sandeep Hurpade
Shri Prakashchand Jain College of Pharmacy & Research, Jamner, Maharashtra, India

Mohammad Irfan
Department of Pharmacy, Faculty of Pharmacy, Integral University, Lucknow, Uttar Pradesh, India

Anas Islam
Department of Pharmacy, Faculty of Pharmacy, Integral University, Lucknow, Uttar Pradesh, India

Manish Kumar Jeengar
Department of Pharmacology, Amrita School of Pharmacy, Amrita Vishwa Vidyapeetham, AIMS Health Sciences Campus, Kochi, Kerala, India

Sushil K. Kashaw
Department of Pharmaceutical Sciences, Dr. Harisingh Gour Vishwavidyalaya, Sagar, Madhya Pradesh, India

Rahul Kaushik
Metro college of Health Sciences and Research, Knowledge Park III, Greater Noida, Uttar Pradesh, India

Kanak Lata
Metro college of Health Sciences and Research, Knowledge Park III, Greater Noida, Uttar Pradesh, India

Devendra Singh Lodhi
Department of Pharmacy, Gyan Ganga Institute of Technology & Sciences, Jabalpur, Madhya Pradesh, India

Nandita Mandal
Department of Health and Family Welfare, Government of West Bengal, India

Purba Mandal
Department of Pharmacy, Faculty of Pharmacy, Integral University, Lucknow, Uttar Pradesh, India

Shubhrat Maheshwari
Bioorganic and Medicinal Chemistry Research Laboratory, Department of Pharmaceutical Sciences, Sam Higginbottom University of Agriculture, Technology and Sciences, Allahabad, Uttar Pradesh, India

Rosaline Mishra
Department of Pharmacy, Metro College of Health Sciences and Research, Uttar Pradesh, India

Sanjay Nagdev
Shri Prakashchand Jain College of Pharmacy & Research, Jamner, Maharashtra, India

Prashant Nikumbh
Shri Prakashchand Jain College of Pharmacy & Research, Jamner, Maharashtra, India

Arijit Nandi
Department of Pharmacy, Sanaka Educational Trust Group of Institutions (SETGOI), Malandighi, Durgapur, West Bengal, India

Dipti Pal
Kalinga University, Kotni, Naya Raipur, Chhattisgarh, India

Arjun Patra
Department of Pharmacy, Guru Ghasidas University, Bilaspur, Chhattisgarh, India

Parveen Rawat
Department of Pharmaceutical Sciences, Dr. Harisingh Gour Vishwavidyalaya, Sagar, Madhya Pradesh, India

Neetu Sachan
Department of Pharmaceutical Sciences, School of Pharmaceutical & Biological Sciences, Harcourt Butler Technical University (HBTU), Kanpur, Uttar Pradesh, India

Himanshu Sharma
Teerthanker Mahaveer College of Pharmacy, Teerthanker Mahaveer University, Moradabad, Uttar Pradesh, India

Shweta Shrivastava
School of Pharmacy, School of Health and Allied Sciences, ARKA JAIN University, Gamharia, Seraikela Kharsawan, Jharkhand, India

Santosh Kumar Singh
School of Pharmacy, School of Health and Allied Sciences, ARKA JAIN University, Gamharia, Seraikela Kharsawan, Jharkhand, India

Puja Sinoriya
Metro college of Health Sciences and Research, Knowledge Park III, Greater Noida, Uttar Pradesh, India

Sakshi Soni
Department of Pharmaceutical Sciences, Dr. Harisingh Gour Vishwavidyalaya, Sagar, Madhya Pradesh, India

Vandana Soni
Department of Pharmaceutical Sciences, Dr. Harisingh Gour Vishwavidyalaya, Sagar, Madhya Pradesh, India

Aditya Singh
Department of Pharmaceutics, Parul Institute of Pharmacy and Research, Parul University, Waghodia, Vadodara, Gujarat, India

Sharad Tayade
Shri Sant Gajanan Maharaj College of Pharmacy, Buldhana, Maharashtra, India

Sandeep Prasad Tiwari
Kalinga University, Kotni, Naya Raipur, Chhattisgarh, India

Varsha Yadav
Department of Pharmacy, Guru Ghasidas University, Bilaspur, Chhattisgarh, India

Mohd Yasir
Department of Pharmacy, College of Health Science, Arsi University, Asella, Oromia region, Ethiopia

Preface

Alzheimer's disease (AD) remains one of the most formidable neurodegenerative disorders, affecting millions worldwide. Despite extensive research, there is still much to uncover about the cellular mechanisms that underlie its pathology and drive disease progression. As the global population ages, the need for effective therapeutic strategies becomes increasingly urgent.

This edited volume, *Cellular and Molecular Mechanisms in Alzheimer's Disease Therapeutics,* aims to provide a detailed and up-to-date exploration of the cellular and molecular processes involved in AD, with an emphasis on their implications for therapeutic interventions. The book brings together contributions from leading experts in the field, offering a comprehensive understanding of how various cellular mechanisms, such as synaptic dysfunction, neuroinflammation, mitochondrial impairment, and protein aggregation contribute to the pathology of AD.

We, as editors, Dr. Prashant Tiwari, Dr. Sunil Kumar Kadiri, and Dr. Kunjbihari Sulakhiya, have curated this volume to facilitate the integration of basic research and clinical application. By focusing on the cellular and molecular insights into AD, we hope to pave the way for novel therapeutic approaches that address the root causes of neurodegeneration. Special attention is given to the role of cellular stress responses, autophagy, tau and amyloid-beta pathology, and the emerging therapeutic potential of gene therapy, precision medicine, and immune modulation.

The contributions in this volume reflect the latest advancements in the field, providing readers with a nuanced understanding of both established and cutting-edge therapeutic strategies aimed at combating AD. It is our hope that this book will serve as a valuable resource for researchers, clinicians, and students dedicated to understanding and treating AD.

We are deeply grateful to all the authors who have contributed to this work, as well as to the team at Apple Academic Press for their support in bringing this project to fruition. Through the collaborative efforts of the scientific community, we remain hopeful that new breakthroughs will emerge that can halt, or even reverse, the devastating effects of AD.

—***Editors***

CHAPTER 1

Exploring the Cellular and Molecular Mechanisms, Pathogenesis, Etiology, and Complications of Alzheimer's Disease

SANDEEP HURPADE[1], SHARAD TAYADE[2], PRASHANT NIKUMBH[1], DEVENDRA SINGH LODHI[3*], PRADEEP GOLANI[3], and SANJAY NAGDEV[1]

[1]*Shri Prakashchand Jain College of Pharmacy & Research, Jamner, Maharashtra, India*

[2]*Sant Shri Gajanan Maharaj College of Pharmacy, Buldhana, Maharashtra, India*

[3]*Department of Pharmacy, Gyan Ganga Institute of Technology & Sciences, Jabalpur, Madhya Pradesh, India*

*Corresponding author

ABSTRACT

In the field of neurodegenerative brain illnesses, Alzheimer's disease (AD) ranks among the most prevalent and catastrophic conditions. It causes dementia by building up amyloid β (Aβ) plaques and neurofibrillary tangles, which in turn cause synaptic dysfunction, neuronal death, and neural disintegration. In terms of dementia types, AD is by far the most common, which is characterized by a wide range of symptoms, such as forgetfulness, an inability to concentrate, and a confused sense of time and place. Because its fundamental mechanism is still a mystery, there is no cure for AD at this time. The involvement of Aβ and tau, as well as glial

Cellular and Molecular Mechanisms in Alzheimer's Disease Therapeutics, Volume 1: Cellular Mechanisms. Prashant Tiwari, Sunil Kumar Kadiri, and Kunjbihari Sulakhiya (Eds.)

DOI: 10.1201/9781779641793-1

contributions to various pharmacological and biological processes in the progression of AD, has been more extensively studied in recent decades. Extensive studies have been conducted on the functions of Aβ and tau in the last several decades, although their involvement in disease remains uncertain. The development of AD has been the subject of numerous suggested mechanisms. Several imaging modalities have revealed distinctive alterations in the brains of AD patients, including structural and functional behavior, through the utilization of magnetic resonance imaging (MRI), positron emission tomography (PET), and CT studies. An additional effort in studying the molecular mechanism, diagnostics, and treatments for AD is required in the future. For the best possible patient management of AD, the clinical analysis includes cutting-edge neuroimaging, genetic profiling, and cognitive evaluations to aid in early detection, disease monitoring, and individualized treatment programs. In this chapter we provide an insight into the cellular and molecular mechanism of AD. Moreover, we have emphasized on imaging techniques with reported clinical evidence.

1.1 INTRODUCTION

In an adult human brain, there are around 100 billion synapses, which are lengthy projections of neurons that connect one neuron to another. Emotional regulation, motor control, language, memory, thought, skill, and sensation are all governed by fast-moving electrochemical nerve impulses that flow between brain circuits [1]. Dementia and other neurodegenerative disorders cause a slow but steady decline in cognitive function and other neurological symptoms due to the loss of brain cells and connections. Dementia severely impacts a person's capacity to do daily tasks. A new instance of dementia is reported every three seconds on a global scale, making it the sixth largest cause of mortality. Dementia now affects 50 million individuals throughout the globe, and experts predict that this number will rise to 152 million by the year 2050 [2]. This includes approximately 23 million in the Asia Pacific region, 8.8 million in Europe, 5.8 million in the United States, and 5.3 million in India. In 2020, a worldwide survey of Alzheimer's disease (AD) reported that the entire expected cost of dementia was one trillion US dollars. By 2030, that number was projected to rise to two trillion US dollars [3]. Modern research has identified AD as a chronic or progressive illness that impairs cognitive function beyond that expected with aging. This impairment impacts many areas of life, including memory, cognition, direction, understanding, language, and judgement. The initial descriptions of the pathophysiology of AD were made more than a century ago. The pathology of AD has been greatly aided by our current understanding of the neurobiological mechanisms at work [4]. Possible explanations for the most striking shifts include the following: the amyloid peptide hypothesis, the cholinergic hypothesis, tau protein function, oxidative stress (OS) involvement, and calcium's role. Depending on the family history, AD can be either early-onset (EOAD) or late-onset (LOAD) and sporadic (SAD). In contrast to SAD, which has a complicated aetiology, including genetic, environmental, metabolic, viral, and other variables, FAD is primarily linked to mutations in the presenilin genes, *PSEN1* and *PSEN2,* and the amyloid β (Aβ) precursor protein (APP) [5]. The most powerful genetic risk factor for SAD is APOE4, one of three isoforms of the polymorphic protein apolipoprotein E (APOE). As far as numbers go, APOE is linked to 40–50% of EOAD and 80% of LOAD. Aβ synthesis, hydrolysis, and removal are all impacted by APOE's significant affinity for the protein. Neurons produced from stem cells are more likely to be injured and have calcium dysregulation issues if they express APOE4 endogenously, which increases the release of phosphorylated tau [6]. Delaying the age of onset and reducing the incidence of AD and Aβ buildup are both protected by the intriguing APOE2 protective factor. According to several studies, expressing APOE4 in mice models enhances the expression of oligomers and plaque deposition, while expressing APOE2 has the opposite effect. Neurodegeneration is linked to aggregation and deposition of Aβ, and both APOE2 and APOE4 isoforms are implicated in this process; however, APOE4 seems to have a substantially larger effect than APOE2 [7]. The only medications that can slow the disease's advancement are those that alleviate symptoms; these largely target Aβ and tau. Various hypotheses are being investigated by scientists in an effort to determine their potential contributions to the onset of AD. Research in these areas has the potential to shed light on the pathogenic mechanisms of AD [8]. Some of the topics covered in this chapter include the various molecular theories surrounding neurofibrillary tangles (NFTs) formed by tau, the processes involved in neuroinflammation and neurodegeneration. The elements that contribute to dementia, and the efficacy and side effects of different medications are all discussed in this chapter. This chapter provides insights for potential perceivers on future therapy in the treatment of AD.

1.2 ETIOLOGY OF ALZHEIMER'S DISEASE

The development of AD is influenced by both hereditary factors and outside factors. Age is the most important deterrent. There is a 3% chance of developing AD at age 65, and that number jumps to more than 30% by age 85. There is little data on the prevalence of AD in those younger than 65 years old, however, rough estimations revealed the number of cases in this age at about 3%. Despite a general upward trend in numbers due to an aging population, several nations have seen a decline in age-specific incidence [9]. Both the onset of symptoms and the presence or absence of inheritance play a role in categorizing AD. There are two types of AD: EOAD and LOAD. EOAD happens before age 65, while LOAD happens beyond 65 and accounts for more than 95% of cases. There is a clear familial link in cases of familial AD, but no such link exists in cases of SAD. The great majority of LOAD occurs randomly, but almost all EOAD are hereditary [10].

1.2.1 Aβ PRECURSOR PROTEIN

The "amyloid hypothesis" states that neuronal death and disease are caused by a cascade of events started by the detrimental buildup of Aβ. A total of 50 different APP mutations have been discovered so far, and they are responsible for about 10% of family cases [11]. According to research, a significant number of mutations induce a rise in Aβ production, or the Aβ 42:40 ratio, which in turn causes an increase in the buildup of amyloid. Promoter mutations or APP duplications can induce AD in extremely uncommon cases [12].

1.2.2 PRESENILINS

PSEN1 and *PSEN2* encode the catalytic elements, which is involved in the digestion of APP. It is believed that between 30 and 50% of family EOAD cases are caused by *PSEN1* variations, the most well-known *Mendelian* gene cause of autosomal dominant AD. The amyloid hypothesis is questioned by research that reveals *PSEN1* and *PSEN2* mutations change Aβ production [13].

1.2.3 OTHER GENETIC RISK FACTORS

The changes in genes, including *PICALM, SORL1, CLU, APOE,* and *TREM2* are linked to an increased risk of AD. The most common hereditary risk factor for AD is the APOE E4 allele, which is present at a frequency of about 13.7% and is a variation of the *apolipoprotein E gene* that is involved in lipid metabolism [14]. The risk is three times higher in people who are heterozygotes for this particular gene. Although less prevalent, another mutation with a comparable impact size is TREM2R47H. A receptor on several immune response cells, *TREM2*, is linked to inflammation, lending credence to the concept that inflammation contributes to the onset of AD [15].

1.2.4 INFLAMMATION

Brain hypoperfusion and inflammation are two of the most prevalent SAD risk factors, which are generally a mix of hereditary and environmental variables. Cognitive impairment, both temporary and permanent, has been associated with inflammation brought on by trauma, sepsis, and infections [16].

1.3 CELLULAR AND MOLECULAR MECHANISMS IN AD

1.3.1 THE MOLECULAR PROCESS IN A HEALTHY BRAIN

Neuronal membranes contain the enormous transmembrane protein APP, which is involved in neuronal growth, survival, and repair. The APP protein extends across the neuronal membrane at both its N- and C-terminal ends. The process of neuronal growth and brain plasticity is aided by the creation of soluble amyloid precursor protein alpha (sAPP'E). An enzymatic process called α-secretase traverses off the end of APP near its C-terminus on the cells' surface, triggering this particular activity [17]. In a non-amyloidogenic pathway, APP is processed normally in the brain by α-secretase and then γ-secretase. The initial enzyme that starts the series of events known as the α-secretase pathway is α-secretase. When neurons are in a normal state, their receptors can easily take in ligands, allowing for the normal transmission of nerve impulses. An important part of neurotransmission between neurons is the binding of muscarinic and nicotinic acetylcholine receptors (nAChRs and mAChRs, respectively) [18].

1.3.2 THE MOLECULAR PROCESS IN A HEALTHY BRAIN AT AD STATE

In the amyloidogenic-APP processing that occurs in the AD state, an enzyme called β-secretase or *βamyloid cleaving enzyme-1 (BACE-1)* takes over from the C-terminal end to produce soluble amyloid precursor protein beta (sAPP β). Caspases (caspase-6) are activated when sAPPβ reaches death receptor-6 on the neuronal membrane. The leftover APP that is attached to the membrane is then identified by an enzyme called γ-secretase, which produces Aβ monomer fragments that are 40–42 amino acids long [19]. The majority of it is Aβ40, with only approximately 10% being Aβ 42. Dense, insoluble oligomers or senile plaques are formed when numerous Aβ peptide monomers assemble. The APP is cleaved from *BACE-1* to create Aβ-peptides according to the amyloid hypothesis. Donor neurons release the misfolded peptides to the extracellular environment in two forms: as naked proteins or as vesicles called exosomes, which recipient neurons collect through receptor-mediated endocytosis. As a result of calcium dysregulation, the Aβ 40/Aβ 42 that has been generated builds up on receptors which in turn hinder synaptic transmission [20]. The development of Aβ plaques may be linked to past cortical tau pathology, however, Aβ independent regulators like as Apolipoprotein-E (ApoE), the metabolism of cholesterol, endocytosis via receptors, and microglial activation can also promote tau pathology. Neuronal Aβ plaques trigger the release of chemokines and cytokines by microglia and astrocytes, leading to the production of ROS [21]. Lipid peroxidation, membrane damage, and neuronal death result from a cascade of events that begins with mitochondrial OS and is triggered by the synthesis of p53, Bad, and Bax. Protein kinases C (PKC), A (PKA), and *extracellular signal-regulated kinases 2 (ERK2)* are activated by ROS produced by Aβ [22]. The Acetylcholinesterase (AchE) enzyme, acts through hydrolytic activity to regulate learning and memory in the brain. An essential neurotransmitter for normal synaptic transmission is acetylcholine. To stop synaptic transmission, AchE breaks down acetylcholine into choline and acetate. An observed process in AD is the degeneration of the cholinergic system, which is caused by the action of AchE, which substantially reduces acetylcholine levels [23].

1.4 PATHOGENESIS OF ALZHEIMER'S DISEASE

The OS theory and the cholinergic hypothesis are two of the many intricate processes that contribute to the formation of AD (Figure 1.1).

Depletion of the neurotransmitter acetylcholine, which is important in memory and cognitive function, is a key component in the cholinergic hypothesis's proposed mechanism of action for the onset and progression of AD. In AD, cholinergic neurons, especially in the basal forebrain, are severely depleted [24]. Memory loss and cognitive decline are symptoms of this condition, which is exacerbated by this loss, which lowers acetylcholine levels in the brain. Although they do not stop the progression of the illness, AchE inhibitors and other drugs that target the cholinergic system are frequently used to treat Alzheimer's symptoms by raising acetylcholine levels. One of the major pathways that contribute to the development of AD is OS. ROS cause damage to cells when they exceed the organism's antioxidant defences [25]. Neuronal proteins, lipids, and nucleic acids are damaged in AD because antioxidant systems are overwhelmed by excessive ROS generation. The malfunction, inflammation, and eventual death of neurons caused by this oxidative damage speeds up the disease's course. In addition to OS and the cholinergic hypothesis, other pathogenic processes in AD include neuroinflammation, synaptic dysfunction, beta-amyloid plaque and tau tangle buildup, and neuroinflammation [26].

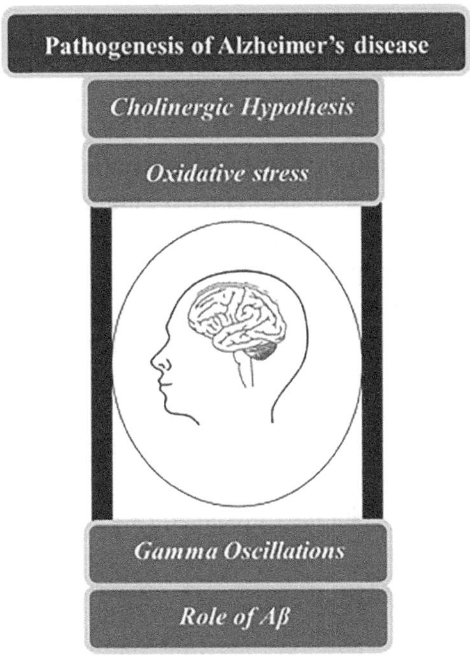

FIGURE 1.1 Pathogenesis of Alzheimer's disease.

1.4.1 INVOLVEMENT OF CHOLINERGIC HYPOTHESIS IN THE PATHOGENESIS OF AD

The cholinergic hypothesis explains the molecular basis of AD pathogenesis and was the initial and most extensively researched method. Memory, learning, concentration, and conscious awareness are all greatly influenced by certain clusters of cholinergic neurons found in areas, such as the medial septum, nucleus basalis, amygdala, frontal cortex, and hippocampus [27]. Degenerative processes in diseases like Alzheimer's are mediated by these neurons, which were first identified more than 30 years ago. The presence of certain markers, such as AchE and acetyltransferase, is linked to the beginning of cognitive decline [28]. The idea that scopolamine-induced cognitive impairment and other non-selective muscarinic antagonists enhance beta-amyloid peptide formation and reduce α-secretase activity is supported by the primary findings. Anamnesia caused by scopolamine can be lessened by taking certain triterpenoid saponins, and both selective and non-selective muscarinic agonists can enhance memory and learning [29]. One important target that connects the main symptoms of AD is selective M1 muscarinic agonists. Researchers are currently trying to pin down the precise molecular mechanisms by which cholinergic medications improve memory and learning, as well as the efficacy of these treatments as a therapeutic intervention. The results of immunohistochemistry, neuroimaging, and other analyses corroborated this hypothesis, showing that nicotinic receptors, primarily of the α4β2 subtype, are less abundant in AD patients, loss of α3, α4, and α7 subunit expression and reduced binding ability of α7 receptors in the hippocampus and α4 receptors in the cortex are observed in both the cortex and the hippocampus [30]. Novel research has demonstrated that the interaction between Aβ and cholinergic receptors impacts the activity of these receptors. The cholinergic theory is supported by the fact that the cholinergic and glutamatergic systems interact considerably during changes in glutamatergic neurotransmission. This is because cholinergic disturbances are associated with AD. According to these assumptions, acetylcholine and its receptors, particularly (α7), are thought to be neuroprotective because they regulate neuronal excitability mediated by glutamate [31].

1.4.2 ROLE OF OXIDATIVE STRESS IN THE PATHOGENESIS OF AD

Indications of oxidative damage, aberrant synthesis of Aβ are observed in the brains of people with AD. The importance of biometals, such as zinc, magnesium, iron, and copper in Aβ and dementia is increasingly demonstrated by growing evidence [32]. The findings lend credence to the notion that zinc and copper bind to high affinity locations. Amyloid plaques include copper, which contributes to the OS observed in AD because copper is a potent generator of the highly reactive hydroxyl radical. The likelihood of producing hydrogen peroxide and other ROS appears to be correlated with the length of the Aβ pieces; Aβ (1–42) is more hazardous than Aβ (1–40) [33]. Furthermore, the neocortex, amygdala, and hippocampus regions of the brain primarily impacted by AD pathology were linked to increased zinc concentrations. The generation of harmful fibrillary Aβ aggregates is caused by the highly structured conformational state of zinc binding. In reaction to non-soluble Aβ plaques, the immune system and inflammation upset the balance of zinc in the body, leading to an unregulated release of zinc in the brain, a characteristic of OS [34]. Consequently, OS and cytotoxicity can be caused by zinc-induced and Aβ-mediated buildup when they are not regulated. Since polyunsaturated fatty acids make up the phospholipids in brain membranes, this organ is especially susceptible to free radical damage. Their ability to form double bonds makes them useful for removing hydrogen ions and increasing lipid peroxidation, the hallmark of AD [35]. The enzymes glutamine synthetase and creatine kinase are particularly vulnerable to oxidative modification; AD brains have significantly lower levels of these enzymes, which reflect changes in glutamate concentrations and increased excitotoxicity; and AD may have impaired energy metabolism due to oxidative impairment of creatine kinase. Protein fibrils and insolubility result from pathologic protein aggregation. The τ protein aggregates undergo hyperphosphorylation to create paired helical filaments, which are known as NFTs [36]. The formation of advanced glycation end products is another consequence of protein oxidation, which occurs after translation and occurs when the amino groups of proteins interact non-enzymatically with monosaccharides. Several other DNA-related effects, such as strand breakage, base modifications, DNA-protein crosslinking, and sister chromatid exchange, can result from brain oxidation [37]. Damage to cell structures and by extension, some diseases and the aging process can be mediated in part by an overproduction of ROS leading to OS.

1.4.3 RECENT DISCOVERIES INTO THE UNDERLYING CAUSES OF AD

1.4.3.1 GAMMA OSCILLATIONS ENHANCE DAMAGE AND COGNITIVE DECLINE IN AD

At 20–50 Hz, gamma oscillations are associated with a wide range of higher-order cognitive processes, including memory formation and attentional selection. There have been reports of gamma oscillation changes in several neurological disorders, including AD [38]. Even in a tau pathology animal model, oscillations in the hippocampus were found to impact spatial memory. The restoration of gamma oscillations may safeguard cognitive performance in AD. Interestingly, research has shown that rescuing learning and memory deficiencies in mice by replacing endogenous mouse APOE4 with the AD-linked human APOE4 gene improved gamma impairments [39]. Optogenetics was used in a study on the hippocampus of 5XFAD mice to treat the visual cortex (VC) in many mouse models without invasive light flickering. Although other frequencies did not reduce AD pathology levels, the results demonstrated that 40-Hz gamma oscillation did. Older animals have a lower plaque load, the tau P301S model has a lower plaque load, and different AD mouse models (including wild-type mice) have lower levels of Aβ 1–40 and Aβ 1–42 isoforms. It is believed that the recruitment of microglial responses played a protective role in decreasing Aβ levels by phagocytosis [40]. Changes in microglial gamma oscillation and how they occur are still unknown. However, these results do point to the possibility that lowering neuroinflammation is a key component in reversing neurodegeneration.

Liu et al. [41] found that modifying gamma oscillations improved cognitive impairment by increasing brain connectivity. Novel studies have shown that high-order cognitive capabilities and the activation of microglia to clear amyloid-β deposits are intimately linked to gamma-band waves of electrical brain activity, specifically oscillations around 40-Hz. In this study, we observed that individuals with probable AD showed a considerable improvement in cognitive performance for up to 8 weeks after receiving 40-Hz high-frequency repeated transcranial magnetic stimulation (rTMS) across the bilateral angular gyrus, as compared to sham stimulation. The following was found after additional testing using TMS-EEG and magnetic resonance imaging (MRI): 40-Hz rTMS had the following effects: (i) it stopped the loss of grey matter volume, (ii) it improved the integration of local and global functions in the bilateral angular gyrus, and (iii) it strengthened the connection between the front and back of the brain. The outcomes of the study show that gamma-band oscillation modulation improves cognitive function in AD patients by increasing synaptic plasticity, long-range connection, and local connectivity [41].

1.4.3.2 THE ROLE OF AB IN AD HIPPOCAMPAL GHRELIN/GHSR1A SIGNALING

The hippocampus is home to GHSR1a, a receptor from the class of G protein-coupled receptor (GPCR) family that is also called the ghrelin receptor. The hedonic, motivational, and learning aspects of food are influenced by the ghrelin/GHSR1a signal in healthy hippocampi. In addition, GHSR1a manages the dopamine receptor D1 (DRD1) to activate Ca^{2+}/calmodulin-dependent protein kinase II (CaMKII) through the non-canonical Gaq-Ca^{2+} signaling pathway, which is involved in hippocampus synaptic physiology and memory preservation [42]. One of the first lesions to manifest in AD is a hippocampal lesion, which impacts cognitive performance and may be associated with GHSR1a. The studies reveal that GHSR1a deficiency causes memory problems and synaptic stress similar to AD. To cure AD, GHSR1a could be a target. Clinical trials of MK0677 in patients with AD, however, did not demonstrate any therapeutic benefit [43]. Patients with AD may have GHSR1a that is resistant to activators, as suggested by these findings. In addition, it has been demonstrated that in young 5XFAD mice, early pathological changes such as amyloid-beta plaque deposition, synaptic dysfunction, and neuroinflammatory responses begin to manifest as early as 2–3 months of age, contributing to progressive cognitive decline and mimicking key features of Alzheimer's disease. Further research is needed to determine if MK0677/SKF81297 is useful for older 5XFAD mice. However, the results of this study suggest that this dual agonist strategy may have a protective impact in AD. Nevertheless, GHSR1a's functions might extend beyond what has been mentioned. Research in 5XFAD mice has demonstrated that ghrelin can reduce hippocampus pathology through neurogenesis [44]. The dentate gyrus of 5XFAD mice undergoes enhanced neurogenesis when GHSR1a and DRD1 are coactivated. GHSR1a dysregulation significantly affects calcium signaling and metabolic processes in the hippocampus and memory loss in the elderly and those with AD. Furthermore, GHSR1a has the potential to influence hypothalamus function and may indirectly cause injury to the hippocampus. Finally, GHSR1a's involvement in AD is too important to dismiss, and it may reveal some interesting treatment targets [45].

1.5 HISTORICAL CONTEXT FOR THE DEVELOPMENT OF THE AD HETEROGENEITY

It has long been noted that AD exhibits phenotypic variability. Two separate subsets of dementia patients were observed in a chronic geriatric hospital setting by McDonald in 1969. He found that some people had problems mainly with visual construction, cortical sensation, and praxis using basic memory, parietal function, and aphasia tests. When they were followed up, they showed more severe progression [46]. The "parietal group" was what he called them. The majority of other patients' symptoms were memory loss, which manifested at a later age and progressed at a slower rate. Something he called "benign memory dysfunction of aging." Despite this early proof of AD heterogeneity, subtype diversity was not widely acknowledged. Future positron emission tomography (PET) investigations would lend more credence to the subtype theory. Similar to how different AD symptoms manifest clinically, different patterns of cerebral hypometabolism were also revealed [47]. There was a consistent correlation between asymmetry in PET imaging of the left hemisphere and increased language impairment in people with AD. On the other hand, people who primarily experienced right parietal hypometabolism were more likely to have visuospatial impairment. The anatomical correlation was good in this diversity. For instance, Gerstmann's syndrome in AD was linked to left angular hypometabolism. Importantly, research using longitudinal follow-up showed that these different syndromes persisted with clinical deterioration of disease and maintained their distinctness across time [48].

1.6 COMPLICATIONS OF ALZHEIMER'S DISEASE

There are various consequences of AD, a neurological disease that worsens with time. Mild forgetfulness to severe impairment affects daily functioning and relationships as the defining symptom of memory loss progresses. Complicating matters further, patients with behavioral and psychiatric symptoms, such as agitation and depression, frequently necessitate specialized care [49]. The need for more help with ADLs grows as a result of functional decline, and health worsens due to physical problems, such as infections and malnutrition. The mental, physical, and financial tolls that carers take on are substantial. To help those living with Alzheimer's and those who care for them, comprehensive management approaches are necessary to tackle the complex issues that this disease presents. Complications can arise at any stage of AD, a neurological ailment that worsens with time [50].

Some of the most prevalent problems that might arise from AD are:

1. ***Memory loss:*** This is a symptom that many people associate with AD. The inability to recall recent events, names, and faces gets worse as the disease advances, impacting the quality of life of the affected person [51].
2. ***Cognitive decline:*** Cognitive faculties, such as reasoning, judgement, and problem-solving are impacted by AD. People may have trouble deciding what to do or even with mundane, everyday activities [52].
3. ***Behavioral changes:*** Personality and behavioral changes are symptoms of AD. Aggression, restlessness, agitation, impatience, and social disengagement are all possible symptoms [53].
4. ***Communication difficulties:*** The inability to express oneself clearly, follow directions, or comprehend others may become more common as the condition advances. Feelings of alienation and frustration may ensue [54].
5. ***Functional decline:*** A decline in autonomy in ADL performance is a hallmark of AD. Things like getting dressed, taking a shower, and feeding oneself fall under this category [55].
6. ***Psychiatric symptoms:*** People suffering from AD often experience feelings of sadness and unease. As the disease develops, mental health issues could become more severe [56].
7. ***Physical decline:*** The physical manifestations of AD include a loss of appetite, weakened muscles, and an accelerated susceptibility to infections as a result of impaired immunological function [57].

8. ***Sleep disturbances:*** Disruptions to the normal sleep-wake cycle are common in AD, causing drowsiness during the day, impaired judgment, and difficulty falling or staying asleep [58].
9. ***Nutritional challenges:*** Malnutrition and dehydration can be the outcomes of swallowing and feeding difficulties. Therefore, Alzheimer's patients run the danger of malnutrition and weight loss [59].
10. ***Increased vulnerability to other illnesses:*** People with AD are at increased risk of contracting infections, including pneumonia and UTIs because their immune systems are compromised [60].

1.7 BRAIN IMAGING TECHNIQUES FOR AD

1.7.1 MRI IN ALZHEIMER'S DISEASE

An essential tool in the diagnosis and therapy of AD is MRI. The MRI helps differentiate AD from other types of dementia by identifying distinct patterns of brain shrinkage. The structural alterations, such as the existence of NFTs and amyloid plaques, that characterize AD can be better understood with the use of these scans. Medical practitioners can monitor the development of a disease with the use of longitudinal MRI, which measures the brain's volume and white matter integrity over time. Moreover, MRI aids in the detection of vascular lesions and other contributors to cognitive impairment in AD patients. In addition to its practical uses, MRI is essential in the ongoing quest to understand the biology of AD and find viable treatments and diagnostic tools [61].

Cai et al. [62] found a way to identify β-amyloid aggregates in AD mice using T1-T2 imaging probes. This study presents the design and preparation of UFNPs@PEG/PZD, which are sub-5 nm ultrasmall ferrite nanoprobes coupled with phenothiazine derivatives, for effective T1-T2 imaging of Aβ plaques. UFNPs@PEG/PZD have remarkable r1 and r2 relaxivities in addition to a strong binding affinity to Aβ plaques. Therefore, the research offers potential ultrasmall nanoplatforms for use in early AD diagnosis as a multimodal imaging probe that targets Aβ [62].

Wang et al. [63] reported the one-pot synthesis of 99mTc-ZW-USIONPs. The stabiliser used was sulfobetains functionalized poly (acrylic acid), while the additives were $Na^{99m}TcO_4$ and $SnCl_2$. Many people are interested in the utilization of ^{99m}Tc and iron oxide nanoparticles together because of the commercialization of SPECT/MRI scanners. One benefit of directly doping radioisotope into nanoparticles is their high radiochemical stability, which allows for almost unlimited surface functionalization. Doping ultrasmall IONPs with ^{99m}Tc without precipitation was particularly difficult due to the complicated technetium chemistry involved. Through precise regulation of radioactivity and reducing agent concentration, we demonstrated the synthesis of ^{99m}Tc doped USIONPs that exhibit desirable water solubility and T1 signal. Because the zwitterionic surface does not require a chelator, it can resist the creation of protein corona, which leads to less RES uptake and greater tumour contrast. In mice bearing the 4T1 tumour, the ^{99m}Tc-ZW-USIONPs showed remarkable tumour SPECT and T1-weighted MR imaging capabilities [63].

1.7.2 SPECT IN ALZHEIMER'S DISEASE

An invaluable diagnostic tool in the management of AD is single-photon emission computed tomography (SPECT) imaging. Brain perfusion patterns can reveal anomalies linked to AD, and SPECT can help evaluate these patterns by providing functional imaging of cerebral blood flow. Due to its reduced cost and accessibility, SPECT imaging is still commonly used, despite PET imaging's superior resolution. In AD, SPECT scans usually show hypoperfusion in the hippocampus and temporal lobes. The neurodegenerative process that characterizes AD is compatible with these perfusion deficiencies, which correlate with neuronal dysfunction [64]. By helping to distinguish AD from other dementias, SPECT imaging can improve prognosis and therapy planning. In addition, SPECT is very useful for tracking how AD patients are responding to medication and how the illness is progressing overall [65]. To better understand the effectiveness of therapeutic interventions, SPECT scans monitor variations in cerebral blood flow over time. This allows physicians to make adjustments to treatment regimens as needed. In clinical evaluations of AD, SPECT is still useful as a supplementary technique, even though it has lesser spatial resolution than PET imaging. Evaluating cerebral perfusion patterns provides supplementary data to structural imaging techniques like MRI, allowing for a more complete picture of the neurobiological alterations happening in AD. In clinical research settings, SPECT imaging has the potential to detect early signs of AD pathology and track the evolution of the disease [66].

Kuwabara et al. [67] compared studies using I-123 IMP and ^{99m}-Tc HMPAO with studies using 0-15 H_2O and F-18 FDG PET. The study included 17 patients, nine of whom had AD, three of whom had Pick's disease, and five of whom had multi-infarct dementia. Studies using IMP and HMPAO SPECT were unable to identify areas with minor effects when contrasted with FDG PET. However, MRI and CT scans revealed mild to moderate brain shrinkage, and decreased blood flow to the bilateral parietal areas in AD [67].

1.7.3 PET IN ALZHEIMER'S DISEASE

Neuroimaging using PET is essential for the detection and treatment of AD. Using PET scans to identify amyloid-beta plaques and tau protein aggregates, researchers can better understand the pathophysiology of AD, which helps with early detection and disease staging. On top of that, PET imaging helps differentiate AD from other types of dementia, which in turn helps with treatment planning [68]. In addition to aiding in diagnosis, PET scans offer important prognostic data by linking the amount of amyloid and tau to how the illness is progressing and the clinical results. Furthermore, PET makes it easier to track how well a patient is responding to treatment, which is very useful in studies testing treatments that alter disease progression. When it comes to understanding the biology of AD and finding possible intervention targets, PET imaging is crucial in the scientific community. When it comes to AD, PET imaging is a game-changer because it helps doctors understand the illness's causes and make better judgements about patient care [69].

An automated system that produces thorough picture presentations and objective diagnostic indices was created by Minoshima et al. Results from fluorine-18-fluorodeoxyglucose PET scans were obtained from 37 individuals with suspected AD, 22 healthy controls, and 5 patients with cerebrovascular illness. After the stereotactic anatomic standardization, the metabolic activity from a person's PET images was projected onto a set of predetermined surface pixels (3D-SSP). These pixels were then utilized in the analysis that followed. The average of the normal subjects' retrieved datasets was used to generate a normal database. For visual inspections, the datasets of individual patients were shown in 3D-SSP views.

Both the unilaterally averaged parietal-temporal-frontal cortex (a diagnostic indicator) and the parietal association cortex (a diagnostic index) demonstrated 100% specificity, and 95 and 97% sensitivity, respectively, when used to distinguish between normal participants and patients with probable AD. No cerebrovascular illness false-positive outcomes were produced by either index. The 3D-SSP allows for accurate stereotactic coordinate localization of metabolic anomalies and quantitative data extraction. To understand functional brain imaging, the suggested method is an encouraging strategy (PET images) [70].

1.8 POTENTIAL AD BIOMARKERS

To narrow down the pool of potential AD drugs for the costly and time-consuming phase 3 clinical trials, biomarkers can be invaluable. In addition to improving the clinical course, biomarkers are critical for proving that a treatment changes the pathophysiology of the disease, which is required to say that the medicine is modifying the disease [71].

1.8.1 GALANTAMINE

Isolation of galantamine from many Amaryllidaceae plants (e.g., *Leucojum* spp., *Narcissus* species, and *Galanthus* spp.) has recently emerged as a crucial treatment option for delaying neuronal deterioration in AD. Galantamine has a low toxicity profile and is safe to use over the long term. In addition to its marketing in Germany and Austria for indications like facial neuralgia, galantamine, sold under the brand name Nivalin, has already received approval in Austria for AD [72].

1.8.2 CANNABINOIDS

One promising new way to treat AD is through the cannabinoid system. Multiple cannabinoid receptors (CBs) make up the compound, with the most well-studied being CB1 and CB2. These effects include lowering damaging beta amyloid peptide activity and tau phosphorylation and enhancing brain's intrinsic repair mechanisms [73].

1.8.3 MULTI-TARGET-DIRECTED LIGAND (MTDL)

Taking into consideration the disease's complicated aetiology, MTDL is a promising technique while searching for new AD medications. Tacrine, donepezil, galantamine, and memantine are just a few examples of the recently discovered compounds that have heterodimeric structures that allow them to interact with a wide variety of targets via pharmacophores. These compounds can be synthetic or natural, and they may or may not have an existing medicinal component [74].

1.8.4 POLY PHENOLIC COMPOUNDS

Polyphenolic chemicals are being investigated as possible natural treatments for several disorders, including neurological issues, due to their powerful anti-oxidant, anti-inflammatory, antibacterial, and anticancer characteristics. In animal models of AD, there is growing evidence that polyphenolic substances can alleviate brain neuropathy and enhance cognitive performance through various pathways [75].

1.9 CASE STUDIES: PRECLINICAL AND CLINICAL EVIDENCE FOR AD

There are many facets to the complex problem that AD poses to people, their families, and medical professionals around the globe. The neurodegenerative disease worsens over time, causing alterations in behavior and personality as well as a decrease in cognitive abilities and memory. To diagnose, intervene with, and effectively manage AD, it is crucial to understand the disease's varied symptoms and progression patterns. It may explore the complex nature of AD, its diagnosis, progression, and care, as well as the preclinical and clinical evidence for the condition, through in-depth case studies [76]. To improve patient outcomes and quality of life, it is crucial to conduct thorough assessments, intervene promptly, and provide holistic care. These case studies are powerful reminders of the immense impact that AD can have on people's lives. Through the analysis of real-life situations, we can uncover important details about the progression of AD, from the early, non-verbal symptoms to the severe cognitive and functional decline seen in the later stages of the disease [77]. We hope that by delving further into this topic, one can improve the ability to diagnose, treat, and provide compassionate care for those affected by AD.

A case report was published by Um et al. detailing the case of a 37-year-old patient with AD who had prominent striatum amyloid retention. Researchers have recently taken a keen interest in the diagnostic potential of amyloid imaging due to its improved capabilities. The patient is a 37-year-old male who presented with a history of progressive cognitive loss, apraxia, disorientation, and sleep problems. The patient's ^{18}F-Florbetaben amyloid scan revealed widespread amyloid retention and strong uptake in the patient's striatum. The clinical value of amyloid imaging in diagnosing EOAD is supported by this finding. Additionally, doctors should pay close attention to this patient because of the significant clinical consequences of the striatal dominant uptake pattern [78].

The neuropathology of AD in humans following amyloid-β peptide vaccination was examined by Nicoll et al. [79]. In the development of AD, the Aβ peptide plays a crucial role. In a transgenic mouse model of AD, immunization with Aβ decreases the buildup of Aβ in the brain as a result of aging1 and the cognitive impairment that comes with it. The initial examination of human neuropathology following Aβ (AN-1792) vaccination is here delivered. Despite definite neuropathological characteristics of AD, the following unexpected traits were seen in the immunized case when compared with seven unimmunized instances of AD: (i) The neocortex was vast and had few Aβ plaques. (ii) The cortex without Aβ plaques had dense tangles, neuropil threads, and cerebral amyloid angiopathy (CAA), which was similar to unimmunized AD. (iii) In some regions without plaques, microglia were associated with Aβ-immunoreactivity. (iv) There was meningoencephalitis caused by T-lymphocytes. (v) Macrophages had infiltrated the cerebral white matter. The results (i–iii) are quite similar to the alterations observed following Aβ immunotherapy in AD1-6 mice models and indicate that this patient's Aβ plaques were cleared by the immune response triggered by the peptide [79].

The amyloidogenic function of the cytokine TGF-β1 in AD and transgenic mice was discussed by Coray et al. [80]. The study found that elderly transgenic mice overexpressing TGF-β1 from astrocytes develop Aβ accumulation in their cerebral blood vessels and meninges. The accumulation of amyloid-β peptide was expedited when TGF-β1 was co-expressed in transgenic mice that overexpressed APP and had symptoms similar to AD. Overexpression of TGF-β1 may be a risk factor for developing AD and experimental models of the disease, since these data suggest that it may initiate or increase amyloidogenesis [80].

1.10 CONCLUSION

Understanding the molecular pathways behind AD has been greatly aided by recent developments in cellular biology. There are a lot of competing ideas on the complex process that is AD, and our current understanding of its molecular underpinnings is inadequate. But none of these hypotheses can explain the illness in its entirety; more research is needed. Some aspects, such as the abnormal production of Aβ and how it impacts neurons and nicotinic acetylcholine receptors, as well as the connection between the interference of cholinergic pathways and the cognitive impairments associated with AD, remain unexplained. Multiple components and changes in behavior of different types of cells within the central nervous system contribute to the aetiology of AD. Microglia and astrocytes can breakdown or eliminate Aβ after it is released to the extracellular space from neurons. Neurons are the primary sites of tau expression, and it is significantly controlled by several PTMs. Tau aggregation and accumulation can be caused by pathogenic tau seeds, abnormal PTMs, or LLPS in numerous ways. During a disease, tau pathology can spread from one location to another, and glial cells are crucial to this process. Neuronal dysfunction, glial activation, and neuroinflammation can be caused by oAβ and other types of Aβ aggregates, as well as tau buildup. Because existing pharmacological treatment is unable to halt the onset or progression of AD, any new information that helps to clarify the molecular pathogenesis of the disease and its relations is vital for the creation of new therapeutic approaches.

KEYWORDS

- **Alzheimer's disease**
- **cellular and molecular mechanism**
- **dementia**
- **etiology**
- **imaging**
- **pathogenesis**

REFERENCES

1. Tzioras, M., et al. (2023). Synaptic degeneration in Alzheimer disease. *Nature Reviews Neurology, 19*(1), 19–38.
2. Wang, C., Song, P., & Niu, Y. (2022). The management of dementia worldwide: A review on policy practices, clinical guidelines, end-of-life care, and challenge along with aging population. *Bioscience Trends, 16*(2), 119–129.
3. Nandi, A., et al. (2022). Global and regional projections of the economic burden of Alzheimer's disease and related dementias from 2019 to 2050: A value of statistical life approach. *EClinicalMedicine, 51*, 101580.
4. Holtzman, D. M., Morris, J. C., & Goate, A. M. (2011). *Alzheimer's disease: The challenge of the second century. Science Translational Medicine, 3*(77), 77sr1.
5. Sepulveda-Falla, D., et al. (2021). A multifactorial model of pathology for age of onset heterogeneity in familial Alzheimer's disease. *Acta Neuropathology, 141*(2), 217–233.
6. Yu, J. T., Tan, L., & Hardy, J. (2014). Apolipoprotein E in Alzheimer's disease: An update. *Annual Review of Neuroscience, 37*, 79–100.
7. Wu, L., & Zhao, L. (2016). ApoE2 and Alzheimer's disease: Time to take a closer look. *Neural Regeneration Research, 11*(3), 412–413.
8. Cruz, C. R., & Smith, R. G. (2008). The growth hormone secretagogue receptor. *Vitamins and Hormones, 77*, 47–88.
9. Reitz, C., Brayne, C., & Mayeux, R. (2011). Epidemiology of Alzheimer disease. *Nature Reviews Neurology, 7*(3), 137–152.
10. Rademakers, R., Cruts, M., & Van Broeckhoven, C. (2003). Genetics of early-onset Alzheimer dementia. *ScientificWorldJournal, 3*, 497–519.
11. Campion, D., et al. (1999). Early-onset autosomal dominant Alzheimer disease: prevalence, genetic heterogeneity, and mutation spectrum. *American Journal of Human Genetics, 65*(3), 664–670.
12. Szaruga, M., et al. (2017). Alzheimer's-causing mutations shift Aβ length by destabilizing γ-Secretase-Aβn interactions. *Cell, 170*(3), 443–456.e14.
13. Vetrivel, K. S., et al. (2006). Pathological and physiological functions of presenilins. *Molecular Neurodegeneration, 1*, 4.
14. Rezazadeh, M., et al. (2019). Genetic discoveries and advances in late-onset Alzheimer's disease. *Journal of Cellular Physiology, 234*(10), 16873–16884.
15. Ulrich, J. D., & Holtzman, D. M. (2016). TREM2 function in Alzheimer's disease and neurodegeneration. *ACS Chemical Neuroscience*, 7(4), 420–427.
16. Armstrong, R. A. (2019). Risk factors for Alzheimer's disease. *Folia Neuropathologica, 57*(2), 87–105.
17. Dawkins, E., & Small, D. H. (2014). Insights into the physiological function of the β-amyloid precursor protein: beyond Alzheimer's disease. *Journal of Neurochemistry, 129*(5), 756–769.
18. Yakel, J. L. (2013). Cholinergic receptors: Functional role of nicotinic ACh receptors in brain circuits and disease. *Pflugers Archiv, 465*(4), 441–450.
19. Chow, V. W., et al. (2010). An overview of APP processing enzymes and products. *Neuromolecular Medicine, 12*(1), 1–12.

20. Jeremic, D., Jiménez-Díaz, L., & Navarro-López, J. D. (2021). Past, present and future of therapeutic strategies against amyloid-β peptides in Alzheimer's disease: A systematic review. *Ageing Research Reviews, 72*, 101496.
21. Johnstone, M., Gearing, A. J., & Miller, K. M. (1999). A central role for astrocytes in the inflammatory response to beta-amyloid; chemokines, cytokines and reactive oxygen species are produced. *Journal of Neuroimmunology, 93*(1-2), 182–193.
22. Rai, S. N., et al. (2019). The role of PI3K/Akt and ERK in neurodegenerative disorders. *Neurotoxicity Research, 35*(3), 775–795.
23. Grisaru, D., et al. (1999). Structural roles of acetylcholinesterase variants in biology and pathology. *European Journal of Biochemistry, 264*(3), 672–686.
24. Teixeira, J. P., et al. (2019). Future therapeutic perspectives into the Alzheimer's disease targeting the oxidative stress hypothesis. *Molecules, 24*(23), 4410.
25. Singh, A., et al. (2019). Oxidative stress: A key modulator in neurodegenerative diseases. *Molecules, 24*(8), 1583.
26. Agostinho, P., Cunha, R. A., & Oliveira, C. (2010). Neuroinflammation, oxidative stress and the pathogenesis of Alzheimer's disease. *Current Pharmaceutical Design, 16*(25), 2766–2778.
27. Hampel, H., et al. (2019). Revisiting the cholinergic hypothesis in Alzheimer's disease: Emerging evidence from translational and clinical research. *The Journal of Preventioanl Alzheimers Disease*, *6*(1), 2–15.
28. Geula, C., et al. (1998). Relationship between plaques, tangles, and loss of cortical cholinergic fibers in Alzheimer disease. *Journal of Neuropathology and Experimental Neurology, 57*(1), 63–75.
29. Moritz, B., et al. (2020). The role of vitamin C in stress-related disorders. *The Journal of Nutritional Biochemistry, 85*, 108459.
30. Paterson, D., & Nordberg, A. (2000). Neuronal nicotinic receptors in the human brain. *Progress in Neurobiology, 61*(1), 75–111.
31. Koukouli, F., & Maskos, U. (2015). The multiple roles of the α7 nicotinic acetylcholine receptor in modulating glutamatergic systems in the normal and diseased nervous system. *Biochemical Pharmacology, 97*(4), 378–387.
32. Chauhan, V., & Chauhan, A. (2006). Oxidative stress in Alzheimer's disease. *Pathophysiology,* 13(3), 195–208.
33. Cheignon, C., et al. (2018). Oxidative stress and the amyloid beta peptide in Alzheimer's disease. *Redox Biology, 14*, 450–464.
34. García-Viñuales, S., et al. (2021). The interplay between lipid and Aβ amyloid homeostasis in Alzheimer's Disease: Risk factors and therapeutic opportunities. *Chemistry and Physics of Lipids, 236*, 105072.
35. Huang, W. J., Zhang, X., & Chen, W. W. (2016). Role of oxidative stress in Alzheimer's disease. *Biomedical Reports, 4*(5), 519–522.
36. Narayanan, S. E., et al. (2020). Exploring the role of aggregated proteomes in the pathogenesis of Alzheimer's disease. *Current Protein & Peptide Science, 21*(12), 1164–1173.
37. Omolaoye, T. S., & du Plessis, S. S. (2020). Male infertility: A proximate look at the advanced glycation end products. *Reproductive Toxicology, 93*, 169–177.

38. Herrmann, C. S., Munk, M. H., & Engel, A. K. (2004). Cognitive functions of gamma-band activity: memory match and utilization. *Trends in Cognitive Sciences, 8*(8), 347–355.
39. Fu, H., et al. (2017). Tau pathology induces excitatory neuron loss, grid cell dysfunction, and spatial memory deficits reminiscent of early Alzheimer's disease. *Neuron, 93*(3), 533–541.e5.
40. Yokoyama, M., et al. (2022). Mouse models of Alzheimer's disease. *Frontiers in Molecular Neuroscience, 15*, 912995.
41. Liu, C., et al. (2022). Modulating gamma oscillations promotes brain connectivity to improve cognitive impairment. *Cerebral Cortex, 32*(12), 2644–2656.
42. Ringuet, M. T., Furness, J. B., & Furness, S. G. B. (2022). G protein-coupled receptor interactions and modification of signalling involving the ghrelin receptor, GHSR1a. *Journal of Neuroendocrinology, 34*(9), e13077.
43. Tian, J., et al. (2019). MK0677, a ghrelin mimetic, improves neurogenesis but fails to prevent hippocampal lesions in a mouse model of Alzheimer's disease pathology. *Journal of Alzheimers Disease, 72*(2), 467–478.
44. Moon, M., Cha, M. Y., & Mook-Jung, I. (2014). Impaired hippocampal neurogenesis and its enhancement with ghrelin in 5XFAD mice. *Journal of Alzheimers Disease, 41*(1), 233–241.
45. Wellman, M., & Abizaid, A. (2015). Growth hormone secretagogue receptor dimers: A new pharmacological target. *eNeuro, 2*(2).
46. Lam, B., et al. (2013). Clinical, imaging, and pathological heterogeneity of the Alzheimer's disease syndrome. *Alzheimer's Research & Therapy, 5*(1), 1.
47. Mehak, S. F., et al. (2023). Apathy in Alzheimer's disease: A neurocircuitry based perspective. *Ageing Research Reviews, 87*, 101891.
48. Sala, A., et al. (2020). Brain metabolic signatures across the Alzheimer's disease spectrum. *European Journal of Nuclear Medicine and Molecular Imaging, 47*(2), 256–269.
49. Ding, X. W., et al. (2020). Nerve growth factor in metabolic complications and Alzheimer's disease: Physiology and therapeutic potential. *Biochimica et Biophysica Acta Molecular Basis of Disease, 1866*(10), 165858.
50. Nayak, A., et al. (2024). Novel ibuprofen prodrug: A possible promising agent for the management of complications of Alzheimer's disease. *Saudi Pharmaceutical Journal, 32*(1), 101888.
51. Gold, C. A., & Budson, A. E. (2008). Memory loss in Alzheimer's disease: implications for development of therapeutics. *Expert Review of Neurotherapeutics, 8*(12), 1879–1891.
52. Wilson, R. S., et al. (2012). The natural history of cognitive decline in Alzheimer's disease. *Psychology & Aging, 27*(4), 1008–1017.
53. Mega, M. S., et al. (1996). The spectrum of behavioral changes in Alzheimer's disease. *Neurology, 46*(1), 130–135.
54. Jones, D. (2015). A family living with Alzheimer's disease: The communicative challenges. *Dementia (London), 14*(5), 555–573.
55. Callahan, C. M., et al. (2017). Targeting functional decline in Alzheimer disease: A randomized trial. *Annals of Internal Medicine, 166*(3), 164–171.

56. Jost, B. C., & Grossberg, G. T. (1996). The evolution of psychiatric symptoms in Alzheimer's disease: a natural history study. *Journal of the American Geriatrics Society, 44*(9), 1078–1081.
57. Brini, S., et al. (2018). Physical activity in preventing Alzheimer's disease and cognitive decline: A narrative review. *Sports Medicine, 48*(1), 29–44.
58. McCurry, S. M., et al. (2000). Treatment of sleep disturbance in Alzheimer's disease. *Sleep Medicine Review, 4*(6), 603–628.
59. Martins, L. B., Malheiros Silveira, A. L., & Teixeira, A. L. (2021). The link between nutrition and Alzheimer's disease: From prevention to treatment. *Neurodegenerative Disease Management, 11*(2), 155–166.
60. Rodrigues, R., Petersen, R. B., & Perry, G. (2014). Parallels between major depressive disorder and Alzheimer's disease: role of oxidative stress and genetic vulnerability. *Cellular and Molecular Neurobiology, 34*(7), 925–949.
61. Filippi, M., et al. (2012). Magnetic resonance imaging in Alzheimer's disease: from diagnosis to monitoring treatment effect. *Current Alzheimer Research, 9*(10), 1198–1209.
62. Cai, J., et al. (2020). Ultrasmall T(1)-T(2) magnetic resonance multimodal imaging nanoprobes for the detection of β-amyloid aggregates in Alzheimer's disease mice. *ACS Applied Materials & Interfaces, 12*(24), 26812–26821.
63. Wang, P., et al. (2021). One pot synthesis of zwitteronic 99mTc doped ultrasmall iron oxide nanoparticles for SPECT/T1-weighted MR dual-modality tumor imaging. *Colloids Surface B Biointerfaces, 197*, 111403.
64. Borroni, B., et al. (2006). Combined 99mTc-ECD SPECT and neuropsychological studies in MCI for the assessment of conversion to AD. *Neurobiology Aging, 27*(1), 24–31.
65. Valotassiou, V., et al. (2015). Clinical evaluation of brain perfusion SPECT with brodmann areas mapping in early diagnosis of Alzheimer's disease. *Journal of Alzheimers Disease, 47*(3), 773–785.
66. Oukoloff, K., et al. (2015). PET and SPECT radiotracers for Alzheimer's disease. *Current Medicinal Chemistry, 22*(28), 3278–3304.
67. Kuwabara, Y., et al. (1990). Comparison of I-123 IMP and Tc-99m HMPAO SPECT studies with PET in dementia. *Annals of Nuclear Medicine, 4*(3), 75–82.
68. Ossenkoppele, R., et al. (2016). Tau PET patterns mirror clinical and neuroanatomical variability in Alzheimer's disease. *Brain, 139*(Pt 5), 1551–1567.
69. Cohen, A. D., et al. (2019). Fluid and PET biomarkers for amyloid pathology in Alzheimer's disease. *Molecular and Cellular Neuroscience, 97*, 3–17.
70. Minoshima, S., et al. (1995). A diagnostic approach in Alzheimer's disease using three-dimensional stereotactic surface projections of fluorine-18-FDG PET. *Journal of Nuclear Medicine, 36*(7), 1238–1248.
71. Han, S. H., & Mook-Jung, I. (2014). Diverse molecular targets for therapeutic strategies in Alzheimer's disease. *Journal of Korean Medical Science, 29*(7), 893–902.
72. Marco, L., & do Carmo Carreiras, M. (2006). Galanthamine, a natural product for the treatment of Alzheimer's disease. *Recent Patents of CNS Drug Discovery, 1*(1), 105–111.
73. Abate, G., Uberti, D., & Tambaro, S. (2021). Potential and limits of cannabinoids in Alzheimer's disease therapy. *Biology (Basel), 10*(6), 542.

74. Bajda, M., et al. (2011). Multi-target-directed ligands in Alzheimer's disease treatment. *Current Medicinal Chemistry, 18*(32), 4949–4975.
75. Reddy, V. P., et al. (2020). Polyphenols in Alzheimer's disease and in the gut-brain axis. *Microorganisms, 8*(2), 199.
76. Mancuso, C., et al. (2012). Natural substances and Alzheimer's disease: From preclinical studies to evidence based medicine. *Biochimca et Biophysica Acta, 1822*(5), 616–624.
77. Banik, A., et al. (2015). Translation of pre-clinical studies into successful clinical trials for Alzheimer's disease: What are the roadblocks and how can they be overcome? *Journal of Alzheimers Disease, 47*(4), 815–843.
78. Um, Y. H., et al. (2017). A case report of a 37-year-old Alzheimer's disease patient with prominent striatum amyloid retention. *Psychiatry Investigation, 14*(4), 521–524.
79. Nicoll, J. A., et al. (2003). Neuropathology of human Alzheimer disease after immunization with amyloid-beta peptide: A case report. *Nature Medicine, 9*(4), 448–452.
80. Wyss-Coray, T., et al. (1997). Amyloidogenic role of cytokine TGF-beta1 in transgenic mice and in Alzheimer's disease. *Nature, 389*(6651), 603–606.

CHAPTER 2

Epigenetic Alterations in Alzheimer's Disease: DNA Methylation and Histone Modifications with Non-coding DNA Involvement

KOMAL BHATI*, RAHUL KAUSHIK, PRAVEEN KUMAR GAUR, KANAK LATA, and PUJA SINORIYA

Metro College of Health Sciences and Research, Knowledge Park III, Greater Noida, Uttar Pradesh, India

*Corresponding author

ABSTRACT

Background: Alzheimer's disease (AD) is a highly complex neurologic degenerative condition distinguished by deterioration of progressive and advanced cognitive ability and a kind of dementia or memory fading conditions. While genetic factors have long been implicated in AD pathology, emerging evidence suggests a significant impact for epigenetic mechanism, with DNA methylation, histone alterations, and non-coding DNA elements, in the progression and development of this prevalent disease.

DNA methylations impact in Alzheimer's disease: An epigenetic modification in which a methyl group is added to cytosine residues in CpG dinucleotides is known as DNA methylation. This process plays a decisive role in the maintenance of DNA daughter cell generation and genetics expression patterns. In AD, modifications in DNA methylation patterns had been recorded in key genes accompanying with neuronal function, synaptic plasticities, and memory enhancing nature or formation. Specifically, the aberrant DNA methylations in promoter regions of genes, such

Cellular and Molecular Mechanisms in Alzheimer's Disease Therapeutics, Volume 1: Cellular Mechanisms. Prashant Tiwari, Sunil Kumar Kadiri, and Kunjbihari Sulakhiya (Eds.)

DOI: 10.1201/9781779641793-2

as *APP* (*amyloid precursor protein*), *BDNF*, and *MAPT* have been linked to dysregulated gene expression and synaptic dysfunction in AD brains. Moreover, genome-wide DNA methylations studies have revealed widespread changes in DNA methylation patterns in AD patients compared to healthy controls. These findings underscore the global dysregulation of epigenetic processes in the framework involved in AD.

The alterations of histone in Alzheimer's disease: Histone changes, including phosphorylation, acetylation, methylation, besides ubiquitination, impart a critical importance in regulating chromatin granule structure and gene expression. In the context of AD, alterations in histone modifications subsidize to the dysregulation of gene countenance systems or their expression involved in synaptic plasticities and memory consolidation. Specifically, dysregulated histone acetylation and methylation patterns have been identified in AD brains. These alterations influence the expression of protein sequence connected with neuro-inflammations, oxidative stress, and the viability of neurons. Furthermore, aberrant histone modifications have found to be associated with the formation of neurofibrillary tangles and amyloid plaques—hallmark neuropathological features of AD.

Non-coding DNA involvement in Alzheimer's disease: Non-coding DNA segments, housing microRNAs besides long uncoding RNAs (lncRNAs), emerge as crucial players in regulating epigenetic mechanisms in AD. The disrupted expression of miRNAs is linked to the modulation of gene expression networks pivotal in AD progression, encompassing amyloid beta metabolism, tau phosphorylation, and synaptic dysfunction. Likewise, deviations in lncRNA expression are correlated with irregular histone modifications and abnormal gene expression in AD brains, hinting at the involvement of non-coding RNAs in orchestrating epigenetic changes within the disease context.

Conclusion: A key finding from this chapter is that epigenetic changes—encompassing DNA methylations, histones alterations, and the involvement of non-coding DNA—are pivotal in the development of AD. Grasping the intricate connections among these epigenetic processes is crucial for comprehending the molecular underpinnings of AD pathology and pinpointing potential treatment targets. Future investigations focused on unraveling the epigenetic terrain for novel diagnostic markers and precise therapeutic strategies in AD offer hope for slowing disease advancement and enhancing patient well-being.

2.1 INTRODUCTION

Alzheimer's disease (AD) is a multifaceted neurodegenerative condition manifested by a progressive cognitive decline and memory impairment, primarily affecting the hippocampus. The pathogenesis of AD originates from an uncontrolled buildup of extra neuronal amyloid-beta (Aβ) peptide in key brain regions, such as the neocortex and temporal lobes [1]. Additionally, the presence of intracellular neurofibrillary tangle (NT) and neurotic plaque further contributes to the progression of the disease. These pathological hallmarks disrupt normal neuronal function and communication, ultimately principal to cognitive decline and memory impairment characteristic of AD [2].

Named after a scientist; Alois Alzheimer, a pioneering German physician or psychiatrist; who first identified the disease's hallmark—amyloid plaques in a patient exhibiting severe memory loss, AD has since been the subject of extensive investigation [3]. This discovery catalyzed significant strides in comprehending, diagnosing, and treating the illness, and has remained a focal point for researchers and medical practitioners alike [4].

The neuropathology of AD is characterized by distinct types of lesions, categorized as positive and negative. Positive lesions involve the accumulation of various deposits within the brain of AD patients, including NT, amyloid plaques, and dystrophic neurites [5]. These pathological features are indicative of the progressive neurodegeneration underlying AD. Conversely, negative lesions signify structural and functional impairments, including neural and neuropil loss, synaptic deficits, and observable atrophy [6, 7]. These alterations in brain morphology and connectivity contribute to the cognitive and behavioral symptoms observed in AD patients.

In addition to these pathological changes, AD diagnosis often uncovers concurrent health issues. For instance, vitamin B12 deficiency and neuronal abnormalities are frequently detected through advanced imaging techniques like magnetic resonance imaging (MRI) [8, 9]. These comorbidities further complicate the clinical picture of AD and may impact disease progression and treatment strategies. Understanding the interplay between these neuropathological features and associated comorbidities is crucial for developing comprehensive approaches to managing AD and improving patient outcomes.

Predominantly diagnosed in the geriatric population aged 65 and above, AD's prevalence is projected to surge significantly by 2050. Clinical staging delineates four progressive phases: preclinical/presymptomatic, early-stage, moderate-stage, and late-stage AD [10–13].

Despite extensive research, the precise pathogenesis of AD remains elusive [14]. However, findings suggest that AD is not merely an inevitable consequence of aging but rather a disruption of neurological processes [15]. Environmental, biological, and genetic factors contribute to molecular and intracellular dysregulation in the AD brain, involving Aβ and tau pathology, inflammatory mechanisms, homeostatic disturbances, and synaptic plasticity alterations [16, 17]. Additionally, emerging evidence indicates a disproportion in epigenetic mechanism, influencing abnormal synaptic plasticities and the dysregulation of memory-associated genes in the AD [18]. Further exploration of these pathways holds promise for unraveling the intricacies of AD pathogenesis and identifying novel therapeutic targets [19–23].

2.2 EPIGENETICS

In 1942, biologist Conrad Hal Waddington introduced the term "epigenetics," defining it as a specific branch of biology concerned with intricate causal communications between genetic materials and their enzyme catalyzed byproducts [24]. Over time, further exploration in this field has not only deepened our understanding of genetic regulation but has also illuminated the mechanisms underlying heritable traits resulting from chromosome changes without alterations to the DNA sequence [25]. Epigenetics, therefore, encompasses about a study of mitotic-ally or meiotic-ally heritable changes in genetic alterations and their function attributed to chromosome alterations rather than modifications to the DNA sequence [26, 27].

While direct alterations to the DNA sequence may not occur, the realm of epigenetics is rich with various modifications that can still take place [28]. These encompass a wide array of chemical changes to DNA centers and structural variations within chromosomal architecture, which directly influence the packaging of DNA. These epigenetic changes hold significant sway over gene expression levels, shaping the intricate dance of biological processes within living organisms [29].

Epigenetic information is not static but dynamic, passed down through cellular generations. This transmission occurs both between mother cells and their own daughter cells, a phenomenon referred as a mitotic inheritance, and across generational lines, considered as meiotic inheritance. Furthermore, epigenetics has emerged as a key player in understanding the vast phenotypic diversity observed across different organisms, shedding light on the molecular mechanisms that underpin these variations [30].

The intersection of epigenetics with neuroscience has unveiled a profound connection, offering insights into a multifaceted interplay between genes and their involvement and environment in shaping neurological processes. This intersection carries significant implications for our understanding of memory formation and allocation, providing a potential substrate for unraveling the mysteries of cognition. Moreover, epigenetic dysregulation has been implicated in a myriad of multifactorial diseases, ranging from neurodegenerative conditions, like AD and Parkinson's disease to autoimmune disorder, for instance amyotrophic lateral sclerosis and multiple sclerosis, as well as neurological disorders like epilepsy [31].

In the context of memory and learning, epigenetic mechanisms are crucial players, orchestrating the intricate molecular processes underlying synaptic plasticity and neuronal connectivity. Perturbations in the epigenetic landscape have been observed in AD, suggesting a probable importance into its pathophysiology. This link between epigenetics and AD underscores the intricate interplay between biological factors, such as aging, lifestyle choices, nutrition, and environmental exposures in shaping disease susceptibility and progression [32]. As our understanding of epigenetics continues to evolve, it promises to unlock new avenues for therapeutic interventions and personalized treatments, offering hope for improved outcomes in the realm of neurological health and beyond.

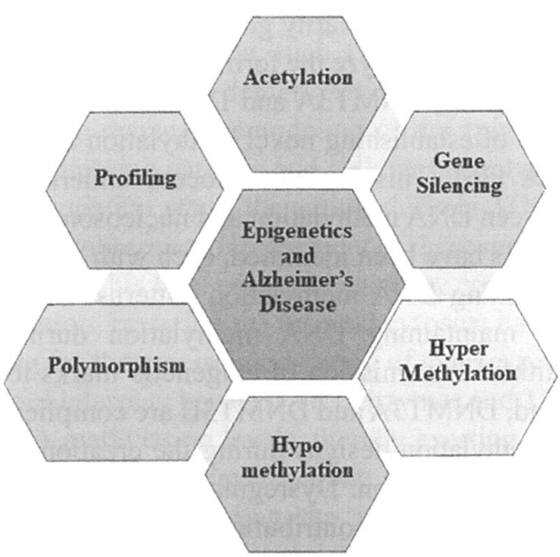

FIGURE 2.1 Representation of computational techniques for epigenetic analysis.

complexities of epigenetic regulation and its impact on health and disease. Understanding the intricate regulation of DNA methylation and its interactions with various factors holds immense promise for elucidating disease mechanisms and developing targeted therapeutic interventions [50].

The delicate process of DNA methylation wove through the genome, silencing some genes while awakening others, orchestrating the symphony of cellular function. In the dim light of the laboratory, scientists marveled at how methylation markers could turn genetic instructions on and off like the flip of a switch. This intricate pattern of DNA methylation held secrets to diseases long thought incurable, a key to unlocking the mysteries of human health. Epigeneticists compared the genome to a piano, with methyl groups serving as the fingers that played its many keys, creating the melody of life. Each cell carried a unique methylation signature, like a biological fingerprint, defining its identity and role within the organism.

DNA methylation was the master regulator, ensuring that skin cells didn't start behaving like brain cells, preserving the harmony of the body's complex systems. As the twin sisters grew older, the differences in their DNA methylation patterns told a story of diverse experiences and environmental exposures. The pharmaceutical breakthrough came when they found a way to reverse abnormal DNA methylation, offering hope for patients with genetic disorders. In the age of personalized medicine, DNA methylation maps were as critical as genetic sequences, guiding doctors in tailoring treatments to individual needs. The environmental toxins had left their mark, subtly altered methylation patterns and thus influencing gene expression and health outcomes.

2.2.3 NOVEL DNA METHYLATION MARKERS IN AD

An advent of deep-sequencing techniques has brought to light a surprising prevalence of 5hmC and 5fC in the brain. However, the precise role of these modifications remains a subject of debate within the scientific community [51, 52]. While some researchers propose that 5hmC and 5fC may serve as new epigenetic marks in their own right, others suggest that they could be intermediate states in the process of DNA demethylation [53].

Beyond CpG dinucleotides, recent studies have revealed the existence of DNA methylation at non-CpG sites, particularly in CpA dinucleotides. This non-CpG DNA methylation phenomenon is particularly pronounced in the brain, accounting for approximately 25% of all cytosine modifications.

Similar to 5mC and 5hmC, non-CpG methylation tends to occur within gene bodies, particularly in regions associated with highly transcribed genes [54, 55]. This intricate landscape of DNA methylation extends our understanding of epigenetic regulation beyond the canonical CpG context, opening new avenues for exploration in the field of neuroepigenetics.

2.2.4 *HISTONE MODIFICATION*

At the core of epigenetic regulation lies the intricate interplay between DNA and histone proteins, which collectively shape chromatin structure and gene expression. Histones serve as fundamental building blocks, around which DNA is wrapped to form nucleosomes, the basic units of chromatin [56, 57]. This arrangement not only compacts DNA but also governs accessibility to the underlying genetic information.

DNA methylation, a common epigenetic modification, influences the positioning of nucleosomes along the DNA strand. Methylation of cytosine residues, typically occurring at CpG dinucleotides, can employ regulatory properties on genome development and gene expression. For example, hyper methylation of promoter regions regularly leads to transcriptional silencing of associated genes, as observed in different cancer inhibiting genes in cancer cells. It is interconnected with different molecular pathways regularly processed in cells, like NF-kB activation pathway, p53 pathway, Wnt-signaling pathway, and different apoptotic mechanisms. Detailed exploration of cancer progression pathways, including NF-kβ signaling pathway, Wnt-B Catenin signaling pathway, JAK-STAT pathway, TNF-α mediated pathway, MAPK/mTor pathway, apoptotic and angiogenic pathway; and effector molecules involved in cancer development have been discussed in this chapter. Critical evaluation of these effector molecules through molecular approaches using phytomolecules can intersect cancer formation and its metastasis. Manipulation in effector molecules like levels of NF-kβ, SOCS, β-catenin, BAX, BAK, VEGF, STAT, Bcl2, p53, caspases, CDKs have played an important role in inhibiting tumor growth and it's spread.

In accumulation to methylation of DNA, posttranslational modifications of histone macro molecules play a pivotal importance in chromatin dynamics. Histones undergo various modifications, including methylation, phosphorylation, acetylation, and ubiquitination, predominantly within their N-terminal tails. These modifications serve as dynamic marks that dictate chromatin structure and gene activity. For instance, histone acetylation, catalyzed by enzymes, such as histone acetyltransferases (HATs), neutralizes a significantly encouraging and +ve charge of histones, thereby slackening chromatin granule structure, in addition promoting genome expression and genes development. In contrast, histone methylation has activating or repressive effects on genetic transcription, conditional on a specific histone residues and the extent of DNA methylation processes [58].

Histone modifications occur in concert with nucleosome positioning, forming a complex regulatory network known as the "histone code." This concept proposes that combinations of histone modifications, akin to a molecular language, convey specific regulatory information to cellular machinery regarding gene activity [59]. For example, the presence of both histone acetylation and methylation marks at specific gene loci can indicate active transcription, while the absence or altered patterns of these marks may signify gene repression or silencing [60].

Furthermore, the organization of histones into nucleosomes and higher-order chromatin structures, such as euchromatin and heterochromatin, further modulates gene accessibility and expression [61]. Euchromatin, characterized by open chromatin structure, is transcriptionally active and allows for efficient gene expression. In contrast, heterochromatin represents densely packed chromatin that is transcriptionally silent, often found in regions of repetitive DNA sequences or inactive genomic regions [62].

Overall, the intricate orchestration of DNA methylations, histone adjustments, and chromatin structure regulates genetic evolutions or genome expression in a dynamic and context-dependent manner. Understanding these epigenetic mechanisms and their impact on gene regulation is essential for unraveling the molecular basis of various physiological and pathological processes, from development and differentiation to disease pathogenesis and therapy.

2.2.5 HISTONE ACETYLATION

Histone modifications, particularly posttranslational modifications, perform a pivotal part in amendable protein sequence expression by influencing chromatin structure and accessibility to DNA. Among these modifications, histone acetylation stands out as a key mechanism associated with increased gene activity [63].

Histone acetylation entails the accumulation of an acetyl group with different lysine deposits within the histone macromolecules of cell, typically occurring at specific sites, such as lysine 9 on histone 3 (H3K9) and lysine

27 on histone 3 (H3K27) in gene promoters, as well as lysine 4 on histone 3, which can be constitutively written as H3K4 and lysines 12 on the molecules of histone 4 (H4K12) in gene bodies. This acetylation process neutralizes the positive charges of histones, thereby weakening their electrostatic interaction with the negatively charged DNA strands. Consequently, chromatin adopts a more relaxed conformation, facilitating transcriptional activation of associated genes [64]. Histone acetylation marks are often enriched into an activator and gene body sections of enthusiastically transcribed amino acid sequences or genetic matters, where they contribute to the formation of open chromatin configurations conducive to gene expression [65].

The regulation of histone acetylation is mediated by HATs, which catalyze the addition of acetyl groups, and histone deacetylases (HDACs), which remove these acetyl groups. Three main connectivities, including HATs—G-NAT, M-YST, and CBP/p300—stand accountable for acetylating lysine residues on histones. The specificity of histone acetylation is determined by the interaction of these HATs with other regulatory proteins, which modulate their activity and substrate specificity [66].

On the flip side, HDACs serve as counterparts to HATs, exerting regulatory control over chromatin dynamics and gene expression. HDACs comprise zinc-dependent class I, II, and IV enzymes, as well as NAD-dependent class III enzymes known as sirtuins. These HDACs function by catalyzing the removal of acetyl groups from histone proteins, thereby reversing the acetylation process initiated by HATs.

Similar to HATs, HDACs display a spectrum of specificity in their actions, a characteristic that is influenced by their interactions with various non-catalytic proteins and protein complexes within the cell. This diversity in specificity allows HDACs to target specific histone residues and modulate chromatin structure in a context-dependent manner.

The dynamic equilibrium between histone acetylation and deacetylation, orchestrated by the delicate interplay of HATs and HDACs, plays a crucial role in fine-tuning chromatin architecture and regulating gene expression profiles. This intricate regulatory mechanism enables cells to respond effectively to diverse cellular signals and environmental stimuli, ultimately influencing cellular processes, such as development, differentiation, and response to stressors. Understanding the nuanced dynamics of HATs and HDACs sheds light on the complex regulatory networks governing gene expression and chromatin remodeling, with implications for various physiological and pathological processes.

The multifaceted regulation of histone acetylation underscores its importance in modulating gene expression programs and cellular processes, from development and differentiation to disease pathogenesis. Elucidating the mechanisms underlying histone acetylation dynamics offers valuable insights into the intricate orchestration of epigenetic regulation and its allegations intended for the health and management of diseases and ailments of living beings.

2.2.6 HISTONE METHYLATION

The impact of histone methylation associated with genes expression varies depending on the specific modification and the targeted amino acid residue. For instance, mono-methylation of lysine 27 on histone 3, which is constitutively known as H3K27me and commonly found into the promoters of actively transcribed genetic matter, while tri-methylations of an amino acid or same amino acid in the H3K27me3 stands often linked to gene repression [65]. This intricate regulatory mechanism underscores the pivotal role of histone methylation dynamics in governing gene expression programs.

Histone methyltransferases, the enzymes responsible for catalyzing histone methylation, display remarkable specificity in targeting specific lysine residues and catalyzing methylation at different levels. These methyltransferases can be broadly categorized into two categories: SET domain-having and non-SET domain-having enzymes. Each class in methyltransferases mediates histone methylations at specific lysines residue, contributing to a diverse array of regulatory outcomes. For instance, SET domain-containing methyltransferases facilitates methylation at lysines 4 on histone 3 (H3K4) and lysines 27 on the histones 3 (H3K27), among others, in mono-, di-, or tri-methylated forms. This diversity in methylation enables precise control over chromatin granule construction and genetic appearance, and their development [66].

Conversely, histone demethylases, known as HDMTs, counteract the activity of methyltransferases by removing methyl groups from histone lysine residues. These demethylases can be broadly classified into two main classes: LSD1- and Jumonji-related enzymes. Each class exhibits specificity in demethylating specific lysine residues and methylation states. For example, LSD1 is capable of demethylating mono- and dimethylation marks on histones 3 lysine 4 (H3K4) and lysines 9 (H3K9), while leaving trimethylation marks unaffected. This selective demethylation activity enables dynamic regulation of histone methylation states and precise modulation of gene expression patterns [66].

The intricate interplay between histone methyltransferases and demethylases, coupled with the specificity of their target residues and methylation states, underscores the complexity of histone methylation regulation. Deciphering the underlying mechanisms of histone methylation dynamics is essential for understanding the epigenetic landscape and its impact on gene regulation in various biological contexts, including development, disease pathology, and cellular homeostasis.

2.2.7 THE HISTONE CODE

"The histone code hypothesis" posited that unique functional outcomes in chromatin regulation are determined by specific combinations of histone modifications. However, emerging evidence from chromatin immunoprecipitation (ChIP) sequencing experiments challenges the conventional view of the histone code's complexity. Recent studies have unveiled a surprising level of redundancy among various histone marks [65]. This redundancy suggests that multiple combinations of histone modifications can elicit similar functional outcomes in terms of chromatin granule structure and gene expression regulation.

Moreover, computational algorithms applied to ChIP sequencing data have identified recurrent patterns of histone modifications, leading to the delineation of distinct chromatin states. These computational analyses have provided valuable insights into the organization and function of chromatin, revealing underlying principles governing the relationship between histone modifications and chromatin structure. By categorizing histone modification patterns into discrete chromatin states, researchers can better understand the regulatory mechanisms that govern gene expression and cellular function. This classification system allows for a more nuanced interpretation of epigenetic data and provides a framework for exploring the functional significance of histone modifications in various biological contexts.

These discrete chromatin states have been found to account for the majority of variability in gene transcription.

The initial idea of the histones code has proposed that different combinations having histone modifications act as molecular signals, encoding specific functional outcomes, such as gene activation or repression. However, the advent of advanced genomic techniques has shed new light on the complexity of chromatin regulation. While individual histone modifications can indeed influence gene expression, their effects appear to be context-dependent and can vary depending on the presence of other histone modifications within the same chromatin region.

Recent ChIP sequencing experiments have revealed that the combinatorial patterns of histone modifications are less diverse than previously envisioned. Instead of a vast array of unique histone codes regulating gene expression, there appears to be a level of redundancy, with certain combinations of histone marks recurring frequently across different genomic loci. Computational analyses have further refined our understanding by identifying specific amalgamations having histone alterations that can be located and connected through distinct chromatin states.

These discrete chromatin states, characterized by their unique combinations of histone modifications, have emerged as key determinants of gene transcriptional activity. By categorizing chromatin into these discrete states, researchers have gained insights into the regulatory mechanisms governing gene expression variability. This paradigm shift highlights the importance of considering the broader context of histone modifications in chromatin regulation and emphasizes the need for integrative approaches to decipher the functional implications of the histone code in gene expression.

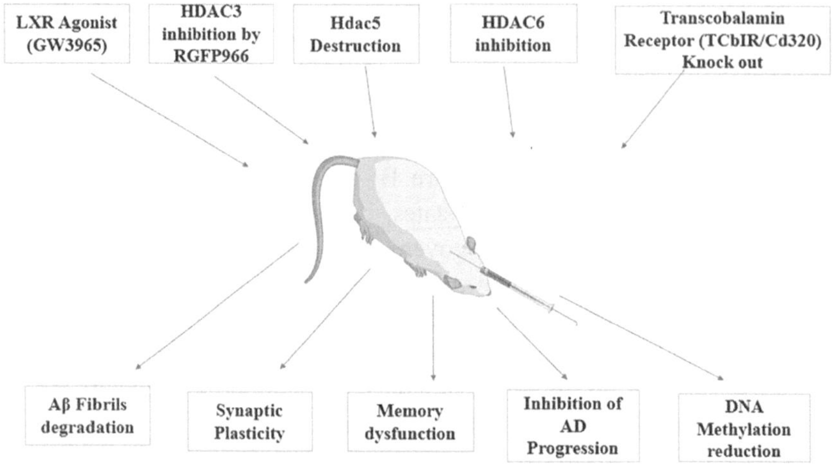

FIGURE 2.2 Figure representation of possible epigenetic modulations and their mechanism of action in improving AD manifestations. LXR, liver X receptor; HDAC, and histone deacetylase.

2.2.8 EPIGENETIC CROSSTALK

Histone modification and DNA methylation are intricately interconnected processes within the realm of epigenetics. Proteins responsible for recognizing and regulating these epigenetic marks often function in multi-protein complex. For instance, proteins like MeCP2 and KAISO are capable of identifying methylated DNA, while also forming associations with HDAC1 and HDAC3, enzymes involved in histone deacetylation [67]. Similarly, other proteins involved in reading DNA methylations, for instance UHRF1, have been found interacting with H3K9 histone methyltransferase G9A, thereby establishing a link between DNA methylation and H3K9 methylation [68]. These interactions illustrate the intricate crosstalk between different epigenetic mechanisms, highlighting the concept of an epigenetic code rather than just a histone code [69, 70].

Therefore, it is imperative to expand the concept of a "histone code" to encompass a broader epigenetic code that extends its influence beyond histone modifications to include DNA methylation and the proteins involved in their regulation. This epigenetic code orchestrates intricate interactions among various epigenetic mechanisms, shaping the chromatin landscape and exerting profound effects on gene expression patterns, particularly within the nervous system [71, 72].

The complexity of the epigenetic code lies in the dynamic interplay between DNA methylation, histone modifications, and the proteins that govern these processes. DNA methylation, mediated by DNMTs, involves the addition of methyl groups to cytosine residues, thereby modulating gene expression. Concurrently, histone modifications, such as acetylation, methylation, phosphorylation, and ubiquitination, dynamically alter chromatin structure and accessibility, influencing gene transcription. The epigenetic code is further modulated by an array of protein complexes that regulate DNA methylation and histone modifications. For instance, methyl-CpG-binding domain proteins (MBDs) and HATs play crucial roles in interpreting and modifying epigenetic marks, thereby influencing gene expression profiles [73]. Similarly, HDACs and DNA demethylases regulate the removal of acetyl groups and methyl groups, respectively, contributing to the dynamic nature of the epigenetic landscape.

In the nervous system, the epigenetic code governs critical processes, such as neuronal development, synaptic plasticity, and memory formation. Dysregulation of the epigenetic code has been implicated in various neurological disorders, including AD, Parkinson's disease, and schizophrenia. Understanding the intricacies of the epigenetic code holds significant promise for elucidating the molecular mechanisms underlying these disorders and identifying novel therapeutic targets. In summary, the epigenetic code represents a complex network of interactions between DNA methylation, histone modifications, and regulatory proteins, with far-reaching implications for gene expression and cellular function. Expanding our understanding of the epigenetic code is essential for unraveling the complexities of gene regulation in health and disease.

2.2.9 NON-CODING RNA

Non-coding RNA (ncRNA) constitutes a class of RNA molecules that do not undergo translation to produce proteins but performs vital functions in the regulation of genetic expressions [74]. Among ncRNAs are dicer-dependent microRNAs (miRNAs) and short-interfering RNAs (siRNAs), which are generated through RNA interference pathways [75].

TABLE 2.2 Localization and Changes of Partial miRNA Expression in Patients with AD and Related Models.

MiRNAs	Locations	Change in AD	References
miR-26b-5p miR-26a-5p miR-107 miR-103a-3p	Parietal lobe	↓	[89]
miRNA-188-5p	Hippocampus	↓	[90]
miRNA-485 miRNA-4723	Hippocampus	↑	[91]
HA-mIR-7 HA-mIR-128 HA-mIR-29c HA-mIR-136	Parietal lobe	↑	[89]
miRNS-7	Superior temporal lobe neocortex	↑	
miRNA-146b-5p miRNA-15b-5p	Blood	↓	[92]
miRNA-483-5p	Plasma	↑	[93]

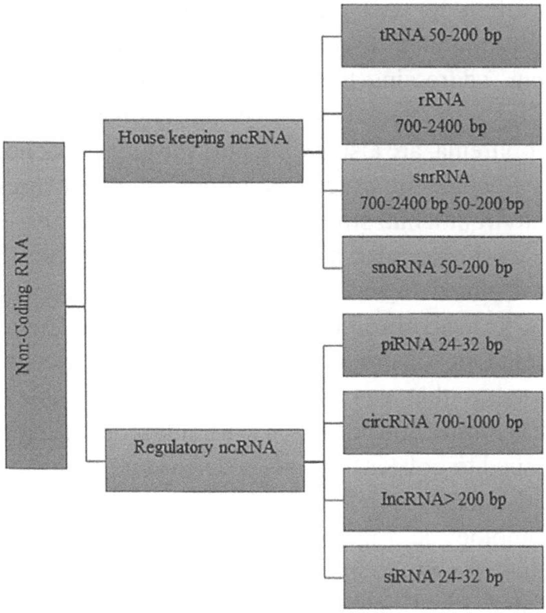

FIGURE 2.3 Flowchart demonstrating the classification of non-coding RNA (ncRNA) and their length in nucleotides.

Recent literature highlights the presence of miRNAs in astrocytes, where they participate in maintaining glutamatergic and inflammatory signaling pathways, ultimately impacting synaptic plasticity [76–82]. SiRNAs typically comprise fewer than 30 nucleotides, whereas long non-coding RNAs (lncRNAs) are characterized by their length, containing more than 200 nucleotides. In addition to its importance in gene silencing, researchers suggested that lncRNAs may be involved in processes like DNA methylations and histone modifications [83–88].

2.2.10 EPIGENETIC THERAPEUTICS BEYOND FOCUS INTO AD

Epigenetic therapies targeting methylation have emerged as promising strategies for treating AD and various other neurodegenerative disorders [94]. The field of epigenetic therapeutics has expanded far beyond its initial focus on AD, opening new avenues for treating a myriad of conditions. Researchers are now exploring how epigenetic modifications can be manipulated to combat cancer, where DNA methylation and histone modifications often go awry, leading to uncontrolled cell growth. In the realm of autoimmune diseases, epigenetic therapies aim to reset the immune system, addressing the underlying genetic miscommunications that cause the body to attack itself. Psychiatric disorders, such as depression and schizophrenia, are also seeing innovative approaches, as scientists investigate how epigenetic drugs can normalize gene expression patterns disrupted by environmental stressors.

Moreover, cardiovascular diseases are being targeted by epigenetic interventions designed to reverse pathological gene expression changes that contribute to heart disease. The promise of epigenetic therapeutics extends to metabolic disorders, where researchers are identifying key epigenetic switches that can be flipped to regulate glucose and lipid metabolism. Even infectious diseases are not left out, as epigenetic strategies are developed to bolster the body's defense mechanisms against viral and bacterial invaders.

As the technology and understanding of epigenetics evolve, personalized medicine becomes increasingly feasible. Tailoring epigenetic treatments to individual genetic and epigenetic profiles could enhance the efficacy and reduce the side effects of these therapies. In oncology, for instance, combining epigenetic drugs with traditional chemotherapies or immunotherapies may offer synergistic effects, improving patient outcomes. The potential of epigenetic therapies in regenerative medicine is also being explored, with the aim of reprogramming cells to regenerate damaged tissues and organs.

The versatility of epigenetic therapeutics is further highlighted by their potential in neurological disorders beyond AD, such as Parkinson's disease and multiple sclerosis, where abnormal gene expression plays a critical role. The future of epigenetic therapeutics is bright, with ongoing research continually uncovering new targets and mechanisms. This burgeoning field promises to revolutionize medicine by offering novel ways to precisely control gene activity, paving the way for treatments that are as dynamic and adaptable as the epigenome itself.

Research examining epigenetic patterns has uncovered similarities in the post-mortem frontal cortex of individuals afflicted with both AD and bipolar disorder. Specifically, genes associated with CpG methylation, DNA methylation, and histone modifications exhibit comparable patterns in both conditions [95]. Notably, both AD and bipolar disorder demonstrate heightened levels of global DNA methylation, increased phosphorylation of histone H3, and elevated methylation at the *BDNF* promoter, alongside decreased methylation of the *COX-2* promoter gene. Furthermore, synaptic markers exhibit CpG methylation in both disorders, although there are nuanced differences [96–98]. For instance, AD shows specific methylation of the synaptophysin promoter, while bipolar disorder displays methylation of the drebin promoter.

In both AD and bipolar disorder, epigenetic alterations are associated with the upregulation of transcription and translation of neuro-inflammatory markers, including microglial activation markers, TNF-α, astrocytic markers, and IL-1β. Additionally, elevated levels of other neuro-inflammation markers, such as IL1β, *CD11b*, *GFAP*, and *COX2* are observed. Conversely, there is a reduction in neurotrophic factors like *BDNF*, as well as decreased levels of presynaptic and postsynaptic markers synaptophysin and drebin [98].

Furthermore, hyper-methylations of *ANK1* have been implicated in AD, Parkinson's disease, and Huntington's disease, suggesting a potential common epigenetic mechanism across these neurodegenerative disorders [99]. These findings underscore the intricate importance of epigenetic dysregulation into neurodegenerative diseases and highlight the probable of different epigenetically derived therapies as novel treatment avenues.

KEYWORDS

- **Alzheimer's disease**
- **DNA methylation**
- **epigenetic alterations**
- **histone modifications**
- **non-coding DNA**

REFERENCES

1. Vlassenko, A. G., Benzinger, T. L. S., & Morris, J. C. (2012). PET amyloid-beta imaging in preclinical Alzheimer's disease. *Biochimica et Biophysica Acta (BBA)—Molecular Basis of Disease, 1822*(3), 370–379. DOI: 10.1016/j.bbadis.2011.11.005.
2. Lane, C. A., Hardy, J., & Schott, J. M. (2018). Alzheimer's disease. *European Journal of Neurology, 25*(1), 59–70. DOI: 10.1111/ene.13439.

3. Burns, A., Iliffe, S. (2009). Alzheimer's disease. *BMJ, 338*, b158–b158. DOI: 10.1136/bmj.b158.
4. Breijyeh, Z., & Karaman, R. (2020). Comprehensive review on Alzheimer's disease: Causes and treatment. *Molecules, 25*(24), 5789. DOI: 10.3390/molecules25245789.
5. Serrano-Pozo, A., Frosch, M. P., Masliah, E., & Hyman, B. T. (2011). Neuropathological alterations in Alzheimer disease. *Cold Spring Harbor Perspectives in Medicine, 1*(1).
6. Singh, S. K., Srivastav, S., Yadav, A. K., Srikrishna, S., & Perry, G. (2016). Overview of Alzheimer's disease and some therapeutic approaches targeting A β by using several synthetic and herbal compounds. *Oxidative Medicine and Cellular Longevity, 2016*, 1–22. DOI: 10.1155/2016/7361613.
7. Spires-Jones, T. L., & Hyman, B. T. (2014). The intersection of amyloid beta and tau at synapses in Alzheimer's disease. *Neuron, 82*(4), 756–771. DOI: 10.1016/j.neuron.2014.05.004.
8. Schachter, A. S., & Davis, K. L. (2000). Alzheimer's disease. *Dialogues in Clinical Neuroscience, 2*(2), 91–100. DOI: 10.31887/DCNS.2000.2.2/asschachter.
9. Jatoi, S., Hafeez, D. A., Riaz, S. U., Ali, A., Ghauri, M. I., & Zehra, M. (2020). Low vitamin B12 levels: An underestimated cause of minimal cognitive impairment and dementia. *Cureus*. DOI: 10.7759/cureus.6976.
10. De-Paula, V. J., Radanovic, M., Diniz, B. S., & Forlenza, O. V. (2012). Alzheimer's disease. *Subcellular Biochemistry, 65*, 329–352.
11. Dubois, B., Hampel, H., Feldman, H. H., Scheltens, P., Aisen, P., Andrieu, S., Bakardjian, H., Benali, H., Bertram, L., Blennow, K. et al. (2016). Preclinical Alzheimer's disease: Definition, natural history, and diagnostic criteria. *Alzheimers Dement, 12*, 292.
12. Kumar, A., Sidhu, J., Goyal, A., & Tsao, J. W. (2022). Alzheimer disease. In *StatPearls* (pp. 1–27) StatPearls Publishing LLC, .
13. Wattmo, C., Minthon, L., & Wallin, Å. K. (2016). Mild versus moderate stages of Alzheimer's disease: Three-year outcomes in a routine clinical setting of cholinesterase inhibitor therapy. *Alzheimer's Research & Therapy, 8*, 7.
14. Gao, X., Chen, Q., Yao, H., Tan, J., Liu, Z., Zhou, Y., & Zou, Z. (2022). Epigenetics in Alzheimer's disease. *Frontiers in Aging Neuroscience, 14*, 911635.
15. Fyfe, I. (2018). Epigenetics links ageing with Alzheimer's disease. *Nature Reviews Neurology, 14*, 254.
16. Hampel, H., Nisticò, R., Seyfried, N. T., Levey, A. I., Modeste, E., Lemercier, P., Baldacci, F., Toschi, N., Garaci, F., Perry, G. et al. (2021). Omics sciences for systems biology in Alzheimer's disease: state-of-the-art of the evidence. *Ageing Research Review, 69*, 101346.
17. Castrillo, J. I., & Oliver, S. G. (2016). Alzheimer's as a systems-level disease involving the interplay of multiple cellular networks. *Methods in Molecular Biology, 1303*, 3–48.
18. Ballard, C., Gauthier, S., Corbett, A., Brayne, C., Aarsland, D., & Jones, E. (2011). Alzheimer's disease. *Lancet, 377*(9770), 1019–1031. DOI: 10.1016/S0140-6736(10)61349-9.
19. Mehler, M. F., & Mattick, J. S. (2007). Non-coding RNAs and RNA editing in brain development, functional diversification, and neurological disease. *Physiological Review, 87*, 799–823.

20. Vecsey, C. G., Hawk, J. D., Lattal, K. M., Stein, J. M., Fabian, S. A., Attner, M. A., Cabrera, S. M., McDonough, C. B., Brindle, P. K., Abel, T., et al. (2007). Histone deacetylase inhibitors enhance memory and synaptic plasticity via CREB: CBP-dependent transcriptional activation. *Journal of Neuroscience, 27*, 6128.
21. Guan, J. S., Haggarty, S. J., Giacometti, E., Dannenberg, J. H., Joseph, N., Gao, J., Nieland, T. J. F., Zhou, Y., Wang, X., Mazitschek, R., et al. (2009). HDAC2 negatively regulates memory formation and synaptic plasticity. *Nature, 459*, 55–60.
22. Michán, S., Li, Y., Chou, M. M. H., Parrella, E., Ge, H., Long, J. M., Allard, J. S., Lewis, K., Miller, M., Xu, W., et al. (2010). SIRT1 is essential for normal cognitive function and synaptic plasticity. *Journal of Neuroscience*, *30*, 9695.
23. Nassar, A., Satarker, S., Gurram, P. C., Upadhya, D., Fayaz, S., & Nampoothiri, M. (2023). Repressor element-1 binding transcription factor (REST) as a possible epigenetic regulator of neurodegeneration and MicroRNA-based therapeutic strategies. *Molecular Neurobiology, 60*, 5557–5577.
24. Henikoff, S. (2023). The epigenetic landscape: An evolving concept. *Frontiers in Epigenetics and Epigenomics, 1*, 1176449.
25. Feil, R., & Fraga, M. F. (2012). Epigenetics and the environment: Emerging patterns and implications. *Nature Reviews Genetics, 2012, 13*(2), 97–109. DOI: 10.1038/nrg3142.
26. Kumari, P., Khan, S., Wani, I. A., Gupta, R., Verma, S., Alam, P., & Alaklabi, A. (2022). Unravelling the role of epigenetic modifications in development and reproduction of angiosperms: A critical appraisal. *Frontiers in Genetics, 13*, 819941.
27. Berger, S. L., Kouzarides, T., Shiekhattar, R., & Shilatifard, A. An operational definition of epigenetics. *Genes & Developments, 23*, 781.
28. Mirbahai, L., & Chipman, J. K. (2014). Epigenetic memory of environmental organisms: A reflection of lifetime stressor exposures. *Mutation Research—Genetic Toxicology and Environmental Mutagenesis,* 764–765, 10–17. DOI: 10.1016/j.mrgentox.2013.10.003.
29. Al Aboud, N. M., Tupper, C., & Jialal, I. (2023). Genetics, epigenetic mechanism. In *StatPearls*. StatPearls Publishing LLC.
30. Jirtle, R. L., & Skinner, M. K. (2007). Environmental epigenomics and disease susceptibility. *Nature Reviews Genetics, 8*, 253.
31. Urdinguio, R. G., Sanchez-Mut, J. V., & Esteller, M. (2009). Epigenetic mechanisms in neurological diseases: Genes, syndromes, and therapies. *Lancet Neurology, 8*, 1056–1072. DOI: 10.1016/S1474-4422(09)70262-5
32. Heyn, H., Li, N., Ferreira, H. J., Moran, S., Pisano, D. G., Gomez, A., et al. (2012). Distinct DNA methylomes of newborns and centenarians. *Proceedings of the. National Academy of Sciences USA, 109*, 10522–10527. DOI: 10.1073/pnas.1120658109
33. Mastroeni, D., Grover, A., Delvaux, E., Whiteside, C., Coleman, P. D., & Rogers, J. (2010). Epigenetic changes in Alzheimer's disease: Decrements in DNA methylation. *Neurobiology of Aging, 31*(12), 2025–2037. DOI: 10.1016/j.neurobiolaging.2008.12.005.
34. Moore, L. D., Le, T., & Fan, G. (2013). Methylation and its basic function. *Neuropsychopharmacology, 38*, 23–38.
35. Zampieri, M., Ciccarone, F., Calabrese, R., Franceschi, C., Burkle, A., & Caiafa, P. (2015). Reconfiguration of DNA methylation in aging. *Mechanisms of Ageing and Development, 151*, 60–70.

36. Cheng, Y., Bernstein, A., Chen, D., & Jin, P. (2015). 5-Hydroxymethylcytosine: A new player in brain disorders? *Experimental Neurology, 268*, 3–9. DOI: 10.1016/j.expneurol.2014.05.008.
37. Vinson, C., & Chatterjee, R. (2012). CG methylation. *Epigenomics, 4*, 655–663. DOI: 10.2217/epi.12.55
38. Rhee, I., Bachman, K. E., Park, B. H., Jair, K. W., Yen, R. W., Schuebel, K. E., et al. (2002). DNMT1 and DNMT3b cooperate to silence genes in human cancer cells. *Nature, 416*, 552–556. DOI: 10.1038/416552a.
39. West, R. L., Lee, J. M., & Maroun, L. E. (1995). Hypomethylation of the amyloid precursor protein gene in the brain of an Alzheimer's disease patient. *Journal of Molecular Neuroscience, 6*(2), 141–146. DOI: 10.1007/BF02736773.
40. Eid, A., Bihaqi, S. W., Renehan, W. E., & Zawia, N. H. (2016). Developmental lead exposure and lifespan alterations in epigenetic regulators and their correspondence to biomarkers of Alzheimer's disease. *Alzheimers & Dementia, 2*, 123–131. DOI: 10.1016/j.dadm.2016.02.002.
41. Wu, J., Basha, M. R., Brock, B., Cox, D. P., Cardozo-Pelaez, F., McPherson, C. A., Harry, J., Rice, D. C., Maloney, B., Chen, D., Lahiri, D. K., & Zawia, N. H. (2008). Alzheimer's disease (AD)-like pathology in aged monkeys after infantile exposure to environmental metal lead (Pb): Evidence for a developmental origin and environmental link for AD. *Journal of Neuroscience, 28*(1), 3–9. DOI: 10.1523/JNEUROSCI.4405-07.2008.
42. Bradley-Whitman, M. A., & Lovell, M. A. (2013). Epigenetic changes in the progression of Alzheimer's disease. *Mechanisms of Ageing and Development, 134*(10), 486–495. DOI: 10.1016/j.mad.2013.08.005.
43. Ellison, E. M., Abner, E. L., & Lovell, M. A. (2017). Multiregional analysis of global 5-methylcytosine and 5-hydroxymethylcytosine throughout the progression of Alzheimer's disease. *Journal of Neurochemistry, 140*(3), 383–394. DOI: 10.1111/jnc.13912.
44. Chouliaras, L., Mastroeni, D., Delvaux, E., Grover, A., Kenis, G., Hof, P. R., Steinbusch, H. W., Coleman, P. D., Rutten, B. P., & van den Hove, D. L. (2013). Consistent decrease in global DNA methylation and hydroxymethylation in the hippocampus of Alzheimer's disease patients. *Neurobiology of Aging 34*(9), 2091–2099. DOI: 10.1016/j.neurobiolaging.2013.02.021.
45. Condliffe, D., Wong, A., Troakes, C., Proitsi, P., Patel, Y., Chouliaras, L., Fernandes, C., Cooper, J., Lovestone, S., Schalkwyk, L., Mill, J., & Lunnon, K. (2014). Cross-region reduction in 5-hydroxymethylcytosine in Alzheimer's disease brain. *Neurobiology of Aging, 35*(8), 1850–1854. DOI: 10.1016/j.neurobiolaging.2014.02.002.
46. Barrachina, M., & Ferrer, I. (2009). DNA methylation of Alzheimer disease and tauopathy-related genes in postmortem brain. *Journal of Neuropathology & Experimental Neurology, 68*(8), 880–891. DOI: 10.1097/NEN.0b013e3181af2e46.
47. Wang, S. C., Oelze, B., & Schumacher, A. (2008). Age-specific epigenetic drift in late-onset Alzheimer's disease. *PLoS One, 3*(7), e2698. DOI: 10.1371/journal.pone.0002698.
48. Fuso, A., Seminara, L., Cavallaro, R. A., D'Anselmi, F., & Scarpa, S. (2005). S-adenosylmethionine/homocysteine cycle alterations modify DNA methylation status with consequent deregulation of PS1 and BACE and beta-amyloid

production. *Molecular and Cellular Neuroscience, 28*(1), 195–204. DOI: 10.1016/j.mcn.2004.09.007.

49. Cao, X. J., Huang, S. H., Wang, M., Chen, J. T., & Ruan, D. Y. (2008). S-adenosyl-L-methionine improves impaired hippocampal long-term potentiation and water maze performance induced by developmental lead exposure in rats. *European Journal of Pharmacology, 595*(1–3), 30–34. DOI: 10.1016/j.ejphar.2008.07.061.
50. Kumar, D., Aggarwal, M., Kaas, G. A., Lewis, J., Wang, J., Ross, D. L., Zhong, C., Kennedy, A., Song, H., & Sweatt, J. D. (2015). Tet1 oxidase regulates neuronal gene transcription, active DNA hydroxy-methylation, object location memory, and threat recognition memory. *Neuroepigenetics, 4*, 12–27. DOI: 10.1016/j.nepig.2015.10.002.. [Ref list]
51. Corder, E. H., Saunders, A. M., Strittmatter. W. J., Schmechel. D. E., Gaskell. P. C., Small, G. W., Roses. A. D., Haines, J. L., & Pericak-Vance. M. A. (1993). Gene dose of apolipoprotein E type 4 allele and the risk of Alzheimer's disease in late onset families. *Science, 261*(5123), 921–923. DOI: 10.1126/science.8346443.
52. Iwata, A., Nagata. K., Hatsuta, H., Takuma, H., Bundo, M., Iwamoto, K., Tamaoka, A., Murayama. S., Saido, T., & Tsuji, S. (2014). Altered CpG methylation in sporadic Alzheimer's disease is associated with APP and MAPT dysregulation. *Human Molecular Genetics, 23*(3), 648–656. DOI: 10.1093/hmg/ddt451.
53. Hahn, M. A., Szabo, P. E., & Pfeifer, G. P. (2014). 5-Hydroxymethylcytosine: A stable or transient DNA modification? *Genomics, 104*, 314–323. DOI: 10.1016/j.ygeno.2014.08.015. [Ref list]
54. Ziller, M. J., Muller, F., Liao, J., Zhang, Y., Gu, H., Bock, C., et al. (2011). Genomic distribution and inter-sample variation of non-CpG methylation across human cell types. *PLoS Genetics, 7*, e1002389. DOI: 10.1371/journal.pgen.1002389
55. Guo, J. U., Su, Y., Shin, J. H., Shin, J., Li, H., Xie, B., et al. (2014). Distribution, recognition and regulation of non-CpG methylation in the adult mammalian brain. *Nature Neuroscience, 17*, 215–222. DOI: 10.1038/nn.3607
56. Kouskouti, A., & Talianidis, I. (2005). Histone modifications defining active genes persist after transcriptional and mitotic inactivation. *EMBO J., 24*(2), 347–357. DOI: 10.1038/sj.emboj.7600516.
57. Clapier, C. R., & Cairns, B. R. (2009). The biology of chromatin remodeling complexes. *Annual Review of Biochemistry, 78*, 273–304.
58. Kouzarides, T. Chromatin modifications and their function. *Cell, 128*, 693–705.
59. Plagg, B., Ehrlich, D., Kniewallner, K. M., Marksteiner, J., & Humpel, C. (2015). Increased acetylation of histone H4 at Lysine 12 (H4K12) in monocytes of transgenic alzheimer's mice and in human patients. *Current Alzheimer Research, 12*(8), 752–760. DOI: 10.2174/1567205012666150710114256.
60. Strahl, B. D., & Allis, C. D. (2000). The language of covalent histone modifications. *Nature, 403*, 41–45.
61. Wen, K. X., Miliç, J., El-Khodor, B., Dhana, K., Nano, J., Pulido, T., Kraja, B., Zaciragic, A., Bramer, W. M., Troup, J., Chowdhury, R., Ikram, M. A., Dehghan, A., Muka, T., & Franco, O. H. (2016). The role of DNA methylation and histone modifications in neurodegenerative diseases: A systematic review. *PLoS One, 11*(12), e0167201. DOI: 10.1371/journal.pone.0167201.

62. Fedorova, E., & Zink, D. (2008). Nuclear architecture and gene regulation. *Biochimica et Biophysica Acta: Molecular Cell Research, 1783*, 2174–2184.
63. Narayan, P. J., Lill, C., Faull, R., Curtis, M. A., & Dragunow, M. (2015). Increased acetyl and total histone levels in post-mortem Alzheimer's disease brain. *Neurobiology of Disease, 74*, 281–294. DOI: 10.1016/j.nbd.2014.11.023.
64. Zhang, K., Schrag, M., Crofton, A., Trivedi, R., Vinters, H., & Kirsch, W. (2012). Targeted proteomics for quantification of histone acetylation in Alzheimer's disease. *Proteomics, 12*(8), 1261–1268. DOI: 10.1002/pmic.201200010.
65. Wang, Z., Zang, C., Rosenfeld, J. A., Schones, D. E., Barski, A., Cuddapah, S., et al. (2008). Combinatorial patterns of histone acetylations and methylations in the human genome. *Nature Genetics, 40*, 897–903. DOI: 10.1038/ng.154.
66. Yang, X. J., & Seto, E. (2007). HATs and HDACs: From structure, function and regulation to novel strategies for therapy and prevention. *Oncogene, 26*, 5310–5318. DOI: 10.1038/sj.onc.1210599.
67. Bannister, A. J., & Kouzarides, T. (2011). Regulation of chromatin by histone modifications. *Cell Research, 21*, 381–395. DOI: 10.1038/cr.2011.22
68. Suzuki, M., Yamada, T., Kihara-Negishi, F., Sakurai, T., & Oikawa, T. (2003). Direct association between PU.1 and MeCP2 that recruits mSin3A-HDAC complex for PU 1-mediated transcriptional repression. *Oncogene, 22*, 8688–8698. DOI: 10.1038/sj.onc.1207182.
69. Rottach, A., Frauer, C., Pichler, G., Bonapace, I. M., Spada, F., & Leonhardt, H. (2010). The multi-domain protein Np95 connects DNA methylation and histone modification. *Nucleic Acids Research, 38*, 1796–1804. DOI: 10.1093/nar/gkp1152
70. Sung, Y. M., Lee, T., Yoon, H., DiBattista, A. M., Song, J. M., Sohn, Y., Moffat, E. I., Turner, R. S., Jung, M., Kim, J., & Hoe, H. S. (2013). Mercaptoacetamide-based class II HDAC inhibitor lowers Aβ levels and improves learning and memory in a mouse model of Alzheimer's disease. *Experimental Neurology, 239*, 192–201. DOI: 10.1016/j.expneurol.2012.10.005.
71. Coppedè, F. (2014). The potential of epigenetic therapies in neurodegenerative diseases. *Frontiers in Genetics, 5*, 220. DOI: 10.3389/fgene.2014.00220.
72. Vanyushin, B. F., Nemirovsky, L. E., Klimenko, V. V., Vasiliev, V. K., Belozersky, A. N. (1973). The 5-methylcytosine in DNA of rats. Tissue and age specificity and the changes induced by hydrocortisone and other agents. *Gerontologia, 19*(3), 138–152. DOI: 10.1159/000211967.
73. Wilson, V. L., Smith, R. A., Ma, S., & Cutler, R. G. (1987). Genomic 5-methyldeoxycytidine decreases with age. *Journal of Biological Chemistry, 262*, 9948–9951.
74. Christopher, M. A., Kyle, S. M., & Katz, D. J. (2017). Neuroepigenetic mechanisms in disease. *Epigenetics & Chromatin, 10*(1), 47. DOI: 10.1186/s13072-017-0150-4.
75. Irier, H. A., & Jin, P. (2012). Dynamics of DNA methylation in aging and Alzheimer's disease. *DNA and Cell Biology, 31*(Suppl 1), S42-S48. DOI: 10.1089/dna.2011.1565.
76. Somel, M., Guo, S., Fu, N., Yan, Z., Hu, H. Y., Xu, Y., et al. (2010). MicroRNA, mRNA, and protein expression link development and aging in human and macaque brain. *Genome Research, 20*, 1207–1218.

77. Nassar, A., Kodi, T., Satarker, S., Chowdari Gurram, P., Upadhya, D., SM, F., Mudgal, J., Nampoothiri, M. (2022). Astrocytic MicroRNAs and transcription factors in Alzheimer's disease and therapeutic interventions. *Cells, 11*, 4111.
78. Morris, K. V., & Mattick, J. S. (2014). The rise of regulatory RNA. *Nature Reviews Genetics, 15*, 423–437. DOI: 10.1038/nrg3722
79. Dahariya, S., Paddibhatla, I., Kumar, S., Raghuwanshi, S., Pallepati, A., & Gutti, R. K. (2019). Long non-coding RNA: Classification, biogenesis and functions in blood cells. *Molecular Immunology, 112*, 82–92. DOI: 10.1016/j.molimm.2019.04.011
80. Chen, Y., Li, Z., Chen, X., & Zhang, S. (2021). Long non-coding RNAs: From disease code to drug role. *Acta Pharmaceutica Sinica B, 11*, 340–354. DOI: 10.1016/j.apsb.2020.10.001
81. Marchese, F. P., Raimondi, I., & Huarte, M. (2017). The multidimensional mechanisms of long noncoding RNA function. *Genome Biology, 18*, 206. DOI: 10.1186/s13059-017-1348-2
82. Fernandes, J. C. R., Acuña, S. M., Aoki, J. I., Floeter-Winter, L. M., & Muxel, S. M. (2019). Long non-coding RNAs in the regulation of gene expression: Physiology and disease. *Noncoding RNA, 5*, 17. DOI: 10.3390/ncrna5010017
83. Maoz, R., Garfinkel, B. P., & Soreq, H. (2017). Alzheimer's disease and ncRNAs. *Advances in Experimental Medicine and Biology, 978*, 337–361. DOI: 10.1007/978-3-319-53889-1_18.
84. Peschansky, V. J., Wahlestedt, C. (2014). Non-coding RNAs as direct and indirect modulators of epigenetic regulation. *Epigenetics, 9*(1), 3–12. DOI: 10.4161/epi.27473.
85. Hébert, S. S., Horré, K., Nicolaï, L., Papadopoulou, A. S., Mandemakers, W., Silahtaroglu, A. N., Kauppinen, S., Delacourte, A., & De Strooper, B. (2008). Loss of microRNA cluster miR-29a/b-1 in sporadic Alzheimer's disease correlates with increased BACE1/beta-secretase expression. *Proceedings of the National Academy of Sciences USA, 105*(17), 6415–6420. DOI: 10.1073/pnas.0710263105.
86. Faghihi, M. A., Modarresi, F., Khalil, A. M., Wood, D. E., Sahagan, B. G., Morgan, T. E., et al. (2008). Expression of a noncoding RNA is elevated in Alzheimer's disease and drives rapid feed-forward regulation of beta-secretase. *Nature Medicine, 14*, 723–730. DOI: 10.1038/nm1784
87. Lin, D., Pestova, T. V., Hellen, C. U., & Tiedge, H. (2008). Translational control by a small RNA: Dendritic BC1 RNA targets the eukaryotic initiation factor 4A helicase mechanism. *Molecular and Cellular Biology, 28*, 3008–3019. DOI: 10.1128/MCB.01800-07
88. Higaki, S., Muramatsu, M., Matsuda, A., Matsumoto, K., Satoh, J., Michikawa, M., et al. (2018). Defensive effect of microRNA-200b/c against amyloid-beta peptide-induced toxicity in Alzheimer's disease models. *PLoS One, 13*, e0196929. DOI: 10.1371/journal.pone.0196929
89. Li, J., Xu, C., Zhang, J., Jin, C., Shi, X., Zhang, C., et al. (2021). Identification of miRNA-target gene pairs in the parietal and frontal lobes of the brain in patients with Alzheimer's disease using bioinformatic analyses. *Neurochemical Research, 46*, 964–979. DOI: 10.1007/s11064-020-03215-8

90. Lee, K., Kim, H., An, K., Kwon, O., Park, S., Cha, J. H., et al. (2016). Replenishment of microRNA-188-5p restores the synaptic and cognitive deficits in 5XFAD mouse model of Alzheimer's disease. *Scientific Reports, 6*, 34433. DOI: 10.1038/srep34433
91. Zolochevska, O., & Taglialatela, G. (2020). Selected microRNAs increase synaptic resilience to the damaging binding of the Alzheimer's disease amyloid beta oligomers. *Molecular Neurobiology, 57*, 2232–2243. DOI: 10.1007/s12035-020-01868-8
92. Wu, H. Z. Y., Thalamuthu, A., Fowler, C., Masters, C. L., Sachdev, P., et al. (2020). Differential blood miRNA expression in brain amyloid imaging-defined Alzheimer's disease and controls. *Alzheimers Res. Ther.* 12:59. DOI: 10.1186/s13195-020-00627-0
93. Sabry, R., El Sharkawy, R. E., & Gad, N. M. (2020). MiRNA -483-5p as a potential noninvasive biomarker for early detection of Alzheimer's disease. *Egyptian Journal of Immunology, 27*, 59–72.
94. Hou, Y., Chen, H., He, Q., Jiang, W., Luo, T., Duan, J., et al. (2013). Changes in methylation patterns of multiple genes from peripheral blood leucocytes of Alzheimer's disease patients. *Acta Neuropsychiatrica, 25*, 66–76. 10.1111/j.1601-5215.2012.00662.x
95. Salcedo-Tacuma, D., Melgarejo, J. D., Mahecha, M. F., Ortega-Rojas, J., Arboleda-Bustos, C. E., Pardo-Turriago, R., et al. (2019). Differential methylation levels in CpGs of the BIN1 gene in individuals with Alzheimer disease. *Alzheimer Disease & Associated Disorders, 33*, 321–326. DOI: 10.1097/WAD.0000000000000329
96. Monti, N., Cavallaro. R. A., Stoccoro, A., Nicolia, V., Scarpa, S., Kovacs, G. G., et al. (2020). CpG and non-CpG Presenilin1 methylation pattern in course of neurodevelopment and neurodegeneration is associated with gene expression in human and murine brain. *Epigenetics, 15*, 781–99. DOI: 10.1080/15592294.2020.1722917
97. Nagata, T., Kobayashi, N., Ishii, J., Shinagawa, S., Nakayama, R., Shibata, N., et al. Association between DNA methylation of the BDNF promoter region and clinical presentation in Alzheimer's disease. *Dementia and Geriatric Cognitive Disorders Extra, 5*, 64–73. DOI: 10.1159/000375367
98. Chang, L., Wang, Y., Ji, H., Dai, D., Xu, X., Jiang, D., et al. (2014). Elevation of peripheral BDNF promoter methylation links to the risk of Alzheimer's disease. *PLoS ONE, 9*, e110773. DOI: 10.1371/journal.pone.0110773
99. Phipps, A. J., Vickers, J. C., Taberlay, P. C., & Woodhouse, A. (2016). Neurofilament-labeled pyramidal neurons and astrocytes are deficient in DNA methylation marks in Alzheimer's disease. *Neurobiology of Aging, 45*, 30–42. DOI: 10.1016/j.neurobiolaging.2016.05.003.

CHAPTER 3

Neuroinflammation and Role of Peripheral Immune Cells in Alzheimer's Disease

SHWETA SHRIVASTAVA[1*], SANTOSH KUMAR SINGH[1], and MANISH KUMAR JEENGAR[2]

[1]*School of Pharmacy, School of Health and Allied Sciences, ARKA JAIN University, Gamharia, Seraikela Kharsawan, Jharkhand, India*

[2]*Department of Pharmacology, Amrita School of Pharmacy, Amrita Vishwa Vidyapeetham, AIMS Health Sciences Campus, Kochi, Kerala, India*

*Corresponding author

ABSTRACT

Neuroinflammation and peripheral immune cells play significant roles in Alzheimer's disease (AD). Neuroinflammation is a form of inflammatory reaction occurring in the brain, involving various immune cells and molecules. In AD, chronic neuroinflammation contributes to neuronal damage and cognitive decline. Immune cells like macrophages and T cells, which are usually found outside central nervous system, can enter brain during neuroinflammation. Microglia, the brain's resident immune cells, also become activated and contribute to the inflammatory reaction seen in AD. The role of peripheral immune cells in AD is multifaceted and intricate. While peripheral immune cells can exacerbate neuroinflammation and contribute to AD pathology through the release of pro-inflammatory cytokines and other molecules, it is essential to recognize that their role in AD is not solely detrimental. Some regulatory T cells may have protective

Cellular and Molecular Mechanisms in Alzheimer's Disease Therapeutics, Volume 1: Cellular Mechanisms. Prashant Tiwari, Sunil Kumar Kadiri, and Kunjbihari Sulakhiya (Eds.)

DOI: 10.1201/9781779641793-3

effects by modulating inflammation and supporting neuronal health as they are immune cells. This chapter focuses on the crosstalk between peripheral immune cells and the brain in the context of systemic inflammation represents a significant avenue of research in understanding the pathogenesis of AD and identifying potential therapeutic targets for intervention.

3.1 INTRODUCTION

Alzheimer's disease (AD) manifests as a neurodegenerative condition marked by initial extracellular accumulation of diffuse and neuritic plaques, comprised of amyloid-β peptides, form initially. This is followed by the development of intracellular neurofibrillary tangles, which are made up of hyperphosphorylated tau protein. Although these aggregates serve as the primary pathological hallmarks of AD, it is often marked by a number of other pathophysiological alterations and processes, including neuroinflammation, metabolic dysregulation, and synaptic dysfunction [1]. Considerable investigation has been undertaken to acquire a deeper understanding of the pathogenesis of the disease. However, a complete understanding of the cause-and-effect relationships underlying the complex biological processes associated with Alzheimer's disease remains elusive. The absence of clear information presents considerable obstacles in the development of effective treatments to managing the AD [2, 3].

Neuroinflammation, driven by both central and peripheral immune cells, plays a pivotal role in the advancement of AD. This triggers the secretion of proinflammatory cytokines like interleukin-1β (IL-1β), IL-6, IL-8, and tumor necrosis factor (TNF), as well as chemokines, complement cytokines, and small molecular messengers such as prostaglandins, nitric oxide (NO), and reactive oxygen species (ROS) [4]. In the brain, acute inflammation serves as a protective response against potential harm to the nervous system, such as infection or other external factors. Over time, the initial inflammation can transition into a chronic state, which then activates the nervous system. Continued inflammation can lead to the release of inflammatory cytokines, which triggers a consistently stronger proinflammatory response compared with the anti-inflammatory response [1, 5]. This exacerbates the inflammation, leading to neuronal damage in AD (Figure 3.1).

This chapter offers a detailed overview of the immune system's involvement, both peripheral and central, which are involved in AD. It also explores their interactions and discusses their potential. These factors have significant implications for gaining a deeper understanding of the development and progression of AD.

3.2 THE CENTRAL NERVOUS SYSTEM

3.2.1 MICROGLIA

Microglia, comprising around 10% of the cellular composition of the nervous system, derive from macrophages and function as the immune cells native to the brain. Microglia, specialist brain cells, promote neurogenesis, axonal growth, synaptic refinement, phagocytosis of neurons undergoing apoptosis, vascular organization, immune monitoring, and response to injury. They have pattern recognition receptors (PPRs) that identify both exogenous pathogen-associated molecular patterns (PAMPs) linked with pathogens and internally derived danger-associated molecular patterns (DAMPs). In this function, cytokines and chemokines are secreted, drawing other cells to uphold brain equilibrium [6, 7].

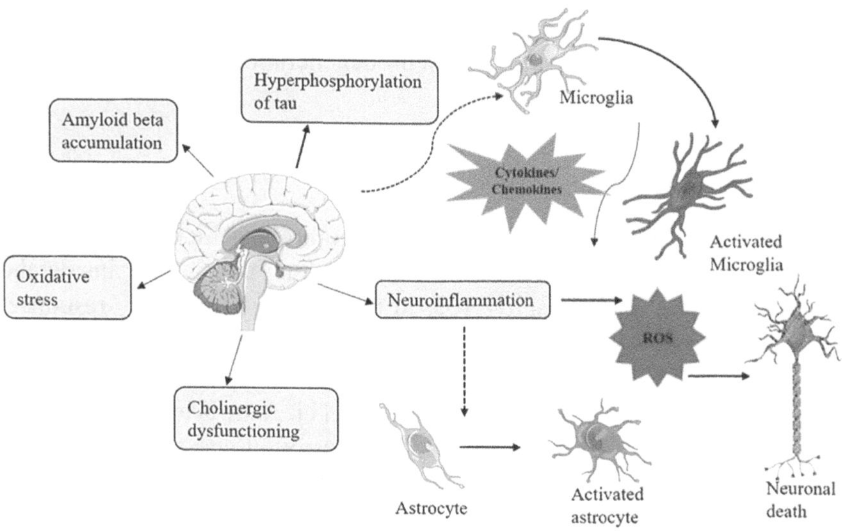

FIGURE 3.1 Neuroinflammation in AD.

In AD, the microglia possess the capability to recognize and attach to Aβ, utilizing several receptors, resulting in their activation. When microglia is stimulated, it enters the activated state, known as M1 microglia phenotype.

This state is characterized by changes in the cytoskeleton, modifications in the release of molecular mediators, and heightened proliferative reactions. This state activation acts as a crucial step in defending and maintaining the brain's equilibrium by facilitating interactions between cytotoxic and neuroprotective factors [8].

The M1 microglial phenotype, also known as the "pro-inflammatory" or "activated" phenotype, is capable of releasing proinflammatory cytokines such as TNF-α, IL-1β, IL-12, and nitric oxide. It can decrease the generation of neurotrophic factors, resulting in heightened inflammation and cytotoxicity.

Alternatively, the M2 phenotype microglia, referred to as the anti-inflammatory phenotype, releases anti-inflammatory cytokines like IL-4, IL-10, and IL-13. As a consequence, there is an increase in the production of neurotrophic factors like brain-derived neurotrophic factor (BDNF) and TGF-β, as well as various signals that are involved in protection and repair processes, while simultaneously reducing inflammatory responses. Neuronal injury during its progression facilitates the subsequent activation of microglia that leads to the release of further proinflammatory cytokines. This persistent neuroinflammatory response that perpetuates itself is frequently deleterious, leading to neurodegeneration and impairment of brain function [9, 10].

3.2.2 ASTROCYTES

Astrocytes are the primary glial cells in the brain, which play a crucial role as essential components of the innate immune system inside the CNS. They also play an active role in the neuroinflammatory response. Astrocytes are a main component of the BBB, astrocytes establish these cells have extensive contact with brain microvessels, providing nearly complete coverage through their processes [11]. Astrocytes in AD have a close relationship with Aβ deposits. They surround and sometimes even penetrate the plaques to protect the nearby healthy tissue. Astrocytes surrounding senile plaques exhibit significant hypertrophy and reactivity, with a notable increase in the expression of the glial fibrillary acidic protein (GFAP). Reactive astrocytes participate in clearing Aβ through the blood–brain barrier (BBB) via Aquaporin 4 (AQP4) and degradation processes. When exposed to Aβ, astrocytes become activated and release various proinflammatory cytokines like IFN-γ, IL-1β, IL6, TNFα, and TGF-β.

They also increase the generation of ROS and NO. These substances have both autocrine and paracrine effects, influencing the activating microglia and contributing to the continued reactive gliosis. Prolonged enhancement of astrocytes causes atrophy and degeneration, leading to decreased coverage of cerebral microvessels by end feet, and alterations in the positioning of AQP4 around blood vessels. The accumulation of Aβ in the brain is caused by the decreased removal of Aβ through the BBB [12, 13].

3.2.3 OLIGODENDROCYTES

Mature oligodendrocytes (OLs) are extensively distributed throughout both the gray and white matter of the CNS. They are responsible for forming myelin sheaths around axons, offering metabolic support to neurons. Recent studies indicate that neurodegenerative diseases are marked by the depletion of oligodendrocytes and demyelination. As a reparative mechanism, this prompts an elevation in the rate of proliferation and differentiation of oligodendrocyte precursor cells (OPCs), resulting in the generation of new OLs. Recent research has unveiled that OLs may possess immune-related capabilities, as evidenced by the expression of diverse immunomodulatory substances, including cytokines/chemokines and their corresponding receptors [14, 15].

3.2.4 ENDOTHELIAL CELLS AND PERICYTES

The neurovascular unit (NVU) forms complex structural and functional connections among neurons, glial cells (including microglia, astrocytes, and oligodendrocytes), and vascular cells (such as pericytes and endothelial cells). The NVU is pivotal in both developing and maintaining the BBB, while controlling ion levels and aiding in the movement of nutrients. Endothelial cells are pivotal components of the NVU, playing a crucial role in its function. Brain endothelial cells display elevated levels of connexins expressed at their intercellular junctions when compared with endothelial cells found in non-neural tissues [16–18]. This characteristic grants them barrier restriction properties, effectively limiting paracellular permeability. Throughout the inflammatory process, peripheral cytokines interact with endothelial cells, resulting in compromised barrier function. Additionally, when endothelial cells become inflamed, they increase the production of adhesion molecules. These molecules help in attracting immune cells and antibodies from the bloodstream, allowing them to cross the barrier. A recent study has discovered a novel pathway through which microvascular endothelial cells can break down myelin debris. This process occurs through the autophagy-lysosome system and has been found to contribute to the development of demyelinating conditions by triggering angiogenesis, inflammation, and the formation of fibrotic scars [19–21]. Pericytes are microvascular wall cells located within basement membrane that encircles the microvasculature. As previously stated, pericytes serve as vital cellular constituents within the neurovascular unit. Pericytes possess the capability to interact with a variety of cell types, including endothelial cells, glial cells, neuronal cells, and perivascular macrophages. These interactions play a crucial role in carrying out their important functions. Pericytes have been found to play a role in both neuro-inflammatory and neurodegenerative diseases in recent years. Pericyte-derived pleiotrophin is believed to play a crucial role as a neurotrophic factor in ensuring the survival of neurons. The absence of pericytes results in the disruption of the BBB, impairment of blood circulation, and neuronal loss in the murine brain [22–25].

3.2.5 BLOOD-BRAIN BARRIER (BBB)

The BBB is established through a delicate network involving brain endothelial cells, astrocytes, pericytes, and various extracellular matrix components. These components, including tight junctions, adherens junctions, and gap junctions, work together to separate the blood from the CNS. BBB is a complex system that controls the exchange of information between neural tissues and the blood and its components. The CNS relies on a specialized barrier to regulate the movement of substances, maintaining neural balance and stability [26, 27].

3.2.5.1 ALTERATION OF THE CENTRAL NERVOUS SYSTEM BARRIERS IN AD

Aβ buildup is a crucial trigger in AD, disrupting the BBB and increasing vascular permeability. This allows toxic molecules to enter or hinders their clearance, triggering inflammatory and immune responses. Chronically activated glial cells release proinflammatory mediators that compromise BBB integrity, leading to neuroinflammation and dysfunction that contribute to AD progression. Reduction in pericytes and vulnerability of oligodendrocytes to neuroinflammation and Aβ further amplify AD-related damage. Additionally, oligodendrocyte progenitor cells may contribute to Aβ secretion [28]. The BBB normally separates the CNS from the immune system in the periphery, but various factors can disrupt this barrier. Proinflammatory molecules like cytokines, chemokines, NO, ROS, and metalloproteinases (MMPs) can lead to change in BBB permeability, compromising the brain's immune protection. Specifically, cerebrovascular endothelial cells, proinflammatory cytokines, and oxidative stress impact tight junctions to preserve barrier integrity, while matrix MMPs break down the endothelial basal lamina, essential for BBB integrity. Furthermore, cerebral amyloid angiopathy (CAA), characterized by the accumulation of beta-amyloid (Aβ) in brain blood vessels, also contributes to BBB disruption. Blood vessels laden with amyloid, often accompanied by degenerated smooth muscle cells, pericytes, and endothelial cells, display diminished integrity, resulting in heightened BBB permeability [29–31].

3.3 PERIPHERAL IMMUNE CELLS

Peripheral immune cells act as crucial element in supporting functionally impaired glial cells within the CNS. Glial cells, encompassing microglia and astrocytes, actively participate in neuroinflammatory processes within the CNS. These responses are orchestrated by both resident CNS glial cells and peripheral immune cells capable of infiltrating the CNS, even breaching the intact BBB. The release of ATP during pathological conditions mediates tissue damage and repair as it directly impacts the integrity along with survival of glial cells. Additionally, the ligands of the glial cell line-derived neurotrophic factor (GDNF) family are expressed not only in neuronal cells but also in glial and peripheral immune system cells, suggesting a potential role in neuroinflammation and neuroprotection [32, 33].

Studies indicate that glial cells have the capacity to modify the phenotypes of other glial cells and modulate neuroinflammatory responses, thereby impacting immune cell targets. Notably, microglia and other glial cells act as a pivotal element in orchestrating immune and damage-repair reactions following brain trauma, thus contributing to neuroinflammation. Furthermore, neurons actively engage in immune responses by regulating both glial cells and infiltrating T cells, underscoring the intricate neuroimmune communication within the CNS [34, 35].

Peripheral immune cells, including macrophages, leukocytes, and T lymphocytes, participate in both peripheral and central pain pathways, highlighting their involvement in regulating glial responses to pain. Activated immune cells and glial cells with immune-like properties have a substantial effect on neuronal function, highlighting the close interaction between the immune and nervous systems. Additionally, infiltrating peripheral immune cells release cytokines that modulate neuroinflammation and the state of glial cells during pathological conditions, emphasizing the cooperative relationship between peripheral immune cells and glial cells within the CNS [36, 37].

In summary, the interaction between peripheral immune cells and glial cells in the CNS plays a critical role in regulating neuroinflammatory responses, tissue repair, and overall CNS homeostasis. This complex interplay underscores the significance of comprehending the functions of both peripheral immune cells and glial cells in sustaining CNS function and addressing pathological conditions.

3.3.1 MONOCYTES/MACROPHAGES

Monocytes and macrophages, acting as peripheral immune cells, play a pivotal role in the neuroinflammatory processes associated with AD. The accumulation of Aβ peptides in the brain triggers a persistent and chronic inflammatory response, leading to hyperphosphorylation of tau, synaptotoxicity, and subsequent neuronal demise, thereby accelerating the gradual decline in cognitive function [38]. These tissues have the capability to infiltrate the CNS and modulate neuroinflammatory responses. Research indicates that the infiltration of peripheral monocytes/macrophages may mitigate the progression of AD pathology [39]. Various studies have elucidated the distinct roles played by microglia, monocytes, and cerebral macrophages in the course of Aβ-related diseases. Perry et al. have conducted a comprehensive analysis highlighting the significance of macrophages and microglia in the context of systemic inflammation and neurodegenerative conditions [40].

Monocytes originate from the monocyte–macrophage dendritic cell progenitor and represent a diverse population of partially differentiated cells that give rise to a heterogeneous mononuclear phagocyte lineage. Both human and mouse monocytes express various clusters of differentiation (CD). In humans, these include CD115, CD11c, CD14, and CD16, while in mice, they consist of CD115, CD11b, and Ly6C. Human and murine monocytes also exhibit varying levels of chemokine receptors, such as CX3CR1 and CCR2. Human monocytes can be classified into three subsets based on their levels of CD14 and CD16 expression: the classical subset (CD14++CD16−), the intermediate subset (CD14++CD16+), and the nonclassical subset (CD14+CD16++). Meanwhile, mouse monocytes can be categorized into two subsets based on chemokine receptors and Ly6C expression levels: the proinflammatory subset (CX3CR1 low CCR2+Ly6C high), which actively migrates to inflamed tissues and participates in inflammatory responses, and the anti-inflammatory subset (CX3CR1 high CCR2−Ly6C low), composed of resident patrolling monocytes that survey the blood vessel lumen and contribute to tissue repair [41–43].

Both monocyte subsets contribute to interactions with the brain in AD, but the anti-inflammatory monocyte subset appears to have a closer functional association with the BBB than the proinflammatory subset. However, the proinflammatory subset primarily engages with the BBB through the process of transmigration, which is essential for reaching the brain parenchyma. For instance, researchers have proved that anti-inflammatory monocytes play a crucial role in monitoring the health of blood vessels by surveilling the endothelium. Numerous reports have underscored the importance of these anti-inflammatory monocytes in AD [44, 45]. Recent research has revealed a reduction in nonclassical CD14+CD16++ monocytes in individuals with Alzheimer's disease compared with those with mild cognitive impairment or healthy individuals of the same age. These monocytes share similarities with mouse anti-inflammatory CX3CR1 high CCR2−Ly6C low monocytes. Moreover, our research group demonstrated the use of two-photon intravital imaging to observe how a specific subset of monocytes adheres to brain vasculature rich in Aβ. These monocytes efficiently remove microglia, internalize and transport Aβ microaggregates from the brain microvasculature into the bloodstream [46–48]. Progenitor cells derived from the bone marrow of Nr4a1−/− mice, a transcription factor involved in the differentiation of anti-inflammatory Ly6Clow monocytes in the bone marrow and their survival, were transplanted into APP/PS1 mice to investigate their contribution to this finding. Notably, the targeted reduction of the anti-inflammatory monocyte subset in APP/PS1 mice led to an accumulation of Aβ within the brain vasculature, resulting in an overall increase in brain Aβ levels and exacerbating cognitive impairment in these mice [49, 50]. Collectively, these observations highlight the significant importance of the interaction between these cells and the brain vasculature in Alzheimer's disease. Overall, the interplay between peripheral immune cells, particularly monocytes and macrophages, and the CNS immune response is critical in the development and progression of neuroinflammation in Alzheimer's disease. Understanding the role of these immune cells in modulating neuroinflammatory processes is essential for developing targeted therapeutic strategies for AD.

3.3.2 LYMPHOCYTES

The adaptive immune system comprises two primary cellular components crucial for immunological responses: T lymphocytes, originating in the thymus, and B lymphocytes, which are antibody-producing cells originating in the bone marrow. After maturation in primary lymphoid organs like the bone marrow and thymus, lymphocytes migrate to secondary lymphoid organs such as the spleen and lymph nodes. In these secondary organs, adaptive immune responses are initiated, influenced by signals from the innate immune system [51]. T cells play a pivotal role in combating infections by directly eliminating infected target cells through the cytotoxic action of CD^{8+} T cells or indirectly via CD^{4+} cells, the predominant form of T cells. CD^{4+} T helper (Th) cells, lacking phagocytic activity, activate other immune cells like Natural Killer cells or CD8 T cells to combat infections indirectly. Peptides presented by the MHC-II complex on macrophages, B cells, and dendritic cells can activate CD^{4+} T lymphocytes. CD^{4+} T-cell receptors (TCRs) bind to peptides presented in conjunction with class II major histocompatibility complexes (MHC-II), including HLA-DQ, HLA-DP, and HLA-DR. Class II MHC molecules' expression on antigen-presenting cells (APCs) is induced by innate immunological stimuli, such as T-cell receptor (TCR) ligands. Upon TCR-induced activation, T cells recognize the peptide–MHC complex on APCs, leading to rapid aggregation of TCR-associated molecules at the interface between

T cells and APCs, forming the "immunological synapse" [51–54]. Th cells are divided into four main subsets: Th1, Th2, Th17, and T-regulatory (Treg) cells, each with distinct functions and cytokine profiles. Th1 cells produce pro-inflammatory cytokines promoting macrophage and CD^{8+} cytotoxic T-cell activation to eliminate intracellular pathogens. Th2 cells produce anti-inflammatory cytokines facilitating B-cell activation and antibody production. Th17 cells release pro-inflammatory cytokines promoting neutrophil and macrophage activation and have implications in autoimmune diseases. Treg cells exhibit immunosuppressive properties, maintaining immune homeostasis and dampening immune responses [55, 56]. CD^{8+} T cells recognize peptide antigens presented via the major histocompatibility complex-I (MHC-I) and exert cytotoxic effects on target cells. Cells of the immune system, including T cells, B cells, dendritic cells, and natural killer (NK) cells, are present in various regions of the CNS, where they play protective or pathological roles depending on the microenvironment [57–59].

Maintaining homeostasis and supporting brain health in the CNS relies on a delicate balance between anti-inflammatory and pro-inflammatory activities. In certain pathological conditions, T cells migrate through the bloodstream and become activated, primarily through the recognition of their T-cell receptors (TCRs) facilitated by major histocompatibility complex (MHC) molecules or antigen-presenting cells (APCs). T-cell activation leads to increased proliferation and numbers through various mechanisms, including direct cell–cell contact, cytokine production, B-cell activation, induction of cell death, or modulation of innate immune cells. Astrocytes and microglia play roles in modulating T-cell activation and differentiation by releasing cytokines and molecules such as IL-1, IL-6, TNF-α, IL-10, and TGF-β, critical for regulatory T cells (Treg) and effector T cells (Teff) differentiation [60, 61].

Teff cells contribute to maintaining a pro-inflammatory microglial phenotype by secreting IFN-γ and IL-17 and releasing granzyme B. Conversely, Treg cells induce neuroprotective responses and maintain antigen-specific immune tolerance by suppressing effector responses against various antigens, including self-antigens. Treg cells suppress Teff activity and proliferation through mechanisms like secretion of immunosuppressive cytokines (TGF-β, IL-10, IL-35) and release of cytotoxic molecules (granzyme B, perforin 1) that induce apoptosis and cytotoxicity in Teff cells. Additionally, Treg cells enhance regenerative activities in tissues such as the kidney, retina, and skin. Depletion of Treg cells leads to increased tissue damage and impaired vascular repair [62–64].

If CD^{4+} and CD^{8+} T cells fail to detect their targets within the brain environment, they exit the CNS via the CNS lymphatic system to migrate to the deep cervical lymph nodes in the periphery, or a local effector immune reaction may be triggered. Neuroinflammatory conditions alter cell trafficking and the secretory capacity of the BBB due to interactions between immune cells and antigens, leading to excessive immune cell migration into brain tissue and exacerbating inflammation. While T lymphocytes are crucial for facilitating molecule movement, cell migration, and neuron functioning, they regulate brain homeostasis through immune signals and secretory molecules without breaching the blood–brain barrier when they fail to detect specific antigens [65, 66].

3.3.3 NEUTROPHILS

Neutrophils, the most abundant type of leukocytes, possess the capability to ensnare microorganisms using structures called neutrophil extracellular traps (NETs). These NETs are elongated formations composed of decondensed chromatin and antimicrobial/granular proteins, enabling neutrophils to effectively eradicate extracellular infections [54].

Limited research has delved into the involvement of neutrophils in AD. Scali et al. conducted a study that unveiled an increase in the CD11b integrin expression in peripheral blood neutrophils of individuals diagnosed with AD. A significant correlation was observed between disease severity and the elevated expression of CD11b in neutrophils, which facilitates their adhesion and migration [67]. In a separate investigation by Bawa et al., 2020 individuals with AD exhibited higher levels of reactive oxygen species in their peripheral blood neutrophils compared with a control group, indicating increased neutrophil activation during AD [67]. Neutrophils in the parenchyma displayed sustained mobility and clustered around amyloid beta (Aβ) plaques in mice with AD, as confirmed by Zenaro et al. in 2015 [68].

Additionally, the study observed heightened levels of neutrophils within the brain parenchyma of both 5XFAD mice and 3xTg-AD mice, another mouse model for AD. Neutrophils displayed increased adherence, attributed to the effects of Aβ on lymphocyte function-associated antigen-1

(LFA-1), a leukocyte integrin. Neutrophil infiltration into the brain parenchyma relied on LFA-1, as evidenced by AD animals with LFA-1 deficiency. Depletion of neutrophils or LFA-1 in brain homogenates of 3xTg-AD mice resulted in enhanced cognitive function, reduced micro-gliosis, and decreased levels of Aβ1–42 [69–71]. Moreover, neutrophil depletion led to decreased levels of phosphorylated tau in AD animals. This study also uncovered elevated neutrophil levels in the cerebral blood vessels and brain parenchyma of individuals with AD compared to age-matched controls, suggesting a potential involvement of neutrophils in AD pathogenesis.

3.3.4 NATURAL KILLER (NK) CELLS

Recent research indicates that Natural Killer (NK) cells play a crucial role in monitoring the decline of cellular immunity associated with aging. As individuals grow older, the reduced function of NK cells becomes associated with an increased vulnerability to infections, the development of cancer, inflammatory conditions, and the buildup of senescent cells. NK cells have the ability to eradicate α-synuclein, alleviate inflammation caused by autologous active T cells, and remove dysfunctional neurons. Studies suggest that NK cells are essential for regulating and suppressing inflammation and abnormal protein aggregation in brain tissue [72]. They have the ability to impact adaptive immune responses as well as regulate immune cell harmony in humans. Additionally, they have the capability to generate cytokines, including IFN-γ, IL-3, IL-5, IL-10, and IL-12 and IL-13 [73].

3.3.5 DENDRITIC CELLS (DCs)

Dendritic cells (DCs) serve as the principal antigen-presenting cells (APCs). Within the central nervous system (CNS), DCs are present under normal conditions, expressing major histocompatibility complex class II (MHCII) and the leukocyte integrin CD11c. These markers are commonly utilized for the detection and localization of DCs within the brain. Numerous studies have demonstrated that during steady states, DCs are distributed throughout various CNS regions, including the meninges, choroid plexus, cerebrospinal fluid, and perivascular spaces.

Through analysis of cell morphology and surface markers, these cells can be identified. The high density of blood vessels in these areas suggests that cerebral DCs originate from the vascular system rather than being indigenous to the brain [74].

Anandasabapathy et al. (2011) reported that DCs present in the meninges and choroidal plexus likely originate from local precursor DCs that migrated into the perivascular region of the brain during early development. This indicates that precursor DCs are derived from the bone marrow instead of monocytes. Chemokines and adhesion molecules facilitate the migration of peripheral DCs to the meninges or choroid plexus, enabling them to encounter antigens and present them to T cells [75]. DCs play a critical role in T-cell activation by providing costimulatory signals, which trigger the activation and proliferation of CD^{8+} cytotoxic T cells (CTLs) and CD^{4+} helper T cells, leading to the production of various proinflammatory cytokines. Under specific circumstances, these proinflammatory effects may contribute to tissue damage. Recent studies have demonstrated that bone marrow-derived DCs can induce neuroprotective regulatory T cells (Tregs), suggesting their potential to confer neuroprotective effects by modulating adaptive immunity [76]. These findings underscore the significance of DCs in the regulation of inflammatory and neurodegenerative conditions.

3.4 THE INFLUENCE OF INFLAMMATORY RESPONSES IN BOTH THE CENTRAL AND PERIPHERAL REGIONS ON AD

The adaptive immune system's role in AD pathogenesis is still being understood, despite the innate immune system's involvement being well-established. Peripheral tissues may be a useful substrate for researching the impact of genetic variables on AD because research suggests that genetic factors linked to the illness are also present in these tissues [77].

The pathogenic mechanisms that contribute to the course of AD, such as the accumulation of Aβ, neuronal dysfunction, and cognitive impairment, are primarily driven by inflammatory processes. Early in the course of AD, glial cell activation may begin before Aβ buildup. Several investigations have revealed increased levels of inflammatory markers in postmortem brain tissues from human AD patients and animal models expressing amyloid precursor protein (APP), including interferon-γ

(IFN-γ), tumor necrosis factor-α (TNF-α), interleukin-1β (IL-1β), and IL-6. Particularly, IFN-γ and TNF-α impair neurons and inhibit the action of insulin-degrading enzymes, which are necessary for the clearance of Aβ. This implies a different mechanism by which inflammation makes Aβ buildup worse [78–80].

Moreover, studies have demonstrated that TNF-α and IFN-γ might increase the synthesis of Aβ and hinder microglial cells' capacity to eliminate it. Systemic and neuroinflammatory inflammation are important factors in the development and course of AD. Even though the brain has historically been thought of as immune-privileged, neuroinflammation can result from cellular injury or an aberrant buildup of inflammatory chemicals, triggering both the innate and adaptive immune systems to respond. Aging-related peripheral inflammation can affect cells in the central nervous system, resulting in oxidative stress and neuroinflammation, both of which exacerbate cognitive impairment in AD [63, 81–83].

When it comes to preserving brain homeostasis or shifting the scales in favor of neurodegeneration in AD, the complex interactions between the innate and adaptive immune systems are critical. Current research highlights the important roles neuroinflammation and the innate immune system play in AD pathogenesis. The peripheral immune system has historically had restricted access to the brain because of BBB, but the central immune system is made up of neurons, glial cells, and other immune cells. On the other hand, recent data point to the possibility that peripheral immune cells may enter the brain and cause neuroinflammation and neurodegeneration. Peripheral cytokines also have the ability to penetrate the blood–brain barrier, where they can directly cause neurotoxicity and activate glial cells [84, 85].

The term "neuroinflammation" describes the various inflammatory responses that arise in the central nervous system as a result of traumas, wounds, or illnesses. Since its first identification, neuroinflammation in AD has been thoroughly studied and is now generally acknowledged as a major factor in the illness's progression. In addition to the well-known clinical characteristics of AD, like the build-up of neurofibrillary tangles (NFTs) and amyloid-beta (Aβ) plaques, neuroinflammation is a key player in the disease's development. Glial cell activation results from neuroinflammatory responses brought on by the presence of Aβ and NFTs in the brain. Note that AD is not the only condition where neuroinflammation priming occurs. Different protein aggregates, referred to as danger-associated molecular patterns (DAMPs), include α-synuclein and TDP-43, and can also trigger comparable inflammatory reactions [58, 86–88].

Studies such as Tejera et al. (2019) have demonstrated the pivotal role that microglia, the principal innate immune cells in the brain, play in the development of AD, namely in the removal of amyloid-beta 89]. Furthermore, modifications to the peripheral immune system have been linked to the pathogenesis of AD and dementia, suggesting that the immune system plays a role in the disease process [90]. There is a substantial interaction between the peripheral and central inflammatory/immune responses in Alzheimer's disease. A common feature of neurodegenerative diseases is the increased susceptibility of particular brain regions to injury. This is probably due to the way neurons react to harmful stimuli, such as misfolded proteins associated with the disease, which eventually causes cell death [91].

According to recent data, the immune system may have a significant role in the onset and course of AD, in addition to the well-established functions of tau and Aβ proteins. Prolonged activation of immune cells within the CNS and increased infiltration of white blood cells across the BBB are linked to neuronal damage in AD. This is consistent with the discovery of a functioning lymphatic system inside the meninges and significant immune cell trafficking via the choroid plexus [29]. These findings suggest that immune cells in the central nervous system and the peripheral have a communication network.

In addition, the innate immune system is a vital line of defense against infections, facilitating the presentation of antigens and the exchange of information between adaptive T and B cells. The immune response's adaptive and innate components can both impact the neuroinflammatory cascade and the development of neurodegenerative diseases like AD and other disorders affecting the central nervous system. Research has indicated that the principal immune cells in the central nervous system, astrocytes and microglia, are deeply implicated in the abnormal molecular pathways linked to AD [10].

Crucial to the brain's innate immune system are microglial cells, which perform functions like tissue surveillance, pathogen removal, and injury response. They also aid in the formation of synapses and the preservation of neuronal life. Microglia display a ramified shape and have restricted antigen-presenting capacities while they are quiescent, mainly because major histocompatibility complex (MHC) molecules are not highly expressed on their surface. Microglia transform morphologically to resemble amoeboid structures and express more MHC and costimulatory molecules when activated, which aids in the presentation of antigens. Interactions with immune cells in the peripheral circulation are made possible by this change [39]. Neurons that are diseased, injured, or inflammed generate aberrant proteins or self-antigens that can activate resting microglia. Microglial activation occurs when harmful stimuli upset the delicate equilibrium between the neuroprotective and neurotoxic processes. One such trigger is Toll-like receptors 4 (TLR4) signaling pathway, which causes microglia to change into a pro-inflammatory type. Morphological changes and the release of pro-inflammatory molecules like IL-1β, IL-6, TNFα, IFN-γ, chemokines, and ROS/RNS are characteristics of this activated state. These chemicals exacerbate the local inflammatory response by facilitating peripheral leukocyte infiltration across the BBB [34, 6]. The major roles of cytokines in AD are summarized in Table 3.1.

Microglia can also activate through a different pathway that entails producing more neurotrophic factors and anti-inflammatory cytokines. This route facilitates neuronal survival and repair by encouraging the phagocytosis of cellular debris. Microglial mechanisms, however, can get caught in a destructive cycle if the inflammatory stimuli continue and the innate immune-related processes are not sufficiently addressed. Pro-inflammatory cytokines are continuously released during this cycle, which sets off a chain reaction of harmful events that finally results in neuronal death [8, 92].

The main immune surveillance regulators in the brain are called microglia. Nevertheless, under some pathological circumstances, external monocytes momentarily become part of the brain's mononuclear phagocyte group, which comprises microglia, and are essential in controlling neuropathological events in the central nervous system [93]. It is worth mentioning that infiltrated monocytes play a significant role in tissue repair, inflammation resolution, and the production of neurotrophic factors. Important cells in the central nervous system called astrocytes release chemicals and other elements required for healthy brain activity and cognitive function, helping to preserve the CNS's equilibrium and support the survival of neurons [94]. They can also regulate the admission of immune cells from the periphery into the brain tissue, and they aid in maintaining the structural integrity and permeability of the blood–brain barrier.

TABLE 3.1 The Major Role of Cytokines in AD.

Molecules	Secreting Cells	Change in AD	Role in the Immune System	References
IL-1β	M1 microglia, monocytes, macrophages	Increased	Proinflammation	[4, 78]
IL-6	Immune cells, endothelial cells, myocytes	Increased/ unchanged	Proinflammation	[4]
IL-10	Monocytes, T lymphocytes, and B lymphocytes; immunosuppressive microglia (M2)	Increased/ increased in the early stage and decreased during disease progression	Anti-inflammation: limiting excessive production of proinflammatory cytokines, ROS	[63, 69, 79]
IL-13	Basophils, eosinophils, Th2 cells, CD4 cells, natural killer T cells, mast cells	Increased	Controversial mechanism: Proinflammation: enhancing MCP-1 expression in monocytes and macrophages; anti-inflammation: induce infiltration to the injured spinal cord and anti-inflammatory polarity of macrophages	[55, 63, 89]
TNF-α	Macrophages, NK cells, T lymphocytes	Increased	Proinflammation	[98]
TGF-β	Macrophages	Increased	Proinflammation, anti-inflammation	[101]
IFN-γ	T cells, NK cells, and macrophages	Increased	Proinflammation	[52, 58, 62]

The importance of the interplay between astrocytes and microglia in understanding pathophysiology has been highlighted by recent research. Certain signal molecules, including complement component 1q (C1q), TNFα, and IL-1α, are released by activated microglia and cause astrocytes to become activated. Similar to activated microglia, these chemicals cause reactive astrocytes. Reactive astrocytes produce cytokines, MHC class II molecules, complement components, and chemokines act as chemoattractants, important for recruiting T cells into CNS [4]. Moreover, it seems that immune cells that penetrate brain tissue play a role in the neuroinflammatory process. Both B and T cells have the capacity to defend against infections and initiate a quick, targeted immune response in cases of repeated infections caused by the same pathogen [56].

T lymphocytes may play a role in the pathogenesis of neurodegenerative diseases since they are found in brain parenchyma during neurodegeneration. The BBB structural integrity and permeability may be compromised by the chronic neuroinflammatory state linked to these disorders. This can result in increased immune cell migration from the periphery into CNS and the diffusion of inflammatory substances through the BBB. These procedures aid in the growth and advancement of lesions in the CNS [95, 96].

The blood–cerebrospinal fluid barrier (BCSFB), which is often thought of as a rigid barrier, actually acts more as a selective entry point for leukocytes entering the CNS. The attraction of immune cells to inflammatory CNS tissue has been suggested to be facilitated by the choroid plexus (CP), a constituent of the BCSFB [96]. When brain parenchymal damage occurs, it has been discovered that neutrophils, monocytes, and T lymphocytes move into the damaged CNS via the BCSFB. Deciphering the processes behind neuroinflammation in neurodegenerative illnesses requires an understanding of the dynamics of immune cell migration across the BCSFB and their interactions within the CNS milieu [97]. Additional investigation in this field could yield understanding of prospective treatment approaches targeted at regulating immune reactions to lessen neurodegeneration.

T cells are divided into two groups: CD^{8+} T cells, also referred to as cytotoxic T cells because of their capacity to eradicate diseased as well as damaged cells, and CD^{4+}T cells, which plays a crucial role in regulating the immune response [66]. Based on varied stimuli, tissue environment, and signaling from antigen-presenting, naïve CD^{4+} T cells (Th0) have the ability to differentiate into distinct T effector cells, like T-helper1 (Th1), T-helper2 (Th2), and T-helper17 (Th17) cells. They may also differentiate into regulatory T cells (Tregs) as well as cytotoxic T lymphocytes (CTLs). The overactivation of Th1 and Th17 cells is linked to neurodegenerative illnesses. These cells release pro-inflammatory cytokines, including IL-17, TNFα, and IFNγ, as well as other inflammatory mediators, which play a direct role in neuroinflammation [98]. However, Th2 cells generate cytokines that reduce inflammation, such as IL-4 making them a possible target for methods of intervention.

Maintaining a healthy environment for the CNS is largely dependent on Th1 and Th2 cells. An imbalance in Th1/Th2 ratio has been found as a potential reason for neurodegeneration [54].

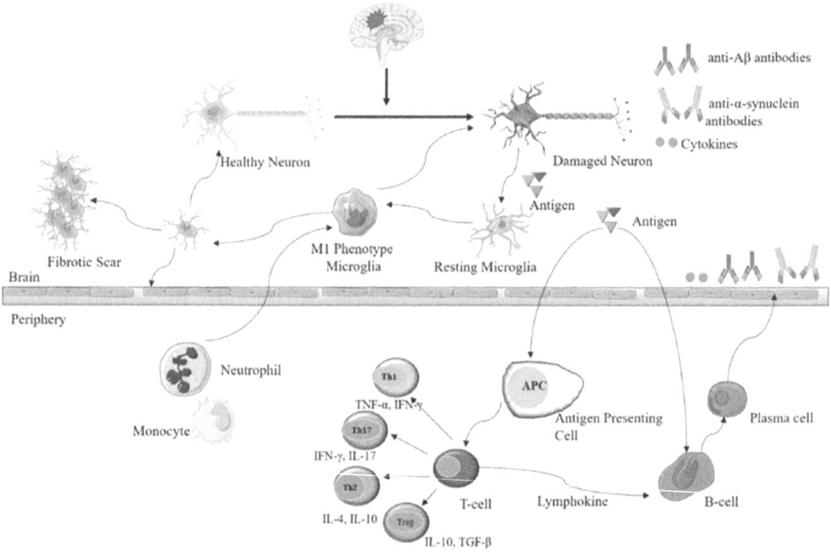

FIGURE 3.2 Crosstalk between central and peripheral immune cells in AD.

Cytotoxic T lymphocytes (CTLs) are specific for antigens that have been linked to pathophysiology of chronic inflammatory diseases, which are linked to neurodegeneration. These CTLs produce cytolysis mediators such as granzymes and perforins, which contribute to inflammation. Conversely, regulatory T (Treg) cells contribute to the reduction of neuroinflammation by enhancing glial cell synthesis of neurotrophic factors and suppressing antigen presentation. Treg cells are immunoregulatory cells that release TGF-β and IL-10, anti-inflammatory factors that are essential for preventing effector T lymphocyte activation [99]. Immune tolerance is developed and maintained in part by this suppression. Treg cells, which have been examined extensively in autoimmune disorders, may also be able to improve outcomes in neurodegenerative disorders due to their immunosuppressive qualities.

Debatable is the intricate role that regulatory T cells (Tregs) play in AD. Tregs may have a positive impact on AD in its early stages by preventing detrimental gliosis, encouraging protective microglia activation, and aiding in leukocyte retrafficking via the choroid plexus (CP). One major cytokine that Tregs produce, TGF-β1, is believed to be important in controlling neuroinflammation in the AD brain [56, 60].

To learn more about whether Tregs can potentially protect neurons during the early phases of amyloid-related neurodegeneration by modulating TGF-β1, preclinical studies utilizing animal models of AD are required. Nevertheless, Tregs may play a more harmful role as the illness advances to later stages. Tregs may impair CP function and reduce the influx of leukocytes that reduce inflammation into the central nervous system in these later phases, which could exacerbate neuroinflammation and accelerate the course of the disease. In general, Tregs play two roles in AD, which emphasizes the complexity of immune responses in neurodegenerative illnesses and the need for more study to completely comprehend the dynamics of Treg activity and the possibilities for therapy in AD [100, 101]. Figure 3.2 highlights the complex interplay between the central and peripheral immune cells in neuroinflammation and AD.

3.5 CONCLUSIONS

The increasing prevalence of AD poses a significant global healthcare challenge, driven by population aging and longer life expectancy. Neuroinflammation is recognized as a significant contributor to the development and progression of AD, alongside Aβ plaques and NFT. In some cases, the failure of antiamyloid therapies highlights the importance of understanding and targeting neuroinflammatory mechanisms. Notably, neuroinflammation in AD involves not only the immune responses within the CNS but also interactions with the peripheral immune system. Immune cells and inflammatory mediators from the peripheral circulation can access the brain parenchyma, contributing to neuroinflammation. This process is facilitated by a functionally impaired BBB, which allows the infiltration of phagocytic tissues and inflammatory cells in the CNS. The interaction between the peripheral and central immune systems underscores the complexity of neuroinflammation in AD. Targeting immune-mediated processes represents a promising avenue for therapeutic intervention in AD. By addressing both central and peripheral immune dysregulation, future treatments may have the ability to modify the disease management and improve patient outcomes.

Currently, there remains a significant gap in the availability of efficacious disease-modifying treatments for AD, highlighting an area of considerable need. Exploring the dysregulated peripheral inflammatory responses emerges as a promising avenue for research. Specifically, interventions targeting the modulation of cytokine/chemokine production, neuro-invasion, monocyte chemotaxis, and Aβ phagocytosis hold potential to attenuate the progression of neurodegeneration associated with AD.

In conclusion, neuroinflammation and engagement of peripheral immune cells is crucial in the development of AD. Understanding the mechanisms underlying peripheral immune cell infiltration and their contribution to neuroinflammation is crucial for developing effective therapeutic interventions for AD. Understanding these immune system dynamics is crucial for developing targeted therapies for AD. Strategies aiming to modulate immune responses, such as immunotherapies that target amyloid-beta or promote anti-inflammatory processes, hold promise in slowing or even reversing disease progression.

KEYWORDS

- **Alzheimer's disease**
- **cytokines**
- **neuroinflammation**
- **peripheral immune cell**

REFERENCES

1. Kumar Thakur, A., Kamboj, P., Goswami, K., & Ahuja, K. (2018, April 26). Pathophysiology and management of Alzheimer's disease: An overview. *Journal of Analytical & Pharmaceutical Research, 7*(2). https://doi.org/10.15406/japlr.2018.07.00230
2. Baazaoui, N., & Iqbal, K. (2022, October 2). Alzheimer's disease: Challenges and a therapeutic opportunity to treat it with a neurotrophic compound. *Biomolecules, 12*(10), 1409. https://doi.org/10.3390/biom12101409
3. Gulisano, W., Maugeri, D., Baltrons, M. A., Fà, M., Amato, A., Palmeri, A., D'Adamio, L., Grassi, C., Devanand, D., Honig, L. S., Puzzo, D., & Arancio, O. (2018, June 12). Role of amyloid-β and tau proteins in Alzheimer's disease: Confuting the amyloid cascade. *Journal of Alzheimer's Disease, 64*(s1), S611–S631. https://doi.org/10.3233/jad-179935
4. Ogunmokun, G., Dewanjee, S., Chakraborty, P., Valupadas, C., Chaudhary, A., Kolli, V., Anand, U., Vallamkondu, J., Goel, P., Paluru, H. P. R., Gill, K. D., Reddy, P. H.,

De Feo, V., & Kandimalla, R. (2021, October 18). The potential role of cytokines and growth factors in the pathogenesis of Alzheimer's disease. *Cells, 10*(10), 2790. https://doi.org/10.3390/cells10102790

5. Calsolaro, V., & Edison, P. (2016, May 11). Neuroinflammation in Alzheimer's disease: Current evidence and future directions. *Alzheimer's & Dementia, 12*(6), 719–732. https://doi.org/10.1016/j.jalz.2016.02.010
6. Sarlus, H., & Heneka, M. T. (2017, September 1). Microglia in Alzheimer's disease. *Journal of Clinical Investigation, 127*(9), 3240–3249. https://doi.org/10.1172/jci90606
7. Villacampa, N., & Heneka, M. T. (2020, March 3). Microglia in Alzheimer's disease: Local heroes! *Journal of Experimental Medicine, 217*(4). https://doi.org/10.1084/jem.20192311
8. McGeer, P. L., & McGeer, E. G. (2014, November 29). Targeting microglia for the treatment of Alzheimer's disease. *Expert Opinion on Therapeutic Targets, 19*(4), 497–506. https://doi.org/10.1517/14728222.2014.988707
9. Wes, P. D., Sayed, F. A., Bard, F., & Gan, L. (2016, April 21). Targeting microglia for the treatment of Alzheimer's disease. *Glia, 64*(10), 1710–1732. https://doi.org/10.1002/glia.22988
10. Fakhoury, M. (2018, May 10). Microglia and Astrocytes in Alzheimer's disease: Implications for Therapy. *Current Neuropharmacology, 16*(5), 508–518. https://doi.org/10.2174/1570159x15666170720095240
11. Rodríguez-Arellano, J., Parpura, V., Zorec, R., & Verkhratsky, A. (2016, May). Astrocytes in physiological aging and Alzheimer's disease. *Neuroscience, 323*, 170–182. https://doi.org/10.1016/j.neuroscience.2015.01.007
12. Chun, H., & Lee, C. J. (2018, January). Reactive astrocytes in Alzheimer's disease: A double-edged sword. *Neuroscience Research, 126*, 44–52. https://doi.org/10.1016/j.neures.2017.11.012
13. Sadick, J. S., O'Dea, M. R., Hasel, P., Dykstra, T., Faustin, A., & Liddelow, S. A. (2022, June). Astrocytes and oligodendrocytes undergo subtype-specific transcriptional changes in Alzheimer's disease. *Neuron, 110*(11), 1788–1805.e10. https://doi.org/10.1016/j.neuron.2022.03.008
14. Cai, Z., & Xiao, M. (2015, July 14). Oligodendrocytes and Alzheimer's disease. *International Journal of Neuroscience, 126*(2), 97–104. https://doi.org/10.3109/00207454.2015.1025778
15. Butt, A. M., De La Rocha, I. C., & Rivera, A. (2019). Oligodendroglial cells in Alzheimer's disease. *Neuroglia in Neurodegenerative Diseases,* 325–333. https://doi.org/10.1007/978-981-13-9913-8_12
16. Sweeney, M. D., Ayyadurai, S., & Zlokovic, B. V. (2016, May 26). Pericytes of the neurovascular unit: Key functions and signaling pathways. *Nature Neuroscience, 19*(6), 771–783. https://doi.org/10.1038/nn.4288
17. Procter, T. V., Williams, A., & Montagne, A. (2021). Interplay between brain pericytes and endothelial cells in dementia. *The American Journal of Pathology, 191*(11), 1917–1931. https://doi.org/10.1016/j.ajpath.2021.07.003
18. J. Alcendor, D. (2020). Interactions between amyloid-β proteins and human brain pericytes: Implications for the pathobiology of Alzheimer's disease. *Journal of Clinical Medicine, 9*(5), 1490. https://doi.org/10.3390/jcm9051490

19. Salmina, A. B., Komleva, Y. K., Lopatina, O. L., & Birbrair, A. (2019). Pericytes in Alzheimer's disease: Novel clues to cerebral amyloid angiopathy pathogenesis. *Advances in Experimental Medicine and Biology*, 147–166. https://doi.org/10.1007/978-3-030-16908-4_7
20. Winkler, E. A., Sagare, A. P., & Zlokovic, B. V. (2014). The pericyte: A forgotten cell type with important implications for Alzheimer's disease? *Brain Pathology, 24*(4), 371-386. https://doi.org/10.1111/bpa.12152
21. ElAli, A., Thériault, P., & Rivest, S. (2014). The role of pericytes in neurovascular unit remodeling in brain disorders. *International Journal of Molecular Sciences, 15*(4), 6453-6474. https://doi.org/10.3390/ijms15046453
22. Uemura, M. T., Maki, T., Ihara, M., Lee, V. M., & Trojanowski, J. Q. (2020). Brain microvascular pericytes in vascular cognitive impairment and dementia. *Frontiers in Aging Neuroscience, 12*, 80.
23. Lendahl, U., Nilsson, P., & Betsholtz, C. (2019). Emerging links between cerebrovascular and neurodegenerative diseases—A special role for pericytes. *EMBO Reports, 20*(11), e48070. https://doi.org/10.15252/embr.201948070
24. Yamazaki, Y., & Kanekiyo, T. (2017). Blood–brain barrier dysfunction and the pathogenesis of Alzheimer's disease. *International Journal of Molecular Sciences, 18*(9), 1965. https://doi.org/10.3390/ijms18091965
25. Huang, Z., Wong, L. W., Su, Y., Huang, X., Wang, N., Chen, H., & Yi, C. (2020). Blood–brain barrier integrity in the pathogenesis of Alzheimer's disease. *Frontiers in Neuroendocrinology, 59*, 100857. https://doi.org/10.1016/j.yfrne.2020.100857
26. Cai, Z., Qiao, P. F., Wan, C. Q., Cai, M., Zhou, N. K., & Li, Q. (2018, May 30). Role of blood–brain barrier in Alzheimer's disease. *Journal of Alzheimer's Disease*, *63*(4), 1223–1234. https://doi.org/10.3233/jad-180098
27. Zenaro, E., Piacentino, G., & Constantin, G. (2017, November). The blood–brain barrier in Alzheimer's disease. *Neurobiology of Disease*, *107*, 41–56. https://doi.org/10.1016/j.nbd.2016.07.007
28. Van der Velpen, V., Teav, T., Gallart-Ayala, H., Mehl, F., Konz, I., Clark, C., Oikonomidi, A., Peyratout, G., Henry, H., Delorenzi, M., Ivanisevic, J., & Popp, J. (2019, November 28). Systemic and central nervous system metabolic alterations in Alzheimer's disease. *Alzheimer's Research & Therapy, 11*(1). https://doi.org/10.1186/s13195-019-0551-7
29. Reinhold, A., & Rittner, H. (2016, December 12). Barrier function in the peripheral and central nervous system—A review. *Pflügers Archiv—European Journal of Physiology*, *469*(1), 123–134. https://doi.org/10.1007/s00424-016-1920-8
30. Anwar, M. M. (2021, October 30). Oxidative stress—A direct bridge to central nervous system homeostatic dysfunction and Alzheimer's disease. *Cell Biochemistry and Function*, *40*(1), 17–27. https://doi.org/10.1002/cbf.3673
31. Storck, S. E., & Pietrzik, C. U. (2018). The blood brain-barrier and its role in Alzheimer's disease. *Neuroforum, 24*(4), A197-A205. https://doi.org/10.1515/nf-2018-A014
32. Cao, W., & Zheng, H. (2018, October 3). Peripheral immune system in aging and Alzheimer's disease. *Molecular Neurodegeneration*, *13*(1). https://doi.org/10.1186/s13024-018-0284-2

33. Prinz, M., & Priller, J. (2017). The role of peripheral immune cells in the CNS in steady state and disease. *Nature Neuroscience, 20*(2), 136-144. https://doi.org/10.1038/nn.4475
34. Dionisio-Santos, D. A., Olschowka, J. A., & O'Banion, M. K. (2019, April 5). Exploiting microglial and peripheral immune cell crosstalk to treat Alzheimer's disease. *Journal of Neuroinflammation, 16*(1). https://doi.org/10.1186/s12974-019-1453-0
35. Greenhalgh, A. D., David, S., & Bennett, F. C. (2020). Immune cell regulation of glia during CNS injury and disease. *Nature Reviews Neuroscience, 21*(3), 139-152. https://doi.org/10.1038/s41583-020-0263-9
36. Le Page, A., Dupuis, G., Frost, E. H., Larbi, A., Pawelec, G., Witkowski, J. M., & Fulop, T. (2018, July). Role of the peripheral innate immune system in the development of Alzheimer's disease. *Experimental Gerontology, 107*, 59–66. https://doi.org/10.1016/j.exger.2017.12.019
37. Bettcher, B. M., Tansey, M. G., Dorothée, G., & Heneka, M. T. (2021, September 14). Peripheral and central immune system crosstalk in Alzheimer disease—A research prospectus. *Nature Reviews Neurology, 17*(11), 689–701. https://doi.org/10.1038/s41582-021-00549-x
38. Dá Mesquita, S., Ferreira, A. C., Sousa, J. C., Correia-Neves, M., Sousa, N., & Marques, F. (2016, September). Insights on the pathophysiology of Alzheimer's disease: The crosstalk between amyloid pathology, neuroinflammation and the peripheral immune system. *Neuroscience & Biobehavioral Reviews, 68*, 547–562. https://doi.org/10.1016/j.neubiorev.2016.06.014
39. Fani Maleki, A., & Rivest, S. (2019, July 31). Innate immune cells: Monocytes, monocyte-derived macrophages and microglia as therapeutic targets for Alzheimer's disease and multiple sclerosis. *Frontiers in Cellular Neuroscience, 13*. https://doi.org/10.3389/fncel.2019.00355
40. Perry, V. H., & Teeling, J. (2013, September). Microglia and macrophages of the central nervous system: The contribution of microglia priming and systemic inflammation to chronic neurodegeneration. In *Seminars in immunopathology* (Vol. 35, pp. 601–612). https://doi.org/10.1007/s00281-013-0382-8
41. Goldeck, D., M. Witkowski, J., Fülop, T., & Pawelec, G. (2016, May 11). Peripheral Immune Signatures in Alzheimer Disease. *Current Alzheimer Research, 13*(7), 739–749. https://doi.org/10.2174/1567205013666160222112444
42. Guilliams, M., Mildner, A., & Yona, S. (2018). Developmental and functional heterogeneity of monocytes. *Immunity, 49*(4), 595-613. https://doi.org/10.1016/j.immuni.2018.10.005
43. Hume, D. A., Irvine, K. M., & Pridans, C. (2019). The mononuclear phagocyte system: The relationship between monocytes and macrophages. *Trends in Immunology, 40*(2), 98–112. https://doi.org/10.1016/j.it.2018.11.007
44. Guo, H., Zhao, Z., Zhang, R., Chen, P., Zhang, X., Cheng, F., & Gou, X. (2019, April 23). Monocytes in the peripheral clearance of amyloid-β and Alzheimer's disease. *Journal of Alzheimer's Disease, 68*(4), 1391–1400. https://doi.org/10.3233/jad-181177
45. Zhao, M., Tuo, H., Wang, S., & Zhao, L. (2020). The roles of monocyte and monocyte-derived macrophages in common brain disorders. *BioMed Research International, 2020*(1), 9396021. https://doi.org/10.1155/2020/9396021

46. Mammana, S., Fagone, P., Cavalli, E., Basile, M. S., Petralia, M. C., Nicoletti, F., et al. (2018). The role of macrophages in neuroinflammatory and neurodegenerative pathways of Alzheimer's disease, amyotrophic lateral sclerosis, and multiple sclerosis: Pathogenetic cellular effectors and potential therapeutic targets. *International Journal of Molecular Sciences, 19*(3), 831. https://doi.org/10.3390/ijms19030831
47. Guedes, J. R., Lao, T., Cardoso, A. L., & El Khoury, J. (2018). Roles of microglial and monocyte chemokines and their receptors in regulating Alzheimer's disease-associated amyloid-β and tau pathologies. *Frontiers in Neurology, 9*, 381108. https://doi.org/10.3389/fneur.2018.00549
48. Zuroff, L., Daley, D., Black, K. L., & Koronyo-Hamaoui, M. (2017). Clearance of cerebral Aβ in Alzheimer's disease: Reassessing the role of microglia and monocytes. *Cellular and Molecular Life Sciences, 74*, 2167-2201. https://doi.org/10.1007/s00018-017-2463-7
49. Mishra, S. K., & Singh, S. K. (2017). Alzheimer's disease and immune response: A brief overview. *Journal of Alzheimer's Disease & Parkinsonism, 7*(5). https://doi.org/10.4172/2161-0460.1000371
50. Spiteri, A. G., Wishart, C. L., Pamphlett, R., Locatelli, G., & King, N. J. (2022). Microglia and monocytes in inflammatory CNS disease: Integrating phenotype and function. *Acta Neuropathologica,* 1–46. https://doi.org/10.1007/s00401-021-02384-2
51. Mietelska-Porowska, A., & Wojda, U. (2017). T lymphocytes and inflammatory mediators in the interplay between brain and blood in Alzheimer's disease: Potential pools of new biomarkers. *Journal of Immunology Research, 2017*(1), 4626540. https://doi.org/10.1155/2017/4626540
52. Carlberg, C., & Velleuer, E. (2022). Cells and tissues of the immune system. *Molecular Immunology,* 1–18. https://doi.org/10.1007/978-3-031-04025-2_1
53. Mathias, C. B. (2019, October 19). Overview of the immune system and its pharmacological targets. *Pharmacology of Immunotherapeutic Drugs,* 1–42. https://doi.org/10.1007/978-3-030-19922-7_
54. Sayed, A., Bahbah, E. I., Kamel, S., Barreto, G. E., Ashraf, G. M., & Elfil, M. (2020). The neutrophil-to-lymphocyte ratio in Alzheimer's disease: Current understanding and potential applications. *Journal of Neuroimmunology, 349*, 577398. https://doi.org/10.1016/j.jneuroim.2020.577398
55. Beňová, K., Hancková, M., Koči, K., Kúdelová, M., & Betáková, T. (2020). T cells and their function in the immune response to viruses. *Acta Virologica, 64*(02), 131–143. https://doi.org/10.4149/av_2020_203
56. Machhi, J., Yeapuri, P., Lu, Y., Foster, E., Chikhale, R., Herskovitz, J., Namminga, K. L., Olson, K. E., Abdelmoaty, M. M., Gao, J., Quadros, R. M., Kiyota, T., Jingjing, L., Kevadiya, B. D., Wang, X., Liu, Y., Poluektova, L. Y., Gurumurthy, C. B., Mosley, R. L., & Gendelman, H. E. (2021, November 19). CD4+ effector T cells accelerate Alzheimer's disease in mice. *Journal of Neuroinflammation, 18*(1). https://doi.org/10.1186/s12974-021-02308-7
57. Chihara, N., Tsuji, A., & Matsumoto, R. (2023). Neuroinflammation and neuroimmunology in Alzheimer's disease: The role of T-lymphocytes in Alzheimer's disease. *Clinical and Experimental Neuroimmunology, 14*(2), 92–99. https://doi.org/10.1111/cen3.12747

58. Heppner, F. L., Ransohoff, R. M., & Becher, B. (2015). Immune attack: The role of inflammation in Alzheimer disease. *Nature Reviews Neuroscience, 16*(6), 358-372. https://doi.org/10.1038/nrn3880
59. Bagyinszky, E., Van Giau, V., Shim, K., Suk, K., An, S. S. A., & Kim, S. (2017). Role of inflammatory molecules in the Alzheimer's disease progression and diagnosis. *Journal of the Neurological Sciences, 376*, 242-254. https://doi.org/10.1016/j.jns.2017.03.031
60. Yang, H., Park, S. Y., Baek, H., Lee, C., Chung, G., Liu, X., Lee, J. H., Kim, B., Kwon, M., Choi, H., Kim, H. J., Kim, J. Y., Kim, Y., Lee, Y. S., Lee, G., Kim, S. K., Kim, J. S., Chang, Y. T., Jung, W. S., et al. (2022). Adoptive therapy with amyloid-β specific regulatory T cells alleviates Alzheimer's disease. *Theranostics, 12*(18), 7668–7680. https://doi.org/10.7150/thno.75965
61. Yeapuri, P., Machhi, J., Lu, Y., Abdelmoaty, M. M., Kadry, R., Patel, M., Bhattarai, S., Lu, E., Namminga, K. L., Olson, K. E., Foster, E. G., Mosley, R. L., & Gendelman, H. E. (2023, December 18). Amyloid-β specific regulatory T cells attenuate Alzheimer's disease pathobiology in APP/PS1 mice. *Molecular Neurodegeneration, 18*(1). https://doi.org/10.1186/s13024-023-00692-7
62. Machhi, J., Kevadiya, B. D., Muhammad, I. K., Herskovitz, J., Olson, K. E., Mosley, R. L., & Gendelman, H. E. (2020, June 5). Harnessing regulatory T cell neuroprotective activities for treatment of neurodegenerative disorders. *Molecular Neurodegeneration, 15*(1). https://doi.org/10.1186/s13024-020-00375-7
63. Princiotta Cariddi, L., Mauri, M., Cosentino, M., Versino, M., & Marino, F. (2022). Alzheimer's disease: From immune homeostasis to neuroinflammatory condition. *International Journal of Molecular Sciences, 23*(21), 13008. https://doi.org/10.3390/ijms232113008
64. Guerriero, F., Sgarlata, C., Francis, M., Maurizi, N., Faragli, A., Perna, S., & Ricevuti, G. (2017). Neuroinflammation, immune system and Alzheimer disease: Searching for the missing link. *Aging Clinical and Experimental Research, 29*, 821–831. https://doi.org/10.1007/s40520-016-0637-z
65. Asamu, M. O., Oladipo, O. O., Abayomi, O. A., & Adebayo, A. A. (2023). Alzheimer's disease: The role of T lymphocytes in neuroinflammation and neurodegeneration. *Brain Research, 1821*, 148589. https://doi.org/10.1016/j.brainres.2023.148589
66. Aries, M. L., & Hensley-McBain, T. (2023, March 3). Neutrophils as a potential therapeutic target in Alzheimer's disease. *Frontiers in Immunology, 14*. https://doi.org/10.3389/fimmu.2023.1123149
67. Bawa, K. K., Krance, S. H., Herrmann, N., Cogo-Moreira, H., Ouk, M., Yu, D., Wu, C. Y., Black, S. E., Lanctôt, K. L., & Swardfager, W. (2020, March 14). A peripheral neutrophil-related inflammatory factor predicts a decline in executive function in mild Alzheimer's disease. *Journal of Neuroinflammation, 17*(1). https://doi.org/10.1186/s12974-020-01750-3
68. Zenaro, E., Pietronigro, E., Bianca, V. D., Piacentino, G., Marongiu, L., Budui, S., Turano, E., Rossi, B., Angiari, S., Dusi, S., Montresor, A., Carlucci, T., Nanì, S., Tosadori, G., Calciano, L., Catalucci, D., Berton, G., Bonetti, B., & Constantin, G. (2015, July 27). Neutrophils promote Alzheimer's disease–like pathology and cognitive decline via LFA-1 integrin. *Nature Medicine, 21*(8), 880–886. https://doi.org/10.1038/nm.3913

69. Pietronigro, E., Zenaro, E., & Constantin, G. (2016, February 15). Imaging of leukocyte trafficking in Alzheimer's disease. *Frontiers in Immunology, 7*. https://doi.org/10.3389/fimmu.2016.00033
70. Dong, Y., Lagarde, J., Xicota, L., Corne, H., Chantran, Y., Chaigneau, T., & Elbim, C. (2018). Neutrophil hyperactivation correlates with Alzheimer's disease progression. *Annals of Neurology, 83*(2), 387–405. https://doi.org/10.1038/nm.3913
71. Smyth, L. C., Murray, H. C., Hill, M., van Leeuwen, E., Highet, B., Magon, N. J., et al. (2022). Neutrophil-vascular interactions drive myeloperoxidase accumulation in the brain in Alzheimer's disease. *Acta Neuropathologica Communications, 10*(1), 38. https://doi.org/10.1186/s40478-022-01347-2
72. Kretzschmar, G. C., Bumiller-Bini, V., Gasparetto Filho, M. A., Zonta, Y. R., Yu, K. S. T., De Souza, R. L. R., et al. (2021). Neutrophil extracellular traps: A perspective of neuroinflammation and complement activation in Alzheimer's disease. *Frontiers in Molecular Biosciences, 8*, 630869. https://doi.org/10.3389/fmolb.2021.630869
73. Solana, C., Tarazona, R., & Solana, R. (2018, November 1). Immunosenescence of natural killer cells, inflammation, and Alzheimer's disease. *International Journal of Alzheimer's Disease, 2018*, 1–9. https://doi.org/10.1155/2018/3128758
74. Ludewig, P., Gallizioli, M., Urra, X., Behr, S., Brait, V. H., Gelderblom, M., Magnus, T., & Planas, A. M. (2016, March). Dendritic cells in brain diseases. *Biochimica Et Biophysica Acta (BBA)—Molecular Basis of Disease, 1862*(3), 352–367. https://doi.org/10.1016/j.bbadis.2015.11.003
75. Anandasabapathy, N., Victora, G. D., Meredith, M., Feder, R., Dong, B., Kluger, C., Yao, K., Dustin, M. L., Nussenzweig, M. C., Steinman, R. M., & Liu, K. (2011, July 25). Flt3L controls the development of radiosensitive dendritic cells in the meninges and choroid plexus of the steady-state mouse brain. *Journal of Experimental Medicine, 208*(8), 1695–1705. https://doi.org/10.1084/jem.20102657
76. Brezovakova, V., Valachova, B., Hanes, J., Novak, M., & Jadhav, S. (2018, June 13). Dendritic cells as an alternate approach for treatment of neurodegenerative disorders. *Cellular and Molecular Neurobiology, 38*(6), 1207–1214. https://doi.org/10.1007/s10571-018-0598-1
77. Rezai-Zadeh, K., Gate, D., & Town, T. (2009, August 11). CNS infiltration of peripheral immune cells: D-day for neurodegenerative disease? *Journal of Neuroimmune Pharmacology, 4*(4), 462–475. https://doi.org/10.1007/s11481-009-9166-2
78. Wang, W. Y., Tan, M. S., Yu, J. T., & Tan, L. (2015). Role of pro-inflammatory cytokines released from microglia in Alzheimer's disease. *Annals of Translational Medicine, 3*(10). https://doi.org/10.3978%2Fj.issn.2305-5839.2015.03.49
79. Uddin, M. S., Kabir, M. T., Jalouli, M., Rahman, M. A., Jeandet, P., Behl, T., & Ashraf, G. M. (2022). Neuroinflammatory signaling in the pathogenesis of Alzheimer's disease. *Current Neuropharmacology, 20*(1), 126. https://doi.org/10.2174%2F1570159X19666210826130210
80. Ghosh, P., Singh, R., Ganeshpurkar, A., Pokle, A. V., bhushan Singh, R., Singh, S. K., & Kumar, A. (2021). Cellular and molecular influencers of neuroinflammation in Alzheimer's disease: Recent concepts & roles. *Neurochemistry International, 151*, 105212. https://doi.org/10.1016/j.neuint.2021.105212
81. Jain, S., Singh, R., Paliwal, S., & Sharma, S. (2023). Targeting neuroinflammation as disease modifying approach to Alzheimer's disease: Potential and challenges. *Mini*

Reviews in Medicinal Chemistry, 23(22), 2097–2116. https://doi.org/10.2174/1389557523666230511122435

82. Bachiller, S., Jiménez-Ferrer, I., Paulus, A., Yang, Y., Swanberg, M., Deierborg, T., & Boza-Serrano, A. (2018). Microglia in neurological diseases: A road map to brain-disease dependent-inflammatory response. *Frontiers in cellular neuroscience, 12*, 488. https://doi.org/10.3389/fncel.2018.00488
83. Gray, S. C., Kinghorn, K. J., & Woodling, N. S. (2020). Shifting equilibriums in Alzheimer's disease: The complex roles of microglia in neuroinflammation, neuronal survival and neurogenesis. *Neural Regeneration Research, 15*(7), 1208–1219. https://doi.org/10.4103/1673-5374.272571
84. Kinney, J. W., Bemiller, S. M., Murtishaw, A. S., Leisgang, A. M., Salazar, A. M., & Lamb, B. T. (2018). Inflammation as a central mechanism in Alzheimer's disease. *Alzheimer's & Dementia: Translational Research & Clinical Interventions, 4*, 575–590. https://doi.org/10.1016/j.trci.2018.06.014
85. Sochocka, M., Diniz, B. S., & Leszek, J. (2017). Inflammatory response in the CNS: Friend or foe? *Molecular Neurobiology, 54*, 8071-8089. https://doi.org/10.1007/s12035-016-0297-1
86. Webers, A., Heneka, M. T., & Gleeson, P. A. (2020). The role of innate immune responses and neuroinflammation in amyloid accumulation and progression of Alzheimer's disease. *Immunology and Cell Biology, 98*(1), 28–41. https://doi.org/10.1111/imcb.12301
87. McCaulley, M. E., & Grush, K. A. (2015). Alzheimer's disease: Exploring the role of inflammation and implications for treatment. *International Journal of Alzheimer's Disease, 2015*(1), 515248. https://doi.org/10.1155/2015/515248
88. Spangenberg, E. E., & Green, K. N. (2017). Inflammation in Alzheimer's disease: Lessons learned from microglia-depletion models. *Brain, behavior, and immunity, 61*, 1–11. https://doi.org/10.1016/j.bbi.2016.07.003
89. Mamsa, R., Prabhavalkar, K. S., & Bhatt, L. K. (2023). Crosstalk between NLRP3 inflammasome and calpain in Alzheimer's disease. *European Journal of Neuroscience, 58*(7), 3719–3731. https://doi.org/10.1111/ejn.16139
90. Busse, M., Michler, E., von Hoff, F., Dobrowolny, H., Hartig, R., Frodl, T., & Busse, S. (2017, June 23). Alterations in the peripheral immune system in dementia. *Journal of Alzheimer's Disease, 58*(4), 1303–1313. https://doi.org/10.3233/jad-161304
91. Leng, F., & Edison, P. (2021). Neuroinflammation and microglial activation in Alzheimer disease: Where do we go from here? *Nature Reviews Neurology, 17*(3), 157-172. https://doi.org/10.2147/NDT.S75546
92. Busse, S., Hoffmann, J., Michler, E., Hartig, R., Frodl, T., & Busse, M. (2021). Dementia-associated changes of immune cell composition within the cerebrospinal fluid. *Brain, Behavior, & Immunity-Health, 14*, 100218. https://doi.org/10.1016/j.bbih.2021.100218
93. Schwabe, T., Srinivasan, K., & Rhinn, H. (2020, September). Shifting paradigms: The central role of microglia in Alzheimer's disease. *Neurobiology of Disease, 143*, 104962. https://doi.org/10.1016/j.nbd.2020.104962
94. Merighi, S., Nigro, M., Travagli, A., & Gessi, S. (2022, October 27). Microglia and Alzheimer's disease. *International Journal of Molecular Sciences, 23*(21), 12990. https://doi.org/10.3390/ijms232112990

95. Tzioras, M., McGeachan, R. I., Durrant, C. S., & Spires-Jones, T. L. (2023). Synaptic degeneration in Alzheimer disease. *Nature Reviews Neurology, 19*(1), 19–38. https://doi.org/10.1038/s41582-022-00749-z
96. Liu, R., Zhang, Z., Chen, Y., Liao, J., Wang, Y., Liu, J., Lin, Z., & Xiao, G. (2022, October 21). Choroid plexus epithelium and its role in neurological diseases. *Frontiers in Molecular Neuroscience, 15*. https://doi.org/10.3389/fnmol.2022.949231
97. Markowicz-Piasecka, M., Markiewicz, A., Darłak, P., Sikora, J., Adla, S. K., Bagina, S., & Huttunen, K. M. (2022, April). Current chemical, biological, and physiological views in the development of successful brain-targeted pharmaceutics. *Neurotherapeutics, 19*(3), 942–976. https://doi.org/10.1007/s13311-022-01228-5
98. Babić Leko, M., Nikolac Perković, M., Klepac, N., Štrac, D. V., Borovečki, F., Pivac, N., Hof, P. R., & Šimić, G. (2020, June 2). IL-1β, IL-6, IL-10, and TNFα single nucleotide polymorphisms in human influence the susceptibility to Alzheimer's disease pathology. *Journal of Alzheimer's Disease, 75*(3), 1029–1047. https://doi.org/10.3233/jad-200056
99. Dai, L., & Shen, Y. (2021, November). Insights into T-cell dysfunction in Alzheimer's disease. *Aging Cell, 20*(12). https://doi.org/10.1111/acel.13511
100. Faridar, A., Thome, A. D., Zhao, W., Thonhoff, J. R., Beers, D. R., Pascual, B., Masdeu, J. C., & Appel, S. H. (2020). Restoring regulatory T-cell dysfunction in Alzheimer's disease through ex vivo expansion. *Brain Communications, 2*(2). https://doi.org/10.1093/braincomms/fcaa112
101. Kapoor, M., & Chinnathambi, S. (2023, March 13). TGF-β1 signalling in Alzheimer's pathology and cytoskeletal reorganization: A specialized Tau perspective. *Journal of Neuroinflammation, 20*(1). https://doi.org/10.1186/s12974-023-02751-8

CHAPTER 4

Blood–Brain Barrier Dysfunction: Relevance to Alzheimer's disease Pathology and Challenges and Opportunities in Drug Discovery

SANDEEP PRASAD TIWARI* and DIPTI PAL

Kalinga University, Kotni, Naya Raipur, Chhattisgarh, India

*Corresponding author

ABSTRACT

Alzheimer's disease (AD), which is marked by a dearth of efficient treatment options, is a serious public health challenge. The breakdown of the blood–brain barrier (BBB), which impairs the brain's capacity to control molecular and cellular exchanges with the circulation, is a major factor in the pathogenesis of AD. This work offers a thorough investigation of BBB failure in AD, illuminating its underlying mechanisms, implications for therapeutic approaches, and related opportunities and problems. In order to clarify the crucial importance of the BBB as a therapeutic target in AD treatment efforts, this chapter will explore the complex relationship between BBB failure and AD progression. Important subjects include the effects of BBB permeability changes on drug delivery to the brain, the escalation of oxidative stress and neuroinflammation, and the promise of novel strategies like tailored medicine and drug delivery systems based on nanotechnology. This research emphasizes therapeutic development opportunities that show promise and emphasizes the value of cross-disciplinary collaboration, even in the face of difficulties connected to BBB. Through the use of recent developments in drug discovery, biomarker

Cellular and Molecular Mechanisms in Alzheimer's Disease Therapeutics, Volume 1: Cellular Mechanisms. Prashant Tiwari, Sunil Kumar Kadiri, and Kunjbihari Sulakhiya (Eds.)

DOI: 10.1201/9781779641793-4

identification, and translational research, there is a noteworthy prospect to transform the treatment of AD and enhance the quality of life for those afflicted with this debilitating illness.

4.1 INTRODUCTION TO ALZHEIMER'S DISEASE AND THE BLOOD-BRAIN BARRIER

Amyloid-beta (Aβ) plaques and neurofibrillary tangles (NFTs) in the brain cause neuronal death and cognitive deterioration in Alzheimer's disease (AD). While much study has been done on these characteristics, current discoveries imply that blood–brain barrier (BBB) disruption plays a major role in AD pathogenesis. Most dementia in elderly persons is caused by AD, a progressive neurological disease. It impairs memory, cognition, and daily life [1]. The disease is characterized by tau protein-based NFTs and Aβ plaques in the brain, which are associated to neuronal malfunction and mortality in affected individuals [2]. The genetic, environmental, and behavioral causes of AD are complicated. Key genetic factors for early-onset AD include mutations in *presenilin-1*, *presenilin-2*, and *amyloid protein precursor* (*APP*), while the *apolipoprotein E (APOE)* ε4 variant is a known risk factor for late-onset AD. Although the exact mechanisms by which these genetic factors affect AD development are unknown, they may affect Aβ production and clearance [3]. Clinical study shows that AD progresses from mild amnesia to severe language, judgment, memory, and decision-making problems. Frequent behavioral shifts including anger, sadness, and apathy can lower patients' quality of life and challenge caretakers. Despite substantial study, AD treatments are scarce and often focus on treating symptoms [4]. Cholinesterase inhibitors including galantamine, rivastigmine, donepezil, and memantine temporarily improve cognition via changing neurotransmitter levels. The US-licensed treatment aducanumab targets Aβ plaques, addressing the disease's root causes. Its acceptance and efficacy have divided the medical community [5]. New studies further emphasize the role of BBB failure in AD pathophysiology, suggesting that BBB integrity may affect Aβ clearance and neuroinflammatory processes, which may affect AD progression. This suggests finding new therapy targets to boost Aβ clearance pathways or BBB function [6]. BBB controls chemical and ion flow between the bloodstream and the brain to maintain brain homeostasis. The BBB, comprised of tight junction proteins, astrocytes, pericytes, and specialized endothelial cells, protects the brain from harmful substances while allowing nutrients and chemicals to enter. BBB alterations in AD may contribute to its origin and development, according to emerging evidence [7]. BBB disruption increases brain parenchyma permeability, allowing immune cells, inflammatory mediators, and toxic substances in. This infiltration exacerbates neuroinflammation, promotes neuronal injury, and accelerates the progression of neurodegenerative processes, particularly in conditions such as Alzheimer's disease and multiple sclerosis. These changes worsen neuronal function, increase neuroinflammation, and destroy neurons, causing neurodegeneration and cognitive impairment [8]. The BBB breaks down in AD due to tight junction protein alterations, oxidative stress, and inflammation. Chronic neuroinflammation, a hallmark of AD, activates endothelial cells and increases adhesion molecules, compromising the BBB. Oxidative stress damages endothelial cells and tight junction proteins, worsening BBB failure and permeability [9]. Aβ peptide buildup at the BBB impairs its function, worsening neuroinflammation and neuronal injury. BBB, a specialized system of capillary endothelial cells, controls the flow of important nutrients and maintains a constant internal environment while protecting the brain from potentially harmful compounds in the blood. Normal brain function depends on this physical and metabolic barrier [10].

BBM Structure: Astrocyte end-feet surround the closely packed endothelial cells that line brain capillaries, which are held up by pericytes and form the BBB [11]. These components form a selective barrier that limits circulatory molecules entering the brain. Tight connections between endothelial cells control what enters the brain's extracellular space through the BBB. These connections prevent paracellular chemical transfer [12].

Function of the BBM: The BBB protects the brain from infections and toxins while allowing ions and nutrients to reach its tissue. Large and hydrophilic molecules are efficiently filtered out, whereas carbon dioxide, oxygen, and small hydrophobic molecules that can traverse cell membranes pass [13]. Important chemicals like glucose and amino acids cannot passively traverse the endothelium and have unique transport routes [14]. The BBB maintains the brain's milieu for proper neuronal activation. However, because of its selective nature, the BBB can limit brain access to therapeutic medicines, making neurological diseases harder to treat [15]. Thus, most research is focused on ways to temporarily open the BBB or improve medicine passage [16]. Understanding how BBB breakdown causes AD pathophysiology will benefit drug development. Pharmaceuticals are less successful because typical medication development ignores the BBB's permeability. Creating drugs that can cross the BBB is difficult, but it allows for imaginative treatment [17]. Focusing on BBB malfunction directly or employing cutting-edge medication delivery technologies to circumvent or change the BBB may improve AD therapy [18]. Advances in PET and MRI allow non-invasive examination of the BBB's integrity and function in AD patients [19]. AD diagnosis and development are greatly aided by these technologies [20]. BBB dysfunction biomarkers such endothelial cell markers and inflammatory cytokines may help detect and prognose AD [21]. Targeting BBB disruption in AD is difficult despite its potential benefits. Complex BBB regulation and its dynamic response to pathological stresses make particular and effective therapies challenging to produce. Drug development is further complicated by AD pathogenesis heterogeneity [22]. Addressing BBB-targeting therapy safety concerns such off-target effects and immune activation is crucial. Recent advances in drug delivery technology may help AD treatments overcome BBB issues [23]. Nanoparticle-based delivery methods, liposomal formulations, and cell-penetrating peptides may improve BBB medication delivery with fewer side effects. Targeted medication delivery techniques that utilize AD pathology's BBB changes could improve efficacy and reduce systemic toxicity [24].

4.2 EFFECTS OF BBB DYSFUNCTION ON AD PATHOLOGY

AD is characterized by increasing development of Aβ plaques, NFTs, and synaptic dysfunction, leading to neuronal death and cognitive impairment [25]. Research on AD has historically concentrated on these features, but new data suggests that BBB disruption is a major cause and course [26] (Figure 4.1) .

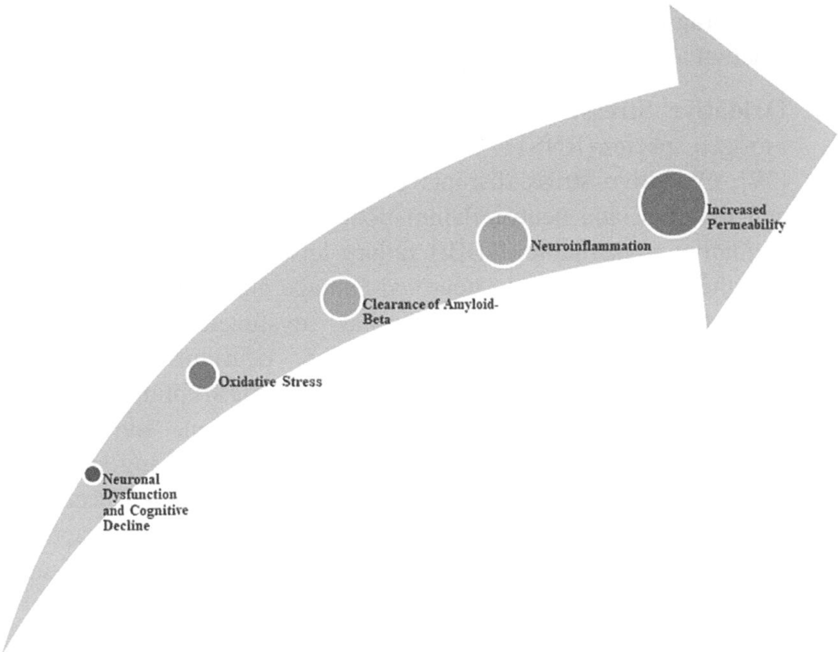

FIGURE 4.1 Blood–brain barrier dysfunction in AD pathology.

Increased permeability is a key sign of BBB failure in Alzheimer's [27]. The BBB closely controls chemical and ion transport between the circulation and brain [28]. In AD, the BBB breaks down, letting inflammatory mediators, neurotoxins, and plasma proteins into the brain [29]. Increased permeability damages and kills neurons [31], disrupts homeostasis, and increases neuroinflammation [30]. Neuroinflammation: BBB dysfunction is closely associated to AD development [32]. Microglia and astrocytes activated upon BBB rupture release pro-inflammatory cytokines, chemokines, and ROS, affecting BBB integrity and perpetuating neuroinflammatory responses [33]. Chronic neuroinflammation can lead to abnormal BBB function, Aβ plaque buildup, tau protein hyperphosphorylation, neuronal degeneration, and cognitive impairment [34, 35]. To eliminate Aβ from the brain, the BBB and glymphatic system, a network of perivascular gaps around cerebral blood vessels, are the main regulators [36]. Breaking the BBB prevents Aβ from being properly removed from the brain, causing it to collect and cluster in the extracellular environment [37]. Aβ directly impacts BBB integrity by altering tight junction proteins and encouraging endothelial cells, leading to a domino effect of BBB malfunction and Aβ buildup [38].

Oxidative Stress: The brain microenvironment produces reactive nitrogen species (RNS) and ROS during AD-related BBB breakdown [39]. Oxidative stress disrupts tight junction proteins, endothelial cell function, and neuroinflammation, worsening BBB failure [40]. Oxidative stress-induced BBB failure lets peripheral immune cells into the brain, worsening neuronal damage and cognitive loss [41]. AD causes widespread neuronal dysfunction and cognitive decline due to BBB dysfunction, neuroinflammation, Aβ accumulation, and oxidative stress [42]. Disrupting the BBB impairs neuronal homeostasis, synaptic function, and neurodegeneration [43–44]. This causes cognitive decline and dementia. BBB disruption, connected to neuroinflammation, Aβ accumulation, oxidative stress, and neuronal dysfunction, greatly impacts AD etiology and course. Targeting the BBB may lead to novel, effective AD treatments that address disease pathophysiology.

Mechanisms of Alzheimer's Disease's BBM Dysfunction: AD is characterized by chronic neuroinflammation, including activation of microglia and astrocytes [45] and release of pro-inflammatory cytokines such TNF-α and IL-6 [46]. Inflammatory mediators disrupt the BBB by activating endothelial cells, upregulating adhesion molecule expression [47], and secreting matrix metalloproteinases (MMPs), which degrade extracellular matrix components and tight junction proteins, as shown in Figure 4.2 [48].

Various Mechanisms of BBB Integrity: Inflammation causes BBB endothelial cells to express additional cell surface adhesion molecules such VCAM-1 and ICAM-1. By allowing immune cells, particularly leukocytes, to cross the BBB, these adhesion molecules cause neuroinflammation and neuronal injury [49].

Tight Junction Disruption: Chronic neuroinflammation alters the expression and distribution of tight junction proteins such occluding, claudins, and ZO-1, which are necessary for BBB integrity. The BBB's ability to prevent pathogens and inflammatory cells into the brain parenchyma is weakened by tight junction disruption, which worsens neuroinflammation and neuronal dysfunction [50]. In response to inflammatory cytokines such TNF-α, IL-1β, and IL-6, endothelial cells and astrocytes release and activate MMPs. The MMP family of proteolytic enzymes breaks down tight junction proteins and extra-cellular matrix components, making the BBB easier to break down. MMP activity opens the BBM, allowing immune cells and inflammatory mediators to enter the brain and worsen neuroinflammation and neurodegeneration [51].

ROS Production: In addition to affecting the BBB, inflammatory mediators produce ROS in the brain microenvironment. ROS like superoxide and hydrogen peroxide damage biological components like BBB endothelial cells. Oxidative stress worsens BBB failure and AD-related inflammation and neuronal damage via activating endothelial cells, tight junctions, and MMPs [52].

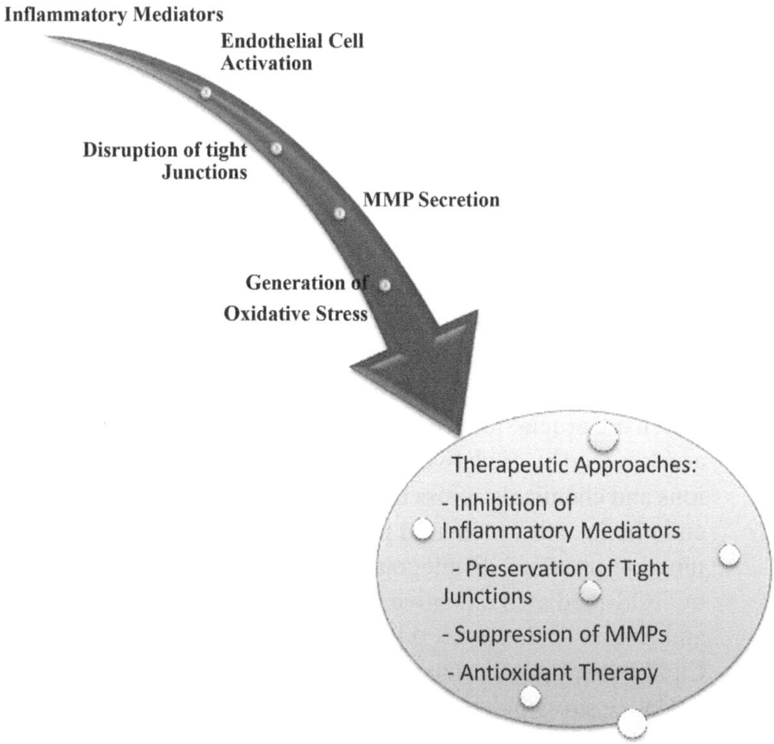

FIGURE 4.2 Mechanism of BBB dysfunction in Alzheimer's disease by inflammatory mediators.

Oxidative Stress: Figure 4.3 shows how important oxidative stress is in breaking down the BBB in AD. Oxidative stress is caused by an imbalance between the production of ROS and antioxidant defense systems. ROS generated in the brain microenvironment damage endothelial cells, disrupt tight junction proteins, and weaken the integrity of the BBB. Moreover, oxidative stress-induced lipid peroxidation and protein modifications lead to alterations in BBB permeability and endothelial cell dysfunction [53].

4.3 OXIDATIVE STRESS INFLUENCES THE INTEGRITY OF THE BLOOD–BRAIN BARRIER

1. ROS Generation: ROS generation and the brain's antioxidant defense systems are out of balance, which leads to oxidative stress. Superoxide, hydrogen peroxide, and hydroxyl radicals are just a few of the biological activities that produce ROS. Other mechanisms include excitotoxicity, inflammation, and mitochondrial respiration. Oxidative damage results from overproduction of ROS, which surpasses the antioxidant capability of the cells [54].
2. Endothelial Cell Damage: ROS produced in the brain microenvironment have the direct ability to harm the BBB's endothelial cells. Because endothelial cells have a high metabolic activity and a high lipid content, they are especially vulnerable to oxidative stress. ROS cause oxidative changes to lipids, proteins, and DNA, which affect cellular function and compromise the integrity of the BBB, ultimately leading to endothelial cell dysfunction [55].
3. Disruption of Tight Junction Proteins: Tight junction proteins, which are crucial for preserving the BBB's selective permeability, are damaged by oxidative stress. Tight junctions control the flow of ions and chemicals across the BBB by sealing the gaps in between endothelial cells. Increased BBB permeability is the result of tight junction proteins' disintegration and changed distribution caused by ROS-mediated oxidative changes, such as occluding, claudins, and zonula occludes (ZO) [56].
4. Lipid Peroxidation: AD-related BBB impairment is also a result of oxidative stress-induced lipid peroxidation. Reactive lipid peroxides and other harmful byproducts are produced when membrane lipids, mainly polyunsaturated fatty acids, undergo oxidative destruction.

This process is known as lipid peroxidation. Lipid peroxidation leads to endothelial cell dysfunction and changes in BBB permeability by upsetting membrane integrity, changing the makeup of lipids, and impairing cellular signaling [57].

5. Protein Modifications: Additionally, post-translational changes to proteins essential for the upkeep and operation of the BBB can be brought on by oxidative stress. The structural integrity and functional characteristics of proteins, such as tight junction proteins, transporters, and BBB receptors, are jeopardized by protein oxidation, nitration, and carboxylation. These protein alterations cause intercellular communication to be disrupted, cellular signaling pathways to be impaired, and BBB dysfunction in AD [58].

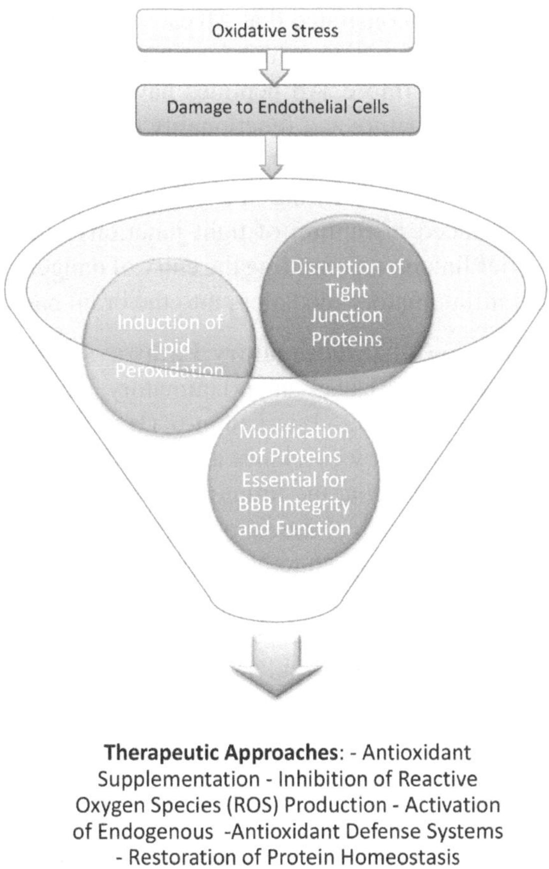

FIGURE 4.3 Mechanism of BBB dysfunction in Alzheimer's disease by oxidative stress.

Amyloid-Beta Accumulation: One of the main pathogenic characteristics of AD is the buildup of Aβ peptides in the brain, which is intimately linked to BBB disruption. By interacting with endothelial cells, rupturing tight junction proteins, and initiating inflammatory pathways, Aβ can have a direct impact on the integrity of the BBM. Additionally, Aβ buildup hinders the BBB's ability to remove Aβ, which causes it to accumulate in the brain parenchyma as shown in Figure 4.4 and exacerbate neuroinflammation and neuronal damage [59].

1. **Contributing to disease pathogenesis in several ways:**

    i. **Direct Effects on BBB Integrity:** Aβ peptides have the ability to directly interact with the endothelial cells that make up the BBB, which can change the integrity of the barrier. It has been demonstrated that Aβ causes endothelial cell activation and dysfunction, which increases the BBB's permeability [60]. Furthermore, Aβ peptides have the ability to interfere with the structure and functionality of tight junction proteins, including claudins, ZO, and occluding, all of which are essential for preserving the BBB's selective permeability. Aβ-induced disruption of tight junctions impairs the BBB's barrier function, permitting the entry of dangerous compounds and inflammatory mediators into the brain parenchyma [61].

    ii. **Activation of Inflammatory Pathways:** A build-up of β at the BBB may cause an inflammatory reaction in the brain microenvironment. Interleukin-6 (IL-6), TNF-α, and other pro-inflammatory cytokines are generated when Aβ peptides activate microglia, the immune cells that dwell in the brain. These inflammatory mediators further erode the BBMs integrity by increasing endothelial cell activity, upregulating the production of adhesion molecules, and permitting immune cell transmigration across the BBB. The subsequent neuroinflammation that results from AD exacerbates neuronal damage and contributes to cognitive decline [62].

    iii. **Impaired Clearance Across the BBB:** Transport systems at the BBB, including the receptor for advanced glycation end products (RAGE) and low-density lipoprotein receptor-related protein 1 (LRP1), are the main regulators of the brain's removal of Aβ. However, changes to these transport systems limit the amount of Aβ that is cleared across the BBB in AD. Aβ peptides build up at the BBB, which causes the brain parenchyma to retain more of them and limit their removal. This build-up accelerates the pathophysiology of AD by exacerbating neuroinflammation, oxidative stress, and neuronal damage [63].

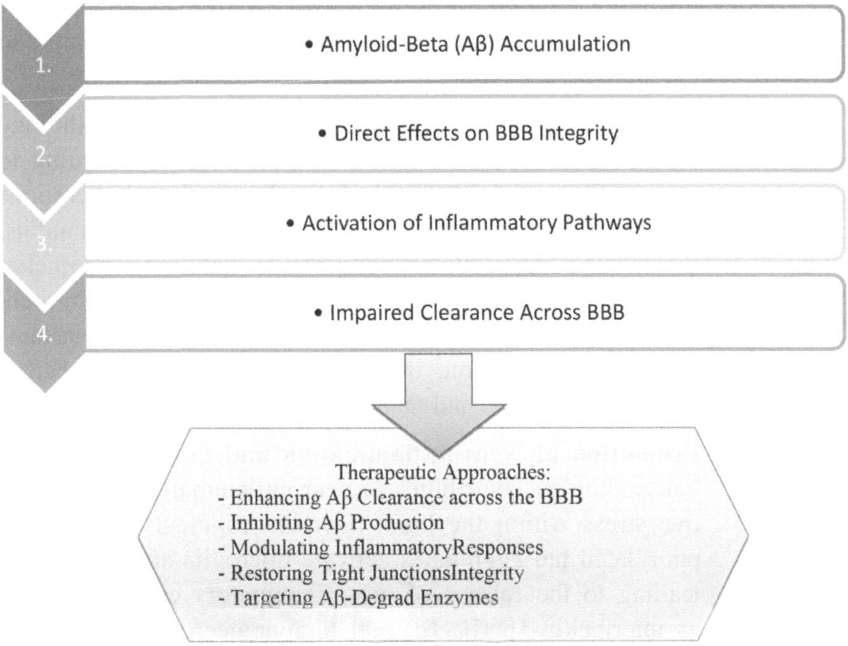

FIGURE 4.4 Mechanism of BBB dysfunction in Alzheimer's disease by amyloid-beta accumulation.

iv. Tau Pathology: Another characteristic of AD pathology that leads to BBB dysfunction is tau protein hyperphosphorylation and aggregation form NFTs. In addition to compromising BBB integrity and causing endothelial cell dysfunction, tau pathology also increases oxidative stress and neuroinflammation. BBB structure and function are further compromised by tau-mediated changes in cytoskeletal dynamics and microtubule stability [64]. Beyond the buildup of Aβ in Figure 4.5, hyperphosphorylation and tau protein aggregation form NFTs are further aspects of the pathophysiology of AD [65].

2. **Exacerbating disease progression through various mechanisms:**
   a. **Disruption of Endothelial Cell Function:** Tau pathology disrupts endothelial cell function at the BBB, impairing its ability to maintain barrier integrity. Hyperphosphorylated tau aggregates can directly interact with endothelial cells, leading to cellular dysfunction and alterations in permeability. This disruption compromises the BBB's ability to regulate the passage of molecules and ions between the bloodstream and the brain parenchyma [66].
   b. **Impaired BBB Integrity:** The accumulation of tau pathology at the BBB compromises its structural integrity, leading to increased permeability and leakage. Tau-mediated alterations in tight junction proteins, such as occluding and claudins, disrupt the intercellular connections between endothelial cells, further compromising BBB integrity. As a result, there is enhanced infiltration of harmful substances and inflammatory mediators into the brain parenchyma, exacerbating neuroinflammation and neuronal injury [67].
   c. **Promotion of Neuroinflammation and Oxidative Stress:** Tau pathology contributes to neuroinflammation and oxidative stress within the brain microenvironment. Hyperphosphorylated tau aggregates activate microglia and astrocytes, leading to the release of pro-inflammatory cytokines, such as interleukin-1β (IL-1β) and tumour necrosis factor-alpha (TNF-α). These inflammatory mediators exacerbate BBB dysfunction by inducing endothelial cell activation and promoting the secretion of MMPs, which degrade tight junction proteins. Additionally, tau-induced oxidative stress further compromises BBB integrity, contributing to neurodegeneration and cognitive decline in AD [68].
   d. **Alterations in Cytoskeletal Dynamics:** Tau pathology disrupts cytoskeletal dynamics and microtubule stability within endothelial cells, affecting BBB structure and function [69]. Hyperphosphorylated tau aggregates impair microtubule assembly and organization, leading to cytoskeletal instability and endothelial cell dysfunction. These alterations compromise the structural integrity of the BBB and its ability to

maintain barrier function, exacerbating BBB permeability alterations and contributing to disease progression in AD [70].

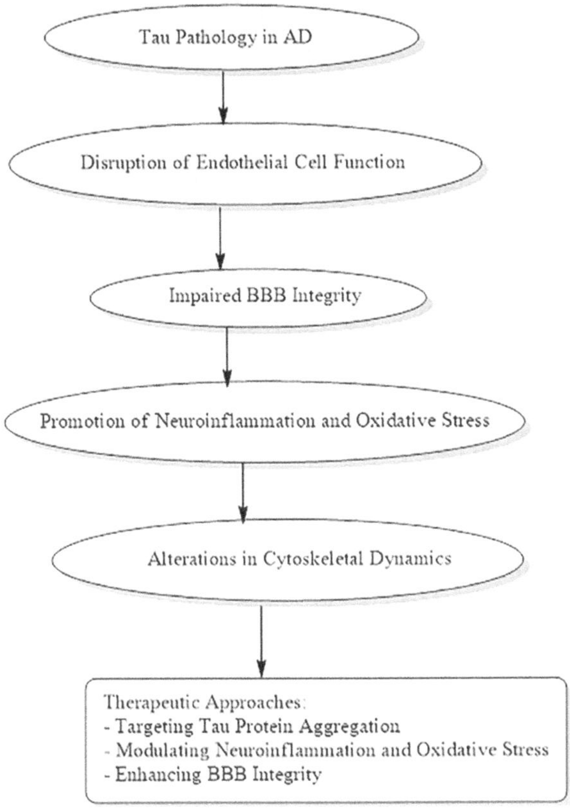

FIGURE 4.5 Mechanism of BBB dysfunction in Alzheimer's disease by tau pathology.

e. **Vascular Dysfunction:** AD is increasingly recognized as a vascular disorder, characterized by cerebrovascular dysfunction and impaired cerebral blood flow regulation [71]. Vascular risk factors, such as hypertension, diabetes, and atherosclerosis, contribute to BBB dysfunction by damaging endothelial cells, compromising tight junction integrity, and promoting neuroinflammation and oxidative stress. Moreover, vascular dysfunction exacerbates cerebral hypoperfusion, leading to neuronal hypoxia, energy depletion, and neuronal injury which is explained in Figure 4.6 [72].

3. **Contributing to BBB dysfunction through various mechanisms:**

   i. **Damage to Endothelial Cells:** Vascular risk factors such as hypertension, diabetes, and atherosclerosis can damage endothelial cells lining cerebral blood vessels. Endothelial cell dysfunction compromises the integrity of the BBB, leading to increased permeability and leakage. Damage to endothelial cells also impairs their ability to regulate cerebral blood flow and maintain vascular tone, further exacerbating cerebral hypoperfusion and neuronal injury [73].

   ii. **Compromised Tight Junction Integrity:** Tight junction proteins, which are essential for preserving the BBB's selective permeability, are broken down by vascular dysfunction. The BBB's integrity is jeopardized when damage is done to tight junction proteins like claudins and occluding, which permits the entry of pathogens and inflammatory mediators into the brain parenchyma. In AD, this disturbance exacerbates neuroinflammation, oxidative stress, and neuronal damage [74].

   iii. **Promotion of Neuroinflammation and Oxidative Stress:** Vascular dysfunction contributes to neuroinflammation and oxidative stress within the brain microenvironment. Hypoperfusion and ischemia brought on by poor regulation of cerebral blood flow produce ROS and inflammatory mediators. Pro-inflammatory cytokines and chemokines are released when these stimuli activate microglia and astrocytes. By stimulating endothelial cell activation and encouraging the release of MMPs, which break down tight junction proteins and extracellular matrix components, the ensuing neuroinflammatory response worsens BBB failure [75].

   iv. **Cerebral Hypoperfusion and Neuronal Injury:** Vascular dysfunction exacerbates cerebral hypoperfusion, leading to neuronal hypoxia, energy depletion, and neuronal injury. Reduced cerebral blood flow deprives neurons of oxygen and nutrients, impairing cellular metabolism and function. Chronic cerebral hypoperfusion contributes to the development of white matter lesions, neuronal loss, and cognitive decline in AD. Additionally, hypoperfusion-induced neuronal injury further compromises BBB integrity, exacerbating neurodegeneration and disease progression [76].

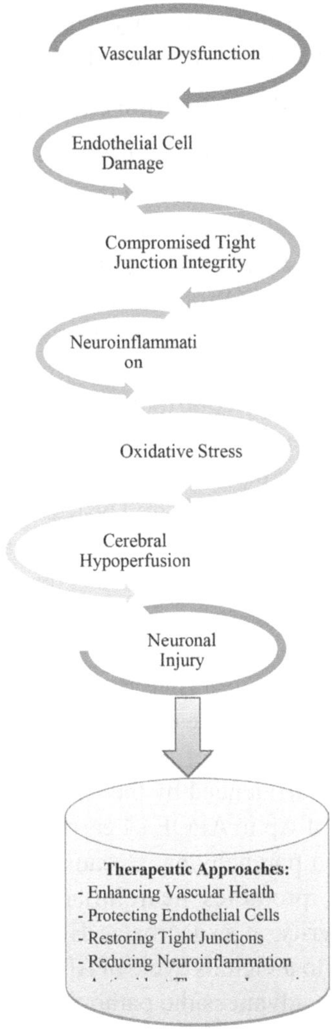

FIGURE 4.6 Mechanism of BBB dysfunction in Alzheimer's disease by vascular dysfunction.

v. **Apolipoprotein E (APOE) Genotype:** APOE, particularly the ε4 allele, is associated with BBB dysfunction and is the major genetic risk factor for late-onset AD. Reduced clearance of Aβ from the brain, decreased endothelial cell function, and increased BBB permeability have all been associated with the APOE ε4 genotype. Lipid metabolism and transport across the BBB depend on APOE. Moreover, APOE ε4 carriers exhibit elevated neuroinflammatory responses and oxidative stress, exacerbating BBB failure in AD, as Figure 4.7 [77] illustrates.

4. **Contributing to disease pathogenesis in several ways:**

    a. Impaired BBB Integrity: In terms of lipid metabolism and transport via the BBB, APOE is essential. Increased permeability and leakage are caused by changes in BBB structure and function linked to the APOE ε4 allele. The BBB integrity is reduced in APOE ε4 carriers, as evidenced by changes in tight junction proteins and dysfunctional endothelial cells. These alterations worsen neuroinflammation and neuronal damage in AD by contributing to the collapse of the BBB and permitting toxic chemicals and inflammatory mediators to enter the brain parenchyma [78].
    b. Altered Lipid Metabolism: Dysregulation of lipid metabolism is seen in APOE ε4 carriers, which may affect BBB function. Because of the APOE ε4 allele's decreased efficiency in lipid transport and clearance, lipoproteins and cholesterol build up at the BBB. AD-related dyslipidemia linked to the APOE ε4 genotype can worsen BBB dysfunction, damage tight junction integrity, and decrease endothelial cell function [79].
    c. Impaired Clearance of Aβ: The removal of Aβ peptides from the brain is influenced by the APOE genotype. Reduced BBB clearance of Aβ in APOE ε4 carriers causes Aβ to accumulate in the brain parenchyma. Because Aβ causes endothelial cell activation, promotes neuroinflammation, and compromises BBB integrity, it exacerbates BBB dysfunction even more. This leads to a vicious cycle of Aβ build-up and BBB malfunction, which advances the pathophysiology of AD [80].
    d. Enhanced Neuroinflammatory Responses and Oxidative Stress: Carriers of APOE ε4 have higher neuroinflammatory responses and intra-brain oxidative stress. The presence of the APOE ε4 allele increases the production of pro-inflammatory cytokines and ROS, which worsens BBB breakdown. When neuroinflammation and oxidative stress are present, endothelial

cell dysfunction, altered tight junctions, and diminished BBB integrity occur, which accelerates the pathogenesis of AD [81].

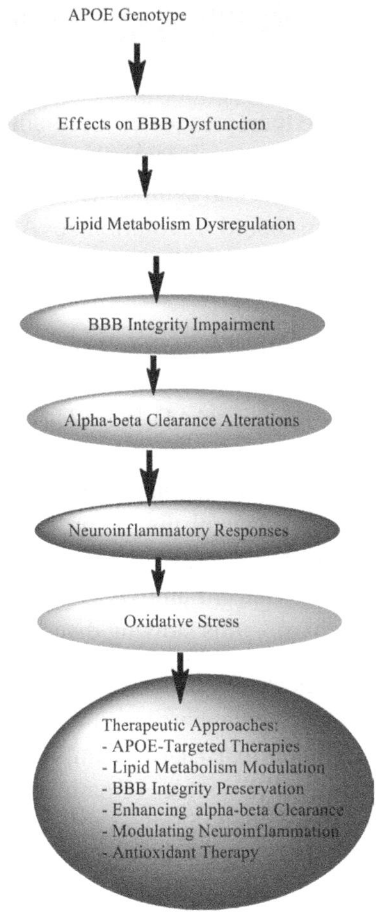

FIGURE 4.7 Mechanism of BBB dysfunction in Alzheimer's disease by apolipoprotein E (APOE) genotype.

4.4 IMPLICATIONS FOR DRUG DISCOVERY: TARGETING BBB DYSFUNCTION

1. **BBB as a Therapeutic Target:** BBB dysfunction plays a crucial role in AD progression. Disruption of the BBB allows harmful

substances and inflammatory mediators into the brain, worsening neuroinflammation, oxidative stress, and neuronal injury [82]. Therapies aiming to restore BBB integrity could mitigate these effects, improving cerebral blood flow regulation and nutrient delivery. By targeting BBB dysfunction, treatments can dampen neuroinflammation and reduce oxidative damage, potentially slowing neurodegeneration. Novel approaches, like small molecules or gene therapies, offer promising strategies to directly address BBB dysfunction, potentially modifying the course of AD rather than just relieving symptoms [83]. Ultimately, targeting BBB dysfunction could revolutionize AD treatment, offering hope for meaningful clinical outcomes and improved quality of life.

2. **Development of BBB-Targeting Therapeutics**: Drug discovery efforts are focused on developing therapeutics that specifically target BBB dysfunction in AD [84]. This therapeutics, including small molecules, antibodies, peptides, gene therapies, and nanoparticle-based delivery systems, aim to modulate BBB permeability, enhance clearance mechanisms, or protect against BBB damage [85]. By restoring BBB integrity, these therapies seek to prevent harmful substances and inflammatory mediators from infiltrating the brain parenchyma, thus slowing disease progression [86].

Small molecules are being investigated for their ability to modulate BBB permeability and enhance barrier function by targeting specific signaling pathways involved in BBB regulation. Monoclonal antibodies are being developed to target molecules or pathways associated with BBB dysfunction, neutralizing or blocking them to restore BBB integrity [87]. Peptide-based therapeutics aim to mimic endogenous BBB-regulating peptides or target specific receptors to promote the clearance of neurotoxic substances [88]. Gene therapies involve delivering genes encoding BBB-regulating proteins to endothelial cells to strengthen barrier integrity. Nanoparticle-based delivery systems encapsulate therapeutics and utilize surface modifications to enhance transport across the BBB, improving efficacy and specificity.

Innovative Drug Delivery Systems: Recent advancements in drug delivery technologies present promising solutions to address BBB-related challenges in AD therapeutics.

4.4.1 INNOVATIVE APPROACHES ARE TRANSFORMING DRUG DELIVERY TO THE BRAIN

1. **Nanoparticle-Based Delivery Systems**: Nanoparticle-based delivery systems employ nanoparticles as carriers to transport therapeutic agents across the BBB. These nanoparticles can encapsulate various drugs, peptides, nucleic acids, or imaging agents, designed to bypass BBB efflux mechanisms and enhance brain penetration. Surface modifications, such as PEGylating or targeting ligands, further facilitate BBB crossing and enhance specificity for brain-targeted delivery. By encapsulating therapeutics within nanoparticles, drug stability can be improved, systemic toxicity reduced, and therapeutic efficacy enhanced [89].
2. **Liposomal Formulations**: Liposomal formulations utilize liposomes, spherical vesicles composed of lipid bilayers, as carriers for drug delivery. Liposomes can encapsulate hydrophilic and hydrophobic drugs, protecting them from degradation and enhancing their solubility and bioavailability. These formulations can be engineered to target specific receptors or transporters expressed at the BBB, facilitating their transport across the barrier and into the brain parenchyma. By encapsulating therapeutics within liposomes, drug delivery to the brain can be improved, leading to enhanced therapeutic efficacy and reduced side effects [90].
3. **Cell-Penetrating Peptides (CPPs)**: CPPs are short peptides with the ability to traverse cell membranes and facilitate the intracellular delivery of cargoes, including drugs, peptides, and nucleic acids [91]. CPPs can be conjugated to therapeutic agents to enhance their uptake into brain cells or to facilitate their transport across the BBB. By exploiting endogenous cellular uptake mechanisms, CPPs offer a versatile and efficient approach for delivering therapeutics to the brain. Additionally, CPPs can be engineered to target specific cell types or subcellular compartments, further enhancing their therapeutic potential in AD [92].

These innovative drug delivery systems offer several advantages for AD therapeutics:

a. Enhanced BBB Penetration: Nanoparticle-based delivery systems, liposomal formulations, and CPPs can bypass BBB efflux mechanisms and enhance the transport of therapeutics across the barrier, enabling their delivery to the brain parenchyma [93].

b. Targeted Delivery: These delivery systems can be engineered to target specific receptors, transporters, or cell types expressed at the BBB or within the brain, allowing for targeted delivery of therapeutics to regions affected by AD pathology [94].
c. Improved Therapeutic Efficacy: By enhancing drug stability, solubility, and brain penetration, these delivery systems can improve the therapeutic efficacy of AD drugs, leading to better treatment outcomes and disease modification [95].
d. Reduced Systemic Toxicity: By minimizing off-target effects and systemic exposure, these delivery systems can reduce the risk of systemic toxicity associated with AD therapeutics, improving patient safety and tolerability [96].

4. **BBB Modulation Strategies:** Novel strategies for modulating BBB permeability offer exciting potential for treating AD. By targeting specific molecular pathways involved in BBB regulation, such as endothelial cell signaling, tight junction dynamics, and transport mechanisms, these approaches aim to enhance drug delivery to the brain while maintaining barrier integrity and minimizing off-target effects [97]. Targeting endothelial cell signaling pathways mediated by growth factors, cytokines, or neurotransmitters could selectively modulate BBB permeability, allowing increased drug delivery to the brain without compromising barrier integrity [98]. Modulating tight junction dynamics through reversible disruption could also facilitate enhanced drug delivery across the BBB, followed by restoration of tight junctions to maintain barrier function [99]. Additionally, strategies to enhance transport mechanisms, such as carrier-mediated transporters and receptor-mediated transcytosis, offer opportunities to improve drug penetration into the brain. Nanotechnology-based approaches utilize engineered nanoparticles to encapsulate drugs and target specific BBB regulation pathways, while bioengineering strategies, such as tissue engineering and 3D organoid models, provide valuable tools for studying BBB physiology and developing innovative therapeutic approaches [100]. Overall, these novel strategies hold promise for developing effective BBB-modulating therapeutics

for AD, potentially revolutionizing treatment approaches for the disease [101].

5. **Combination Therapies:** Combining BBB-targeting therapies with existing AD treatments shows great promise for enhancing therapeutic efficacy and overcoming treatment resistance. These combination approaches offer synergistic benefits by addressing both the underlying pathology of AD and the associated BBB dysfunction [102]. By targeting multiple aspects of the disease simultaneously, combination therapies may have a greater impact on slowing disease progression and improving cognitive outcomes compared to monotherapy approaches. Many existing AD treatments have shown limited efficacy in clinical trials due to the complexity of the disease and BBB dysfunction, which limits drug delivery to the brain parenchyma. BBB-targeting therapies could enhance drug delivery across the BBB, overcoming treatment resistance and improving the response to existing AD treatments [103].

   Combination therapies could provide a more comprehensive approach to treating AD by targeting multiple pathways implicated in BBB dysfunction. For example, combining therapies that enhance BBB integrity, reduce neuroinflammation, and promote Aβ clearance could address different aspects of AD pathology and synergistically improve therapeutic outcomes. These combination therapies could be personalized based on individual patients' disease stage, genetic profile, and BBB status. Biomarkers associated with BBB dysfunction could be used to identify patients most likely to benefit from combination therapies, optimizing therapeutic outcomes and minimizing potential adverse effects [104].

   Future research efforts should focus on identifying optimal combinations of BBB-targeting therapies and existing AD treatments, as well as elucidating the mechanisms underlying their synergistic effects. Preclinical studies and clinical trials are needed to evaluate the safety, efficacy, and long-term benefits of combination therapies in AD. Additionally, innovative drug delivery technologies and biomarker-based approaches could facilitate the development and implementation of personalized combination therapies for AD [105]. Combination therapies that combine BBB-targeting therapies with existing AD treatments offer a promising approach for enhancing therapeutic efficacy, overcoming treatment resistance, and providing a more comprehensive treatment strategy for AD. By addressing both the underlying pathology of AD and the associated BBB dysfunction, combination therapies hold great potential for improving clinical outcomes and quality of life for patients with AD [106].

6. **Personalized Medicine Approaches:** Personalized medicine approaches hold significant promise for optimizing the effectiveness of BBB-targeting therapies in AD treatment. Genetic profiling can identify individuals most likely to benefit from BBB-targeting therapies based on their genetic risk profile, such as APOE ε4 carriers who may be more responsive to therapies aimed at restoring BBB integrity [107]. Additionally, considering the disease stage is crucial, with early-stage AD patients potentially benefiting from preserving BBB integrity and later-stage patients requiring more aggressive interventions to restore barrier function. Biomarkers associated with BBB dysfunction can assess individual BBB status, guiding treatment strategies to enhance barrier function or reduce permeability as needed. Incorporating biomarker-based monitoring allows clinicians to assess treatment response and adjust strategies over time, optimizing therapeutic outcomes and minimizing adverse effects [108]. Ultimately, personalized medicine approaches offer tailored treatment strategies that improve clinical care for AD patients by optimizing BBB-targeting therapies to individual patient needs.

4.5 CHALLENGES IN DEVELOPING THERAPEUTICS FOR BBB DYSFUNCTION IN ALZHEIMER'S DISEASE

a. BBB Complexity: The BBB regulates molecular exchange between the circulation and brain parenchyma through a highly specialized and dynamic interface. Therapy that penetrates the BBB while retaining barrier integrity is difficult to develop. Targeted therapeutics are difficult to design since the BBB has endothelial cells, tight junctions, pericytes, astrocytes, and basement membrane components [109].
b. Ensuring selectivity and specificity in BBB-targeting therapies is crucial to reduce off-target effects and toxicity. Therapeutics must selectively target BBB-related pathways or mechanisms

implicated in AD pathogenesis without affecting normal BBB function or other physiological processes. Therapeutic agents must be designed and optimized to achieve this equilibrium [110].

c. Drug Delivery: Addressing brain delivery challenges is crucial for developing BBB-targeting therapies for AD. Size, charge, hydrophilicity, and efflux transporter interactions make many therapeutic compounds poorly brain penetrant. Drug delivery systems that penetrate the BBB, circumvent efflux processes, and release therapies in the brain parenchyma are difficult to design [111].
d. Controlled and reversible BBB permeability modulation is essential for safe and successful AD treatment. Effective BBB permeability therapies must avoid severe barrier disruption, which can cause neuroinflammation, edema, and neurotoxicity. Modulating BBB permeability precisely while reducing off-target effects is difficult [112].
e. Identifying biomarkers for BBB failure in AD is crucial for patient classification, therapy monitoring, and efficacy assessment. Validated indicators for BBB state or function *in vivo* are lacking. Developing sensitive and precise biomarkers for AD BBB impairment requires investigation and confirmation [113].
f. Limited Preclinical Models: BBB-targeting therapies for AD have challenges in development. Numerous animal models fail to replicate human AD's complicated pathophysiology, including BBB failure. For treatment efficacy and safety evaluation, more physiologically relevant and predictive preclinical models that accurately mimic AD BBB impairment are needed [114]. Achieving clinical translation of promising preclinical results into viable AD medications is a serious hurdle. Emerging issues prevent many BBB-targeting therapies from proving efficacy or safety in clinical trials. To overcome these obstacles and successfully translate preclinical research into clinical practice, academia, industry, regulatory agencies, and healthcare providers must collaborate [115].

4.6 CHALLENGES IN BBB ASSESSMENT

Exploring the challenges in BBB assessment reveals the complexities inherent in accurately evaluating BBB integrity and function. Current methods face limitations that hinder diagnostic accuracy and reliability.

Addressing these challenges requires innovative strategies and advancements in BBB assessment techniques (Table 4.1).

TABLE 4.1 Overview of the Challenges and Potential Solutions.

S. No.	Problem	Solution
1.	Limited Accessibility: One major challenge in BBB assessment is the limited accessibility of the barrier for direct measurement. Traditional methods, such as cerebrospinal fluid sampling or brain tissue biopsy, are invasive and carry risks for patients. Non-invasive imaging techniques, such as magnetic resonance imaging (MRI) or positron emission tomography (PET), offer alternatives but may lack spatial resolution or sensitivity to detect subtle BBB alterations.	Advancements in imaging technology, such as higher-resolution MRI sequences or novel PET tracers targeting BBB-specific markers, could improve the sensitivity and specificity of non-invasive BBB assessment. Additionally, developing blood-based biomarkers that reflect BBB integrity could provide accessible and less invasive alternatives for assessing BBB function.
2.	Complexity of BBB Dynamics: The BBB is a dynamic interface influenced by multiple factors, including neuroinflammation, oxidative stress, and vascular dysfunction. Assessing BBB integrity and function requires consideration of these complex interactions, which may vary spatially and temporally across different brain regions and disease stages.	Integrating multimodal imaging techniques, biomarker panels, and computational modeling approaches could provide a more comprehensive understanding of BBB dynamics in health and disease. By combining data from multiple sources, researchers can elucidate the underlying mechanisms driving BBB dysfunction and develop more accurate diagnostic tools.
3.	Lack of Standardization: There is a lack of standardized protocols and criteria for assessing BBB integrity and function, leading to variability and inconsistency across studies. Differences in experimental conditions, imaging parameters, and analytical methods can confound results and hinder reproducibility.	Establishing standardized protocols and guidelines for BBB assessment could improve consistency and comparability across studies. Collaborative efforts between researchers, clinicians, and regulatory agencies are needed to define best practices for BBB assessment and ensure the reliability of diagnostic tests and research findings.
	Clinical Translation Challenges: Translating preclinical findings into clinical applications poses challenges due to differences in species, disease models, and patient populations. Biomarkers and imaging techniques validated in preclinical studies may not always translate to humans, limiting their clinical utility.	Conducting longitudinal studies in human cohorts and validating findings across different patient populations are critical steps for clinical translation. Additionally, leveraging advances in personalized medicine and precision imaging techniques could enhance the relevance and applicability of BBB assessment methods in clinical practice.

4.7 CONCLUSION

Despite the formidable challenge posed by the BBB in AD therapeutics, recent advancements in drug discovery and innovative approaches provide promising avenues for overcoming these obstacles. By delving into the intricate mechanisms of BBB dysfunction in AD pathology, researchers can develop targeted therapeutics aimed at restoring barrier integrity, enhancing drug delivery to the brain, and mitigating neuroinflammation and neuronal injury. Innovative strategies such as nanotechnology-based drug delivery systems, BBB-targeting ligands, peptide-based therapeutics, and personalized medicine approaches offer significant potential for improving treatment outcomes. Moving forward, emphasis should be placed on advancing preclinical research, validating biomarkers, and fostering collaboration between academia, industry, regulatory agencies, and healthcare providers to accelerate the development and implementation of BBB-targeting therapeutics for AD. Despite challenges, the opportunities and innovations in drug discovery offer hope for bringing us closer to effective disease-modifying treatments for AD, thereby enhancing the lives of patients and families affected by this devastating neurodegenerative disorder.

KEYWORDS

- **Alzheimer's disease**
- **blood–brain barrier dysfunction**
- **drug discovery**
- **nanotechnology**
- **personalized medicine**
- **therapeutic challenges**

REFERENCES

1. Sahana, S., Kumar, R., Nag, S., Paul, R., Chatterjee, I., & Guha, N. (2020). A review on Alzheimer disease and future prospects.
2. Cummings, J., Lee, G., Zhong, K., Fonseca, J., & Taghva, K. (2021). Alzheimer's disease drug development pipeline: 2021. *Alzheimer's & Dementia: Translational Research & Clinical Interventions*, *7*(1), e12179.

3. Simrén, J., Leuzy, A., Karikari, T. K., Hye, A., Benedet, A. L., Lantero-Rodriguez, J., Mattsson-Carlgren, N., Schöll, M., Mecocci, P., Vellas, B., & Tsolaki, M. (2021). The diagnostic and prognostic capabilities of plasma biomarkers in Alzheimer's disease. *Alzheimer's & Dementia, 17*(7), 1145–1156.
4. Sheppard, O., & Coleman, M. (2020). *Alzheimer's disease: Etiology, neuropathology and pathogenesis*. Exon Publications, 1–21.
5. Lutz, M. W., Sprague, D., Barrera, J., & Chiba-Falek, O. (2020). Shared genetic etiology underlying Alzheimer's disease and major depressive disorder. *Translational Psychiatry, 10*(1), 88.
6. Ma, C., Hong, F., & Yang, S. (2022). Amyloidosis in Alzheimer's disease: Pathogeny, etiology, and related therapeutic directions. *Molecules, 27*(4), 1210.
7. Bossaerts, L., Cacace, R., & Van Broeckhoven, C. (2022). The role of ATP-binding cassette subfamily A in the etiology of Alzheimer's disease. *Molecular Neurodegeneration, 17*(1), 31.
8. Bellenguez, C., Küçükali, F., Jansen, I., Andrade, V., Moreno-Grau, S., Amin, N., Naj, A. C., Grenier-Boley, B., Campos-Martin, R., Holmans, P. A., & Boland, A. (2020). New insights on the genetic etiology of Alzheimer's and related dementia. *MedRxiv*, 2020–2010.
9. Kurz, C., Walker, L., Rauchmann, B. S., & Perneczky, R. (2022). Dysfunction of the blood–brain barrier in Alzheimer's disease: Evidence from human studies. *Neuropathology and Applied Neurobiology, 48*(3), e12782.
10. Sharma, C., Woo, H., & Kim, S. R. (2022). Addressing blood–brain barrier impairment in Alzheimer's disease. *Biomedicines, 10*(4), 742.
11. Nehra, G., Bauer, B., & Hartz, A. M. (2022). Blood–brain barrier leakage in Alzheimer's disease: From discovery to clinical relevance. *Pharmacology & Therapeutics, 234*, 108119.
12. Alahmari, A. (2021). Blood–brain barrier overview: Structural and functional correlation. *Neural Plasticity, 2021* .
13. Kadry, H., Noorani, B., & Cucullo, L. (2020). A blood–brain barrier overview on structure, function, impairment, and biomarkers of integrity. *Fluids and Barriers of the CNS, 17*, 1–24.
14. Lochhead, J. J., Yang, J., Ronaldson, P. T., & Davis, T. P. (2020). Structure, function, and regulation of the blood–brain barrier tight junction in central nervous system disorders. *Frontiers in Physiology, 11*, 558491.
15. Pandit, R., Chen, L., & Götz, J. (2020). The blood–brain barrier: Physiology and strategies for drug delivery. *Advanced Drug Delivery Reviews, 165*, 1–4.
16. Andjelkovic, A. V., Situ, M., Citalan-Madrid, A. F., Stamatovic, S. M., Xiang, J., & Keep, R. F. (2023). Blood–brain barrier dysfunction in normal aging and neurodegeneration: mechanisms, impact, and treatments. *Stroke, 54*(3), 661–672.
17. Pandit, R., Chen, L., & Götz, J. (2020). The blood–brain barrier: Physiology and strategies for drug delivery. *Advanced Drug Delivery Reviews, 165*, 1–4.
18. Hussain, B., Fang, C., & Chang, J. (2021). Blood–brain barrier breakdown: An emerging biomarker of cognitive impairment in normal aging and dementia. *Frontiers in Neuroscience, 15*, 688090.

19. Van Oostveen, W. M., & de Lange, E. C. (2021). Imaging techniques in Alzheimer's disease: a review of applications in early diagnosis and longitudinal monitoring. *International Journal of Molecular Sciences, 22*(4), 2110.
20. Hameed, S., Fuh, J. L., Senanarong, V., Ebenezer, E. G., Looi, I., Dominguez, J. C., Park, K. W., Karanam, A. K., & Simon, O. (2020). Role of fluid biomarkers and PET imaging in early diagnosis and its clinical implication in the management of Alzheimer's disease. *Journal of Alzheimer's Disease Reports, 4*(1), 21–37.
21. Villa, C., Lavitrano, M., Salvatore, E., & Combi, R. (2020). Molecular and imaging biomarkers in Alzheimer's disease: A focus on recent insights. *Journal of Personalized Medicine, 10*(3), 61.
22. Barisano, G., Montagne, A., Kisler, K., Schneider, J. A., Wardlaw, J. M., & Zlokovic, B. V. (2022). Blood–brain barrier link to human cognitive impairment and Alzheimer's disease. *Nature Cardiovascular Research, 1*(2), 108–115.
23. Banks, W. A., Reed, M. J., Logsdon, A. F., Rhea, E. M., & Erickson, M. A. (2021). Healthy aging and the blood–brain barrier. *Nature Aging, 1*(3), 243–254.
24. Pardridge, W. M. (2020). Treatment of Alzheimer's disease and blood–brain barrier drug delivery. *Pharmaceuticals, 13*(11), 394.
25. Xie, J., Van Hoecke, L., & Vandenbroucke, R. E. (2022). The impact of systemic inflammation on Alzheimer's disease pathology. *Frontiers in Immunology, 12*, 796867.
26. Ju, Y., & Tam, K. Y. (2022). Pathological mechanisms and therapeutic strategies for Alzheimer's disease. *Neural Regeneration Research, 17*(3), 543–549.
27. Kurz, C., Walker, L., Rauchmann, B. S., & Perneczky, R. (2022). Dysfunction of the blood–brain barrier in Alzheimer's disease: evidence from human studies. *Neuropathology and Applied Neurobiology, 48*(3), e12782.
28. He, J. T., Zhao, X., Xu, L., & Mao, C. Y. (2020). Vascular risk factors and Alzheimer's disease: blood–brain barrier disruption, metabolic syndromes, and molecular links. *Journal of Alzheimer's Disease, 73*(1), 39–58.
29. Uprety, A., Kang, Y., & Kim, S. Y. (2021). Blood–brain barrier dysfunction as a potential therapeutic target for neurodegenerative disorders. *Archives of Pharmacal Research, 44*(5), 487–498.
30. Lin, Z., Sur, S., Liu, P., Li, Y., Jiang, D., Hou, X., Darrow, J., Pillai, J. J., Yasar, S., Rosenberg, P., & Albert, M. (2021). Blood–brain barrier breakdown in relationship to Alzheimer and vascular disease. *Annals of Neurology, 90*(2), 227–238.
31. Barisano, G., Montagne, A., Kisler, K., Schneider, J. A., Wardlaw, J. M., & Zlokovic, B. V. (2022). Blood–brain barrier link to human cognitive impairment and Alzheimer's disease. *Nature Cardiovascular Research, 1*(2), 108–115.
32. Onyango, I. G., Jauregui, G. V., Čarná, M., Bennett, Jr J. P., & Stokin, G. B. (2021). Neuroinflammation in Alzheimer's disease. *Biomedicines, 9*(5), 524.
33. Kurz, C., Walker, L., Rauchmann, B. S., & Perneczky, R. (2022). Dysfunction of the blood–brain barrier in Alzheimer's disease: evidence from human studies. *Neuropathology and Applied Neurobiology, 48*(3), e12782.
34. Kloske, C. M., & Wilcock, D. M. (2020). The important interface between apolipoprotein E and neuroinflammation in Alzheimer's disease. *Frontiers in Immunology, 11*, 532040.

35. Leng, F., & Edison, P. (2021). Neuroinflammation and microglial activation in Alzheimer disease: where do we go from here? *Nature Reviews Neurology, 17*(3), 157–172.
36. Chachaj, A., Gąsiorowski, K., Szuba, A., Sieradzki, A., & Leszek, J. (2023). The lymphatic system in the brain clearance mechanisms-new therapeutic perspectives for Alzheimer's disease. *Current Neuropharmacology, 21*(2), 380.
37. Chachaj, A., Gąsiorowski, K., Szuba, A., Sieradzki, A., & Leszek, J. (2023). The lymphatic system in the brain clearance mechanisms-new therapeutic perspectives for Alzheimer's disease. *Current Neuropharmacology, 21*(2), 380.
38. Liu, J., Guo, Y., Zhang, C., Zeng, Y., Luo, Y., & Wang, G. (2022). Clearance systems in the brain, from structure to function. *Frontiers in Cellular Neuroscience, 15*, 729706.
39. Song, K., Li, Y., Zhang, H., An, N., Wei, Y., Wang, L., Tian, C., Yuan, M., Sun, Y., Xing, Y., & Gao, Y. (2020). Oxidative stress-mediated blood–brain barrier (BBB) disruption in neurological diseases. *Oxidative Medicine and Cellular Longevity, 2020*, 1–27.
40. Singh, A., Kukreti, R., & Kukreti, S. (2022). Oxidative stress and cellular dysfunction in neurodegenerative disease. *Neurodegenerative Diseases Biomarkers: Towards Translating Research to Clinical Practice, 2022*, 121–154.
41. Merelli, A., Repetto, M., Lazarowski, A., & Auzmendi, J. (2021). Hypoxia, oxidative stress, and inflammation: three faces of neurodegenerative diseases. *Journal of Alzheimer's Disease, 82*(s1), S109–126.
42. Montagne, A., Nation, D. A., Sagare, A. P., Barisano, G., Sweeney, M. D., Chakhoyan, A., Pachicano, M., Joe, E., Nelson, A. R., D'Orazio, L. M., & Buennagel, D. P. (2020). APOE4 leads to blood–brain barrier dysfunction predicting cognitive decline. *Nature, 581*(7806), 71–76.
43. Nakamura, T., Zou, K., Shibuya, Y., Michikawa, M. (2021). Oral dysfunctions and cognitive impairment/dementia. *Journal of Neuroscience Research, 99*(2), 518–528.
44. Viggiano, D., Wagner, C. A., Martino, G., Nedergaard, M., Zoccali, C., Unwin, R., & Capasso, G. (2020). Mechanisms of cognitive dysfunction in CKD. *Nature Reviews Nephrology, 16*(8), 452–469.
45. Hampel, H., Caraci, F., Cuello, A. C., Caruso, G., Nisticò, R., Corbo, M., Baldacci, F., Toschi, N., Garaci, F., Chiesa, P. A., & Verdooner, S. R. (2020). A path toward precision medicine for neuroinflammatory mechanisms in Alzheimer's disease. *Frontiers in Immunology, 11*, 456.
46. Uddin, M. S., Kabir, M. T., Jalouli, M., Rahman, M. A., Jeandet, P., Behl, T., Alexiou, A., Albadrani, G. M., Abdel-Daim, M. M., Perveen, A., & Ashraf, G. M. (2022). Neuroinflammatory signalling in the pathogenesis of Alzheimer's disease. *Current Neuropharmacology, 20*(1), 126.
47. Rather, M. A., Khan, A., Alshahrani, S., Rashid, H., Qadri, M., Rashid, S., Alsaffar, R. M., Kamal, M. A., & Rehman, M. U. (2021). Inflammation and Alzheimer's disease: Mechanisms and therapeutic implications by natural products. *Mediators of Inflammation, 2021*, 1–21.
48. Rauf, A., Badoni, H., Abu-Izneid, T., Olatunde, A., Rahman, M. M., Painuli, S., Semwal, P., Wilairatana, P., & Mubarak, M. S. (2022). Neuroinflammatory markers:

key indicators in the pathology of neurodegenerative diseases. *Molecules, 27*(10), 3194.
49. Guerra-Espinosa, C., Jiménez-Fernández, M., Sánchez-Madrid, F., & Serrador, J. M. (2024). ICAMs in immunity, intercellular adhesion and communication. *Cells, 13*(4), 339.
50. Greene, C., Hanley, N., & Campbell, M. (2020). Blood–brain barrier associated tight junction disruption is a hallmark feature of major psychiatric disorders. *Translational Psychiatry, 10*(1), 373.
51. Luchian, I., Goriuc, A., Sandu, D., & Covasa, M. (2022). The role of matrix metalloproteinases (MMP-8, MMP-9, MMP-13) in periodontal and peri-implant pathological processes. *International Journal of Molecular Sciences, 23*(3), 1806.
52. Takata, F., Nakagawa, S., Matsumoto, J., & Dohgu, S. (2021). Blood–brain barrier dysfunction amplifies the development of neuroinflammation: Understanding of cellular events in brain microvascular endothelial cells for prevention and treatment of BBB dysfunction. *Frontiers in Cellular Neuroscience, 15*, 661838.
53. Song, K., Li, Y., Zhang, H., An, N., Wei, Y., Wang, L., Tian, C., Yuan, M., Sun, Y., Xing, Y., & Gao, Y. (2020). Oxidative stress-mediated blood–brain barrier (BBB) disruption in neurological diseases. *Oxidative Medicine and Cellular Longevity, 2020*, 1–27.
54. Aborode, A. T., Pustake, M., Awuah, W. A., Alwerdani, M., Shah, P., Yarlagadda, R., Ahmad, S., Silva Correia, I. F., & Chandra, A., Nansubuga, E. P., & Abdul-Rahman, T. (2022). Targeting oxidative stress mechanisms to treat Alzheimer's and Parkinson's disease: A critical review. *Oxidative Medicine and Cellular Longevity, 2022.*
55. Anwar, M. M. (2022). Oxidative stress-a direct bridge to central nervous system homeostatic dysfunction and Alzheimer's disease. *Cell Biochemistry and Function, 40*(1), 17–27.
56. Teleanu, D. M., Niculescu, A. G., Lungu, I. I., Radu, C. I., Vladâcenco, O., Roza, E., Costăchescu, B., Grumezescu, A. M., & Teleanu, R. I. (2022). An overview of oxidative stress, neuroinflammation, and neurodegenerative diseases. *International Journal of Molecular Sciences, 23*(11), 5938.
57. Feng, J., Zheng, Y., Guo, M., Ares, I., Martínez, M., Lopez-Torres, B., Martínez-Larrañaga, M. R., Wang, X., Anadón, A., & Martínez, M. A. (2023). Oxidative stress, the blood–brain barrier and neurodegenerative diseases: The critical beneficial role of dietary antioxidants. *Acta Pharmaceutica Sinica B, 13*(10), 3988–4024.
58. Buccellato, F. R., D'Anca, M., Fenoglio, C., Scarpini, E., & Galimberti, D. (2021). Role of oxidative damage in Alzheimer's disease and neurodegeneration: From pathogenic mechanisms to biomarker discovery. *Antioxidants, 10*(9), 1353.
59. Selkoe, D. J. (2001). Alzheimer's disease results from the cerebral accumulation and cytotoxicity of amyloid\beta-protein. *Journal of Alzheimer's Disease, 3*(1), 75–82.
60. Gandy, S. (2005). The role of cerebral amyloid β accumulation in common forms of Alzheimer disease. *The Journal of Clinical Investigation, 115*(5), 1121–1129.
61. Rao, C. V., Asch, A. S., Carr, D. J., & Yamada, H. Y. (2020). "Amyloid-beta accumulation cycle" as a prevention and/or therapy target for Alzheimer's disease. *Aging Cell, 19*(3), e13109.
62. Glabe, C. (2001). Intracellular mechanisms of amyloid accumulation and pathogenesis in Alzheimer's disease. *Journal of Molecular Neuroscience, 17*, 137–145.

63. Bayer, T. A., & Wirths, O. (2010). Intracellular accumulation of amyloid-Beta-a predictor for synaptic dysfunction and neuron loss in Alzheimer's disease. *Frontiers in Aging Neuroscience*, *2*, 1359.
64. Gao, Y., Tan, L., Yu, J. T., & Tan, L. (2018). Tau in Alzheimer's disease: mechanisms and therapeutic strategies. *Current Alzheimer Research*, *15*(3), 283–300.
65. Chu, D., & Liu, F. (2018). Pathological changes of tau related to Alzheimer's disease. *ACS Chemical Neuroscience*, *10*(2), 931–944.
66. Michalicova, A., Majerova, P., & Kovac, A. (2020). Tau protein and its role in blood–brain barrier dysfunction. *Frontiers in Molecular Neuroscience*, *13*, 570045.
67. Song, L., Wells, E. A., & Robinson, A. S. (2021). Critical molecular and cellular contributors to tau pathology. *Biomedicines*, *9*(2), 190.
68. Zhang, H., Wei, W., Zhao, M., Ma, L., Jiang, X., Pei, H., Cao, Y., & Li, H. (2021). Interaction between Aβ and tau in the pathogenesis of Alzheimer's disease. *International Journal of Biological Sciences*, *17*(9), 2181.
69. Rawat, P., Sehar, U., Bisht, J., Selman, A., Culberson, J., & Reddy, P. H. (2022). Phosphorylated tau in Alzheimer's disease and other tauopathies. *International Journal of Molecular Sciences*, *23*(21), 12841.
70. Sinsky, J., Pichlerova, K., & Hanes J. (2021). Tau protein interaction partners and their roles in Alzheimer's disease and other tauopathies. *International Journal of Molecular Sciences*, *22*(17), 9207.
71. Klohs, J. (2020). An integrated view on vascular dysfunction in Alzheimer's disease. *Neurodegenerative Diseases*, *19*(3–4), 109–127.
72. Kelleher, R. J., & Soiza, R. L. (2013). Evidence of endothelial dysfunction in the development of Alzheimer's disease: Is Alzheimer'sa vascular disorder? *American Journal of Cardiovascular Disease*, *3*(4), 197.
73. Dickstein, D. L., Walsh, J., Brautigam, H., Stockton, Jr S. D., Gandy, S., & Hof, P. R. (2010). Role of vascular risk factors and vascular dysfunction in Alzheimer's disease. *Mount Sinai Journal of Medicine: A Journal of Translational and Personalized Medicine: A Journal of Translational and Personalized Medicine*, *77*(1), 82–102.
74. Solis, Jr E., Hascup, K. N., & Hascup, E. R. (2020). Alzheimer's disease: the link between amyloid-β and neurovascular dysfunction. *Journal of Alzheimer's Disease*, *76*(4), 1179–1198.
75. Govindpani, K., McNamara, L. G., Smith, N. R., Vinnakota, C., Waldvogel, H. J., Faull, R. L., & Kwakowsky, A. (2019). Vascular dysfunction in Alzheimer's disease: A prelude to the pathological process or a consequence of it? *Journal of Clinical Medicine*, *8*(5), 651.
76. Shabir, O., Berwick, J., & Francis, S. E. (2018). Neurovascular dysfunction in vascular dementia, Alzheimer's and atherosclerosis. *BMC Neuroscience*, *19*, 1–6.
77. Yamazaki, Y., Zhao, N., Caulfield, T. R., Liu, C. C., & Bu, G. (2019). Apolipoprotein E and Alzheimer disease: pathobiology and targeting strategies. *Nature Reviews Neurology*, *15*(9), 501–518.
78. Yu, J. T., Tan, L., & Hardy, J. (2014). Apolipoprotein E in Alzheimer's disease: An update. *Annual Review of Neuroscience*, *37*, 79–100.
79. Roses, MD A. D. (1996). Apolipoprotein E alleles as risk factors in Alzheimer's disease. *Annual Review of Medicine*, *47*(1), 387–400.

80. Liu, C. C., Kanekiyo, T., Xu, H., & Bu, G. (2013) Apolipoprotein E and Alzheimer disease: Risk, mechanisms and therapy. *Nature Reviews Neurology*, *9*(2), 106–118.
81. Troutwine, B. R., Hamid, L., Lysaker, C. R., Strope, T. A., & Wilkins, H. M. (2022). Apolipoprotein E and Alzheimer's disease. *Acta Pharmaceutica Sinica B*, *12*(2), 496–510.
82. Pardridge, W. M. (2003). Blood–brain barrier drug targeting: the future of brain drug development. *Molecular Interventions*, *3*(2), 90.
83. Uprety, A., Kang, Y., & Kim, S. Y. (2021). Blood–brain barrier dysfunction as a potential therapeutic target for neurodegenerative disorders. *Archives of Pharmacal Research*, *44*(5), 487–498.
84. Pardridge, W. M. (2015). Blood–brain barrier endogenous transporters as therapeutic targets: A new model for small molecule CNS drug discovery. *Expert Opinion on Therapeutic Targets*, *19*(8), 1059–1072.
85. Pardridge, W. M. (2009). Alzheimer's disease drug development and the problem of the blood–brain barrier. *Alzheimer's & Dementia*, *5*(5), 427–432.
86. Uprety, A., Kang, Y., & Kim, S. Y. (2021). Blood–brain barrier dysfunction as a potential therapeutic target for neurodegenerative disorders. *Archives of Pharmacal Research*, *44*(5), 487–498.
87. Jakki, S. L., Senthil, V., Yasam, V. R., Chandrasekar, M. J., & Vijayaraghavan, C. (2018). The blood brain barrier and its role in Alzheimer's therapy: An overview. *Current Drug Targets*, *19*(2), 155–169.
88. Yamazaki, Y., & Kanekiyo, T. (2017). Blood–brain barrier dysfunction and the pathogenesis of Alzheimer's disease. *International Journal of Molecular Sciences*, *18*(9), 1965.
89. Sharma, H. S., Castellani, R. J., Smith, M. A., & Sharma, A. (2012). The blood–brain barrier in Alzheimer's disease: novel therapeutic targets and nanodrug delivery. *International Review of Neurobiology*, *102*, 47–90.
90. Erickson, M. A., & Banks, W. A. (2013). Blood–brain barrier dysfunction as a cause and consequence of Alzheimer's disease. *Journal of Cerebral Blood Flow & Metabolism*, *33*(10), 1500–1513.
91. Li, J., Zheng, M., Shimoni, O., Banks, W. A., Bush, A. I., Gamble, J. R., & Shi, B. (2021). Development of novel therapeutics targeting the blood–brain barrier: From barrier to carrier. *Advanced Science*, *8*(16), 2101090.
92. Sethi, B., Kumar, V., Mahato, K., Coulter, D. W., & Mahato, R. I. (2022). Recent advances in drug delivery and targeting to the brain. *Journal of Controlled Release*, *350*, 668–687.
93. Poudel, P., & Park, S. (2022). Recent advances in the treatment of Alzheimer's disease using nanoparticle-based drug delivery systems. *Pharmaceutics*, *14*(4), 835.
94. Anselmo, A. C., & Mitragotri, S. (2014). An overview of clinical and commercial impact of drug delivery systems. *Journal of Controlled Release*, *190*, 15–28.
95. Choonara, Y. E., Kumar, P., Modi, G., & Pillay, V. (2016). Improving drug delivery technology for treating neurodegenerative diseases. *Expert Opinion on Drug Delivery*, *13*(7), 1029–1043.
96. Umar, M., Rehman, Y., Ambreen, S., Mumtaz, S. M., Shaququzzaman, M., Alam, M. M., & Ali, R. (2024). Innovative approaches to Alzheimer's therapy: Harnessing the

power of heterocycles, oxidative stress management, and nanomaterial drug delivery system. *Ageing Research Reviews*, *10*, 102298.

97. Li, J., Zheng, M., Shimoni, O., Banks, W. A., Bush, A. I., Gamble, J. R., & Shi, B. (2021). Development of novel therapeutics targeting the blood–brain barrier: From barrier to carrier. *Advanced Science*, *8*(16), 2101090.
98. Cao, J., Hou, J., Ping, J., & Cai, D. (2018). Advances in developing novel therapeutic strategies for Alzheimer's disease. *Molecular Neurodegeneration*, *13*, 1–20.
99. Nguyen, T. T., Nguyen, T. T., Nguyen, T. K., Vo, T. K. (2021). Advances in developing therapeutic strategies for Alzheimer's disease. *Biomedicine & Pharmacotherapy*, *139*, 111623.
100. Xiong, B., Wang, Y., Chen, Y., Xing, S., Liao, Q., Chen, Y., Li, Q., Li, W., & Sun, H. (2021). Strategies for structural modification of small molecules to improve blood–brain barrier penetration: A recent perspective. *Journal of Medicinal Chemistry*, *64*(18), 13152–13173.
101. Liu, Z., Liu, Q., Zhang, B., Liu, Q., Fang, L., & Gou, S. (2021). Blood–brain barrier permeable and no-releasing multifunctional nanoparticles for Alzheimer's disease treatment: Targeting NO/cGMP/CREB signalling pathways. *Journal of Medicinal Chemistry*, *64*(18), 13853–13872.
102. Kanwar, J. R., Sriramoju, B., & Kanwar, R. K. (2012). Neurological disorders and therapeutics targeted to surmount the blood–brain barrier. *International Journal of Nanomedicine*, *9*, 3259–3278.
103. Xie, J., Shen, Z., Anraku, Y., Kataoka, K., & Chen X. (2019). Nanomaterial-based blood–brain-barrier (BBB) crossing strategies. *Biomaterials, 224*, 119491.
104. Kanwar, J. R., Sriramoju, B., & Kanwar, R. K. (2012). Neurological disorders and therapeutics targeted to surmount the blood–brain barrier. *International Journal of Nanomedicine, 9*, 3259–3278.
105. Cooper, C. G., Kafetzis, K. N., Patabendige, A., & Tagalakis, A. D. (2024). Blood–brain barrier disruption in dementia: Nano-solutions as new treatment options. *European Journal of Neuroscience*, *59*(6), 1359–1385.
106. Teleanu, R. I., Preda, M. D., Niculescu, A. G., Vladâcenco, O., Radu, C. I., Grumezescu, A. M., & Teleanu, D. M. (2022). Current strategies to enhance delivery of drugs across the blood–brain barrier. *Pharmaceutics*, *14*(5), 987.
107. Marini, B. L., Benitez, L. L., Zureick, A. H., Salloum, R., Gauthier, A. C., Brown, J., Wu, Y. M., Robinson, D. R., Kumar, C., Lonigro, R., & Vats, P. (2017). Blood–brain barrier–adapted precision medicine therapy for pediatric brain tumors. *Translational Research*, *188*, 27–e1.
108. Ghosh, S., Lalani, R., Patel, V., Bhowmick, S., & Misra, A. (2019). Surface engineered liposomal delivery of therapeutics across the blood brain barrier: Recent advances, challenges and opportunities. *Expert Opinion on Drug Delivery, 16*(12), 1287–1311.
109. Cacabelos, R., Cacabelos, P., & Torrellas, C. (2014). Personalized medicine of Alzheimer's disease. *Handbook of Pharmacogenomics and Stratified Medicine*, *2014*, 563.
110. Moorthy, H., & Govindaraju, T. (2021). Dendrimer architectonics to treat cancer and neurodegenerative diseases with implications in theranostics and personalized medicine. *ACS Applied Bio Materials*, *4*(2), 1115–1139.

111. Rana, D., Salave, S., Khunt, D., & Benival, D. (2024). Blood brain barrier and its significance in drug delivery to brain in Alzheimer disease. In *Alzheimer's disease and advanced drug delivery strategies* (pp. 47–60). Academic Press.
112. Wong, K. H., Riaz, M. K., Xie, Y., Zhang, X., Liu, Q., Chen, H., Bian, Z., Chen, X., Lu, A., & Yang, Z. (2019). Review of current strategies for delivering Alzheimer's disease drugs across the blood–brain barrier. *International Journal of Molecular Sciences*, *20*(2), 381.
113. Montagne, A., Zhao, Z., & Zlokovic, B. V. (2017). Alzheimer's disease: A matter of blood–brain barrier dysfunction? *Journal of Experimental Medicine*, *214*(11), 3151–3169.
114. Jakki, S. L., Senthil, V., Yasam, V. R., Chandrasekar, M. J., & Vijayaraghavan, C. (2018). The blood brain barrier and its role in Alzheimer's therapy: An overview. *Current Drug Targets*, *19*(2), 155–169.
115. Banks, W. A. (2012). Drug delivery to the brain in Alzheimer's disease: Consideration of the blood–brain barrier. *Advanced Drug Delivery Reviews*, *64*(7), 629–639.

CHAPTER 5

Interdisciplinary Methods: The Integration of Neurology, Genetics, and Systems Biology Approaches to Comprehend Alzheimer's Disease

PHOOL CHANDRA[1*], HIMANSHU SHARMA[1], and NEETU SACHAN[2]

[1]*Teerthanker Mahaveer College of Pharmacy, Teerthanker Mahaveer University, Moradabad, Uttar Pradesh, India*

[2]*Department of Pharmaceutical Sciences, School of Pharmaceutical & Biological Sciences, Harcourt Butler Technical University (HBTU), Kanpur, Uttar Pradesh, India*

*Corresponding author

ABSTRACT

Alzheimer's disease (AD) is a highly common neurological condition that is exceedingly difficult to cure. Healthcare systems, families, and people are all significantly impacted. Even after extensive research, the precise cause and course of AD remain unknown. Wide approaches that combine ideas from the disciplines of systems biology, inheritance, and neuroscience have developed feasible strategies for handling that difficult issue and provide light on the complex and multidimensional characteristics of AD. A thorough understanding of neurobiology is required to appreciate the behavioral issues, memory loss, and cognitive decline associated with AD. Researchers may now be able to see both functional and structural changes in the brain areas of sufferers of Alzheimer's as a result of enhancements in imaging techniques like PET and magnetic resonance imaging (MRI). The above methods offer light on the origins and course

Cellular and Molecular Mechanisms in Alzheimer's Disease Therapeutics, Volume 1: Cellular Mechanisms. Prashant Tiwari, Sunil Kumar Kadiri, and Kunjbihari Sulakhiya (Eds.)

DOI: 10.1201/9781779641793-5

of AD. Additionally, longitudinal studies make it possible to monitor how an illness develops, which raises the possibility of early diagnosis and therapy. It has been shown that mutations in the PSEN1, PSEN2, and APP chromosomes constitute substantial inheritable risk factors for AD. The vulnerability to late-onset AD is conferred by common variations that have been identified by genome-wide association studies (GWAS). These discoveries have shed light on the underlying chemical reactions that contribute to AD pathogenesis. Owing to more studies on AD in families, uncommon and highly penetrant mutations have been found, offering novel mechanistic insights into the biology of AD. Using high-throughput computational simulations, systems biology techniques provide insight into the complex networks that underlie AD development.

5.1 INTRODUCTION

It is possibly difficult to grasp all the different aspects and complexity of Alzheimer's disease (AD), a neurodegenerative disorder. To unravel its mysteries, skills in the sciences of systems biology, genetics, and neurology may be merged with interdisciplinary methods. Understanding the basic structure, functions, and changes in the brain associated with AD may be made easier by studying neurology [1]. Despite the use of revolutionary neuroimaging methods like magnetic resonance imaging (MRIs) and PET scans, investigators can study the anatomical and cognitive abnormalities in the brain linked to the advancement of AD. Understanding the processes in which these behavioral changes appear is further aided by cognitive evaluations. When determining the molecular mechanisms and risk factors linked to AD, genetics is a crucial component to consider. According to GWAS, aware genetic variations are linked to an increased risk of the illness [2]. It is possible to identify possible therapy targets and create individualized treatment plans by clarifying these hereditary features. To provide an extensive understanding of system biology, data gathered from proteins, cells, and tissues combine with information from multiple states' physiological levels. We can recreate the intricate relationships between proteins, genes, and biological processes that are the foundation of AD appreciations to computational models [3]. An illness networking is useful for a couple of reasons: One is to identify important nodes and the other is to forecast how medical care targeting these areas may impact the evolution of the illness [4]. This is made possible by systems biology techniques.

A thorough knowledge of AD, including its molecular causes and clinical symptoms, is made possible by the integration of these multidisciplinary approaches. Brain imaging investigations have the potential to disclose structural alterations in particular brain areas that are linked to genetic risk alleles identified by GWAS. The impacts on neuronal function and mental health could change in relevant regions of the brain [5]. Moreover, by bringing together specialists with varying specialties and viewpoints, interdisciplinary cooperation promotes creativity [6]. Researchers may create new treatment targets, diagnostic tools, and therapies for AD by fusing knowledge from neurology, genetics, and systems biology (see Figure 5.1).

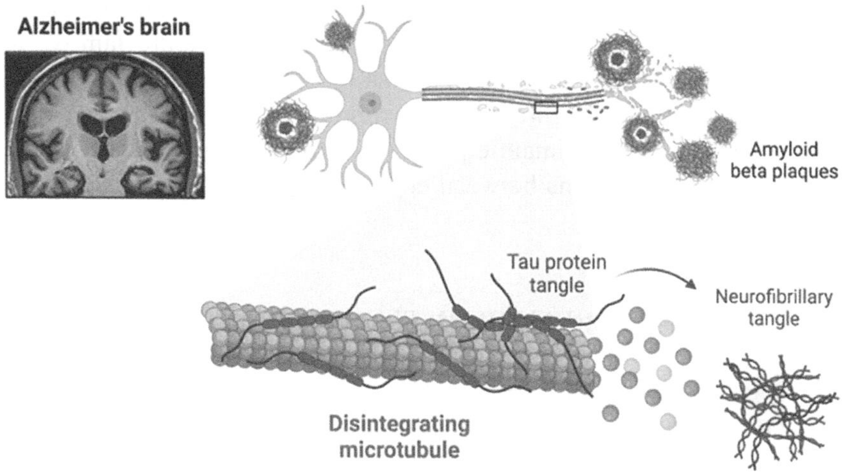

FIGURE 5.1 Alzheimer's illness and microtubule disintegration.

Decreases in cognitive function and aberrant protein deposits in the brain are hallmarks of AD, the most widespread kind of Alzheimer [7]. The condition known as Alzheimer's is a progressive neurodegenerative ailment. Dr. Alois Alzheimer first wrote about it in 1906 after he saw unique brain damage in a patient who had passed away and showed a noteworthy decline in memory and other cognitive impairments. Extracellular amyloid-beta plaques and intracellular neurofibrillary tangles are two of the disease's most characteristic neuropathological traits [8]. Amyloid-beta plaques are an outcome of the oxidation of APP into Alzheimer-beta polypeptide bundles. These plaques lead to neurodegeneration and death of cells in addition to neurotoxicity and impaired neuronal transmission. The protein responsible for creating tangled neurofibrillary cells is hyperphosphorylated tau, which bends filamentous constructions inside neurons, disrupts regular activity, and finally triggers the death of cells [9]. Phases of Alzheimer's disease may range from moderate anxiety and memory loss to severe mental decline and the inability to perform daily duties. People may also become nonverbal, lose their motor skills, and behave strangely as the disease progresses. Age is the single biggest risk factor for AD, with those over 65 having a far higher incidence of the disorder. It is essential to recognize, therefore, that AD is not a normal part of aging [10]. While aging is associated with an increased risk, environmental factors, intrinsic characteristics, and lifestyle choices all have a significant impact on the development of disease. Few cases less than 5% are family, and those caused by gene mutations such as *APP*, *PSEN1*, and *PSEN2* often manifest symptoms before reaching reproductive age. However, most Alzheimer's cases are unanticipated since the disease is largely caused by complex interactions between environmental factors and hereditary predisposition [11].

5.2 NEUROLOGY IN ALZHEIMER'S DISEASE RESEARCH

Neurology is a crucial area of inquiry in science because it highlights the morphological and physiological alterations to the neural network that occur with the advancing stages of AD. Neurology uses an assortment of strategies that improve our awareness of the disease, diagnosis, and possible therapy options for AD [12]. One of the primary ways that neurology has advanced the area of cognitive impairment studies is via the establishment of sophisticated neuroscience methods. Utilizing this sort of image processing, AD-related structural modifications in the neural network, such as cortical atrophy and hippocampal distortion, may be was observed. These structural alterations help identify Alzheimer's from other forms of cognition and serve as helpful markers of the disease's growth. Using PET imaging, tau tangles, and amyloid-beta plaques are two of the main neuropathological features of AD [13]. To give information on the biology of the illness and aid in the development of more specialized therapies, PET scans may identify the presence and extent of tau and amyloid pathology. A kind of magnetic resonance imaging (fMRI) that monitors variations in the flow of blood to the brain and neuronal activity can be applied by researchers to examine abnormalities in connectivity between functional networks associated with AD [14]. The neurobiological underpinnings of cognitive loss in patients with Alzheimer's have been significantly improved upon by resting-state fMRI searches, which have shown discrepancies in neural connections related to memory, attention, and executive function. Neurological tests, including neurological assessments and cognitive testing, play a major role in the diagnosis and follow-up of AD. Medical practitioners analyze executive function, memory, and language difficulties that may be symptomatic of the illness using standardized cognitive evaluations including the MMSE and the AD Assessment Scale-Cognitive subscale (ADAS-Cog). Neurosurgery is involved in both the therapeutic and symptomatic medication study and evaluation process for AD. Expertise in neurology is used in the development of protocols, assessment of treatment efficacy, and monitoring of potential side effects in clinical trial subjects [15].

5.2.1 CLINICAL MANIFESTATIONS OF ALZHEIMER'S DISEASE

Language impairment is characterized by a person's inability to find words, communicate thoughts, or comprehend spoken language [16]. This leads to communication problems. Disorientated people often question where they are, when they are, and even who they know. Mood swings, agitation, despair, or social alienation are examples of behavioral modifications that may affect a person's relationships and alter their personality [17]. The onset of neurological disorders may have a detrimental effect on balance, coordination, and fine motor ability [18]. Due to its impairment in spatial perception and directions, tactile disorder may make it impossible to calculate distances or identify objects. Executive dysfunction impairs the ability to plan, organize, and multitask, aggravating difficulties with daily functioning [19]. Insomnia and excessive fatigue during the day are common sleep problems. People may also struggle with personal grooming, meal skipping, and hygiene and self-care (see Table 5.1).

TABLE 5.1 Clinical Manifestations Commonly Observed in Alzheimer's Disease.

Clinical Manifestation	Description	Examples	References
Memory loss	Difficulty recalling recent events or information	Forgetting recent conversations or appointments	[20]
Cognitive decline	Progressive deterioration in cognitive abilities	Difficulty in problem-solving or decision-making	[20]
Language impairment	Difficulty finding words or understanding speech	Struggling to follow conversations or express thoughts	[20]
Disorientation	Confusion regarding time, place, or people	Getting lost in familiar places or forgetting the day	[20]
Behavioral changes	Alterations in mood, personality, or behavior	Increased agitation, depression, or social withdrawal	[21]
Motor impairment	Decline in motor skills and coordination	Difficulty with walking, balance, or fine motor tasks	[21]
Visuospatial dysfunction	Problems with spatial perception or navigation	Difficulty judging distances or recognizing objects	[22]
Executive dysfunction	Impaired ability to plan and organize tasks	Struggling with multitasking or problem-solving	[22]
Sleep disturbances	Disruptions in sleep patterns	Difficulty falling or staying asleep, daytime fatigue	[23]
Hygiene and self-care issues	Neglect of personal hygiene and self-care	Forgetting to bathe, dress, or maintain nutrition	[23]

5.2.2 NEUROIMAGING TECHNIQUES

Neuroimaging techniques are crucial to the study of AD and the development of new treatment approaches because they allow researchers to see and quantify the structural, functional, and cellular alterations that take place in the brain [24]. Numerous advanced neuroimaging methods have been employed to follow the disease growth and development, diagnose, and assist prognosis in addition to studying the mechanism of the condition. High-resolution images of the physical makeup of the brain may be obtained using MRI without the need for invasive treatments [25]. In the context of Alzheimer's investigations, localized brain reduction is detected and assessed using MRI imaging), particularly in brain regions susceptible to neurodegeneration such as the entorhinal cortex and hippocampal regions. While fMRI tracks alterations in functional connectivity networks associated with AD and monitors the activity of the brain during cognitive tasks, diffusion MRI can additionally assess white matter integrity [26]. PET imaging allows one to observe and quantify the chemical processes that take place within the brain by connecting radiotracers to specific locations. PET scans are widely utilized in AD investigations to assess and identify tau and amyloid-beta pathology, affording insight into the location and advancement of neuropathological changes. PET imaging also assesses brain breakdown of glucose and may detect the patterns of regional a lack of metabolism characteristic of AD [27]. Another nuclear medicine method used to assess cerebral blood flow and brain activity is SPECT imaging. SPECT scans may be used in AD studies to pinpoint regions of lack of perfusion or hypometabolism that may indicate neuronal dysfunction and disease progression. The neurochemical alterations associated with Alzheimer's illness may also be emphasized by SPECT imaging using radiotracers that target various neurotransmitter systems, such as acetylcholine. Amyloid imaging techniques, such as PET scans using amyloid-specific radiotracers (like Pittsburgh compound B), allow for the visualization and measurement of amyloid-beta buildup in the brain [28]. These imaging markers help in early Alzheimer's pathology identification as well as in distinguishing between Alzheimer's disease and various other forms of dementia. Furthermore, tracking therapy regimens intended to lower the formation of amyloid-beta is made less complicated in clinical trials utilizing amyloid imaging [29].

5.2.3 LONGITUDINAL STUDIES AND EARLY DIAGNOSIS

To better understand how dementia progresses, highlight risk factors, and assess the beneficial effects of treatments over time, long-term studies are extremely useful in the discipline of AD research [30]. From preclinical phases to clinical appearance as well as beyond, these ongoing follow-up studies permit experts to monitor fluctuations in metabolite levels of difficulty, neural design, and mental function. In addition to offering possibilities for early identification and treatment, longitudinal studies provide unique perspectives into the usual course of AD [31]. To initiate treatment on time and enhance results for patients, early diagnosis of the illness is essential. Years before clinical symptoms appear, researchers could identify early markers of AD via long-term investigations. These markers may include anomalies in neuroimaging, changes in mental performance, and biomarker concentrations (such as tau and amyloid-beta). Through ongoing surveillance of these indicators, scientists may be able to create risk and prediction model scores that will enable them to group people as to their risk of AD [32]. This allows to obtain specialist treatment and therapies that are more focused. In the preclinical and prodromal phases of AD, long-term studies are also helpful in assessing possible disease-modifying treatments. Researchers may evaluate how medicines impact the progression of a disease and choose the most efficient course of therapy via tracking changes over time in biomarkers, cognitive function, and clinical outcomes. In addition, follow-up investigations provide valuable insights into the long-term safety, effectiveness, and tolerability of therapies. These insights may help direct future efforts in therapeutic research and clinical practice [33]. Moreover, longitudinal research might aid in our understanding of the complicated relationships involving genetic, environmental, and personal factors that influence the onset and development of AD. Through the prospective collection of details on individuals' genetic profiles, medical histories, lifestyle decisions, and exposure to pollutants, professionals could potentially be able to find safeguards and modifiable risk factors associated with the possibility of AD [34]. The data gathered could facilitate public health programs that promote emotional wellness and lower the overall risk of dementia [35].

5.3 GENETICS OF ALZHEIMER'S DISEASE

A significant portion of AD's etiology, associated risks, and therapeutic targets are mediated by genetics. While some cases of AD are hereditary and consequently linked to heredity, the majority have complicated relationships with both genetic and environmental variables. This section examines the genetic factors associated with the beginning of AD in instances that are sporadic as well as familial [36]. Familial Alzheimer's disease (FAD) accounts for less than 5% of instances of AD and is identifiable by an autosomal dominant line of inheritance. *APP*, *PSEN1*, and

PSEN2 are three genes that have been linked to FAD [37]. These mutations promote the production or aggregation of amyloid-beta peptides, which leads to an accumulation of amyloid deposits that harm the nervous system. Individuals suffering from FAD frequently exhibit increasing signs of their ailment between their 30s and 60s. Most individuals are impacted by this erratic illness, and the chance of developing AD is heightened by many factors, both genetic and environmental [38]. The ε4 allele of the *APOE* gene is the most well-established genetic risk factor for late-onset AD (LOAD), the most common kind of dementia. With homozygous carriers being much more sensitive than those with heterozygous or non-carriers, the probability of AD becomes greater among *APOE ε4* carriers [39]. Due to its involvement in amyloid-beta metabolism, tau phosphorylation, and neural inflammation, *APOE ε4* plays a role in the development of AD. GWAS has connected an assortment of genetic mutations to an elevated risk of AD. These include modifications related to immunity, lipid metabolism, and synaptic function, as well as *APP*, *PSEN1*, *PSEN2*, and *APOE*. These findings highlight the complexity and polygenicity of AD, whereby several genetic differences interact with environmental variables to modify the chance of developing the condition [40].

5.3.1 GENETIC RISK FACTORS

An elevated risk of AD is linked to the *APOE ε4* allele; homozygous carriers (*APOE ε4/ε4*) have a much greater risk than heterozygous carriers (*APOE ε3/ε4*) or noncarriers (*APOE ε3/ε3*). The etiology of AD is believed to be altered by *APOE ε4* in terms of tau phosphorylation, neuroinflammation, and amyloid-beta aggregation. Rare mutations in the *APP*, *PSEN1*, and *PSEN2* genes are the cause of FAD [41]. The generation of amyloid plaques and neuronal toxicity may be brought on by these mutations, which enhance the formation or aggregation of amyloid-beta peptides [42]. FAD is passed down in an autosomal dominant manner and usually manifests at a younger age. Many common genetic variations have been linked to an elevated risk of LOAD by GWAS research. Genes related to immunological response (e.g., *CLU*, *CR1*), lipid metabolism (e.g., *ABCA7*), synaptic function (e.g., *BIN1*), and neural signaling (e.g., *SORL1*) have these variations [43]. Although the risk of AD is somewhat increased by each of these genetic variations alone, the polygenic morality of AD susceptibility is facilitated by their simultaneous existence (see Figure 5.2).

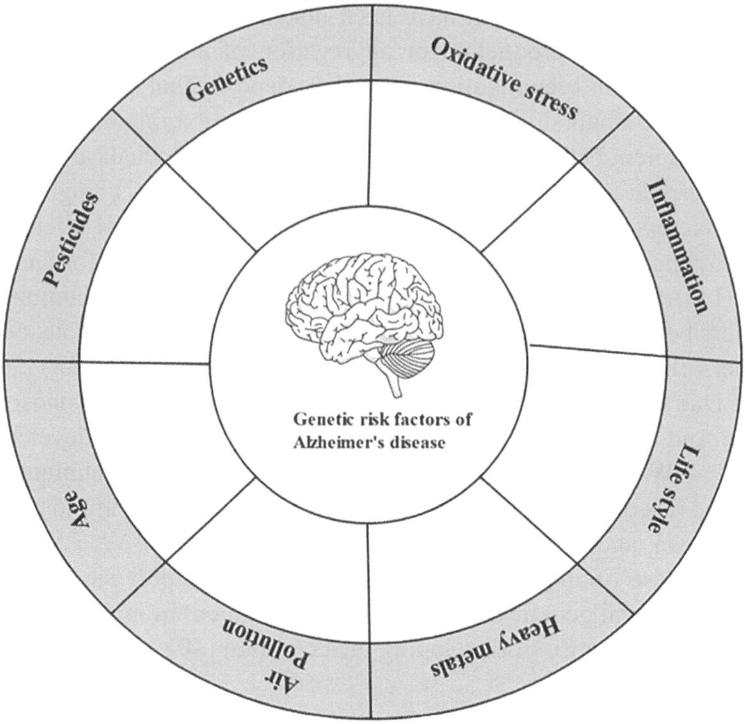

FIGURE 5.2 Genetic risk factors of Alzheimer's disease.

5.3.2 GENOME-WIDE ASSOCIATION STUDIES (GWAS)

Thanks to GWAS, we now have a far better understanding of the genetic basis of complex illnesses like AD. Genetic variation associated with the risk of illness may be recognized through GWAS, a highly effective method that assesses several hundred thousand to a billion SNPs throughout the entirety of a genome in large cohorts of individuals with and without the condition [44]. Several genetic variants have been related to an elevated probability of LOAD, which is the most common kind of Alzheimer's illness, according to the GWAS [45]. These mutations affect, among several other biological functions, genes associated with brain signaling, lipid metabolism, defence against infection, and synaptic transmission. One of the most well-established genetic risk factors for LOAD identified by GWAS is the ε4 allele of the *APOE* gene. It is thought that *APOE ε4* influences amyloid-beta metabolism, neuroinflammation, and tau phosphorylation, all of which have an impact on the etiology of AD. It is associated with a higher chance of developing AD. Besides *APOE*, GWAS has revealed many more genetic loci associated with LOAD risk [46]. For example, alterations in complement receptor 1 (CR1), clusterin (CLU), phosphatidylinositol binding clathrin assembly protein (PICALM), and CLU have been connected to immunological response and synaptic function [47]. Variations in the *ATP-binding cassette transporter A7* (*ABCA7*) gene modify the elimination of amyloid-beta particles and the metabolism of lipids. Other genes associated with brain signaling pathways, endocytosis, and synaptic function found in the GWAS database include *BIN1*, *CD2AP*, and *SORL1*. While any of these mutations alone could a little the risk of AD, their cumulative effect accounts for the majority of the polygenic feature of AD susceptibility. The unique insights into the molecular pathways underpinning the pathogenesis of AD afforded by the GWAS data encouraged further study into potential treatment targets and procedures [48].

5.3.3 FAMILIAL ALZHEIMER'S DISEASE MUTATIONS

FAD, which often appears at a younger age and is marked by an autosomal dominant pattern of inheritance, is a rarer form of the illness than spontaneous AD [49]. FAD is caused by polymorphisms in three genes: *APP*, *PSEN1*, and *PSEN2*. A portion of cases of AD in families are brought about by polymorphisms in the *APP* gene. Enzymes hydrolyze the *APP*-encoded amyloid precursor protein to produce amyloid-beta peptides. A mutation in the *APP* gene conduces to the altered synthesis of the amyloid precursor protein [50]. Amyloid-beta peptide production or aggregation increases as a consequence, leading to the formation of amyloid plaques in the brain. The majority of individuals who have such mutations experience early-onset FAD, often starting to show abnormalities in their 30s or 50s. *PSEN1* gene abnormalities, which account for over 70% of circumstances, are an extremely prevalent manifestation of familial AD [51]. A crucial component of the gamma-secretase complex that cleaves amyloid precursor protein is the presenilin-1 protein, which is produced by *PSEN1*. *PSEN1* mutations change how amyloid precursor protein is processed, which speeds up the manufacture of amyloid-beta peptides and the buildup of amyloid plaques. FAD typically manifests in persons between the ages of 30 and 50, and *PSEN1* mutations often cause the disease to manifest early. Less common, *PSEN2* gene mutations can cause familial AD, albeit they contribute simply to a very tiny percentage of cases [52]. The presenilin-2 protein, which is encoded by *PSEN2*, is an additional component of the gamma-secretase complex. Amyloid-beta peptide production is enhanced and protein amyloid precursor processing suffers when *PSEN2* mutations occur, triggering the emergence of amyloid plaques. FAD is common in persons in their 40s to 60s and is frequently brought on by *PSEN2* mutations [53].

5.4 SYSTEMS BIOLOGY APPROACHES

A solid basis for understanding the complex biological pathways behind AD has been offered by systems biology techniques [54]. Molecular, cellular, and physiological data, along with information from multiple states biological levels, are brought together in systems biology to provide insights into the complex chains and interactions that drive AD pathogenesis (see Table 5.2).

TABLE 5.2 Applications of Various Biological Techniques to Alzheimer's Illness.

Systems Biology Approach	Description	Application in Alzheimer's Disease	References
Network analysis	Analysis of molecular interaction networks to identify key genes, proteins, and pathways implicated in AD	Identifying biomarkers, therapeutic targets, and elucidating AD pathophysiology	[55]
Multi-omics integration	Integration of data from genomics, transcriptomics, proteomics, and metabolomics to comprehensively study AD	Identifying molecular signatures, subtypes of AD, and personalized treatment strategies	[55]
Computational modeling	Simulation of biological systems affected by AD to understand dynamic interactions and predict disease outcomes	Predicting the effects of genetic variations, environmental factors, and therapeutic interventions on AD	[55]
Systems pharmacology	Modeling drug-target interactions within biological networks to predict drug efficacy and safety for AD	Identifying potential drug candidates, optimizing treatment regimens, and personalized medicine for AD	[56]

5.4.1 COMPUTATIONAL MODELING

Multidisciplinary techniques have significantly increased our comprehension of AD via modeling on computers that incorporate information from the biology of systems, genetics, and neurology [57]. The complex interplay between genetic, molecular, and cellular components that underpin the onset and development of AD can potentially be understood by scientists using a holistic approach, leading to novel discoveries and potential treatment modalities [58]. About the clinical manifestations of AD, including memory loss, neurodegeneration, including cognitive impairment, a vital understanding of neurology is also accessible. Thanks to neuroimaging techniques such as PET and MRI, mathematical representations are achievable to investigate morphological and functional changes in the brain that are linked to the deterioration of AD. In addition to helping identify indicators for early diagnosis and enabling tailored treatment strategies, these models track the progression of a disease over time [59]. Genetic factors play an important part in understanding the genetic risk factors that lead to AD susceptibility. GWAS have shown there is a link between an increased risk of AD and many genetic variants, most notably the *APOE* gene. Computer approaches to modeling like as quantitative genetics and machine learning help us better understand the complex genetic architecture of AD and prioritize potential genes for further experimental validation. It is simpler to find potential therapeutic targets when genetic data is integrated with neuroimaging and clinical data since this enhances computational model predictiveness [60]. An in-depth knowledge of the pathophysiology of AD is possible via the use of systems biology, which models the intricate network of molecular interactions that undergrow the illness. Computerized models are used for simulating the shifting patterns of gene transcription, chemical reactions, and protein signaling related to AD pathophysiology [61]. According to these hypotheses, protein aggregation, synaptic dysfunction, and neuronal death all characteristic features of the disease are brought on by changes in cellular homeostasis resulting from genetic differences. Through the integration of feedback loops and nonlinear interactions with multi-omics data (genomics, transcriptomics, proteomics, etc.), computational models can capture the novel features of complicated organisms and identify crucial nodes for medical therapy [62].

5.4.2 INTEGRATION OF SEVERAL OMICS

A powerful method in biological research is called "multi-omic integration," which synthesizes information from several molecular levels (transcriptomics, proteomics, metabolomics, and epigenomics, for example) to fully understand complex biological systems [63]. By merging different omics information sets, analysts can gain novel perspectives into the molecular concepts underlying life processes, develop markers for the diagnosis and prediction of illnesses, and identify targeted therapies. In essence, multi-omic integration requires combining data from many chemical levels to more thoroughly depict biological processes [64]. This method recognizes that different chemical states and complex interactions across biological systems may lead to a limited understanding of complex biological processes when studied in isolation [65]. By combining data from several omics technologies, scientists may reveal regulatory networks, uncovered connections, and the basic molecular principles underlying disease and cellular activity. Multi-omic biosynthesis has several vital uses, including the research of human health and disorders. Researchers may discover molecular markers connected to a variety of illnesses, including cancer, heart troubles, and disorders of the brain, by analyzing omics data extracted from patient fluids like blood, urine, or tissues. Integrating transcriptomics, proteomics, and genomics data might enable researchers to recognize variations in genes, alterations in gene expression, and protein modifications that influence the underlying cause of illness [66]. By identifying biomarkers for disease diagnosis, prognosis, and therapy response prediction, this comprehensive approach may pave the way for patient-specific personalized medicine techniques. In cancer research, multi-omic integration has altered our knowledge of the pathways beneath resistance to therapy and tumor heterogeneity [67]. Through the integration of transcriptomic, epigenomic, and genome-wide data obtained from cancer samples, scientists may identify driver mutations, epigenetic alterations, and gene expression variations associated with the initiation and progression of illness. By identifying promising therapeutic targets and developing specialist medications that are tailored to the unique biological traits in specific tumors, this comprehensive strategy improves the lives of patients and increases the efficacy of cancer therapy [68].

5.4.3 NETWORK ANALYSIS

The structure, dynamics, and function of complex systems are investigated using the analysis of networks, a potent technique that is used in many disciplines, including computer science, epidemiology, biology, and sociology [69]. To analyze relationship patterns, highlight essential nodes or clusters, and unearth emergent system features, ranging it requires describing transactions or links among entities as the edges and nodes in a system's network. Here, using a 300-phrase structure shall be discussed the core concepts and uses of analyzing networks [70]. In essence, network analysis offers a graphical depiction of complex systems, with edges standing in for the connections or interactions (e.g., physical interactions, co-expression, social links) and nodes for separate entities (e.g., genes, proteins, persons). Through the visualization and analysis of networking topology, scholars may get an improved comprehension of the basic frame and organization of the system [71]. Studying biological systems, such as protein-protein interaction networks, gene regulatory networks, and physiological networks, is one of the largest applications of network analysis. Researchers may determine functioning components or pathways, clarify the chemical processes behind the functions of cells, and forecast the outcomes of genetic perturbations or medication treatments by examining the connections between genes, proteins, and metabolites [72]. Finding disease-associated genes, locating potential therapeutic targets, and comprehending the intricate interactions between variations in genes in complex illnesses like cancer and disorders of the brain have all been made possible by network-based techniques. Network analysis is a tool used in the fields of psychology and sociology to examine collaborative networks, communication patterns, and social networks [73]. Through the examination of social network structure, scholars may pinpoint significant persons or factions, describe the dissemination of knowledge and its impact, and evaluate the adaptability of social networks to disruptions or actions. Network-based techniques have been used to investigate the transmission of contagious illnesses, examine the dissemination of knowledge in virtual networks of friends, and to analyze the dynamics of group behavior in human cultures [74].

5.5 INTEGRATION OF NEUROLOGY, GENETICS, AND SYSTEMS BIOLOGY

Because to the integration of neurology, genetics, and biochemistry, our awareness of complex diseases of the brain such as AD, personality disorder

(PD), and ASD has greatly progressed. Finding novel biomarkers, developing customized medication regimens, and grasping the intricate cellular processes underlying illness pathogenesis are all achievable with the use of a collaborative strategy [75]. Important clinical and neuroimaging data that characterize the phenotypic manifestations of neurological illnesses are provided by neurology. With the use of cutting-edge imaging techniques like PET and MRI, experts may see morphological and functional changes in the brain related to the progression of sickness [76]. Neurological testing, including tests for cognition and cerebral examinations, provides insight into the patient's psychomotor and cognitive issues. By combining the fields of neurology, genetics, and systems biology, scientists may link clinical symptoms to the underlying cellular and genetic factors. This facilitates the identification of illness subgroups and the creation of personalized treatment regimens [77]. One of genetics' main goals is discovering the inherited causes of diseases of the brain. Using next-generation sequencing technology and GWAS, several genetic variants that are linked to a higher probability of disease have been uncovered. By amalgamating genomic data with neuroimaging and clinical information, researchers may be able to make sense of how hereditary characteristics impact treatment response, disease progression, and susceptibility [78]. In addition to studies in genetics, systems biology methods explain the complex interactions across genes, proteins, and chemical reactions that contribute to the etiology of disease. Systems biology uses the integration of multi-omics data (e.g., genomes, transcriptomics, proteomics) and the construction of molecular networks to shed light on the biological processes behind brain diseases and identify potential therapeutic targets [79]. Neurology, genetics, and the biology of systems together may be useful in personalized healthcare approaches to neurological illnesses. Scientists may divide patients into distinct groups based on their molecular traits and underlying illness processes by combining genetic, clinical, and biological information [80]. This categorization enables the development of personalized medications for every patient, improving patient outcomes and treatment efficacy (see Figure 5.3).

5.5.1 SYNTHESIZING DIVERSE DATA SETS

In interdisciplinary fields like biomedicine, where synthesizing diverse data sets is necessary, integrating information collected from various places

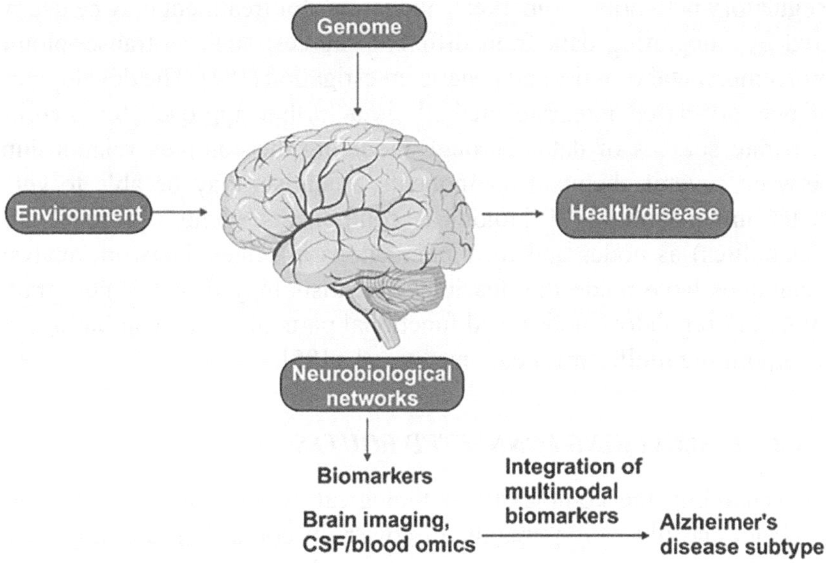

FIGURE 5.3 The methodology of systems biology both genetic and environmental variables have an impact on neural networks.

and modalities is a crucial aspect of current research toward a comprehensive knowledge of complex biological systems [81]. Combining data from many experimental platforms, methodologies, and inputs results in a cohesive and comprehensive conceptualization of biological processes or occurrences. We will look at the basic concepts and methods of merging multiple sources of information in a 300-word context [82]. The ultimate aim of the process of synthesizing several data sets that integrate and harmonize different data types such as genomic, transcriptomic, proteomic, metabolomic, and clinical data is a deeper knowledge of biological systems. This strategy requires a thorough assessment of normalization, data integrity, as well as integration techniques to ensure consistency and interoperability across various sources of data. Data fusion, integrating systems, and machine learning algorithms are a few of the statistical and computational techniques used to combine disparate data sources and extract insightful information [83]. A common strategy for merging multiple sets of data unreported phrase together into a single is to build multi-omics analysis pipelines. These pipelines fuse data from many omics' technologies, such as mass spectrometry, RNA sequencing, and genome sequencing, to clarify the genetic origins of biological events or illnesses.

Regulatory networks, biomarkers, and targets for treatment may be uncovered by integrating data from different sources, such as transcriptome, proteome, metabolomic, and genetic investigations [84]. The development of network-based integrate methods is a further approach for merging disparate sources of data. Through modeling the complex relationships between several chemical components, scientists may be able to get a better understanding of biological components (genes, proteins, and metabolites) as nodes and their interactions as edges. Network analysis techniques have made this feasible. Scientists may discover new traits, important regulatory nodes, and functional parts of systems in biology by incorporating multi-omics data in networks [85].

5.5.2 DISCOVERING CONNECTED ROUTES

Understanding the complexity of biological systems and their function in health and illness requires the discovery of connecting pathways [86]. Carefully controlling gene regulation, metabolism, signal transmission, and other cellular functions is the job of biological pathways, a network of interconnected cellular processes. By elucidating the connections and interplay among distinct pathways, scientists might identify novel therapeutic targets and acquire enhanced comprehension of the fundamental origins of illnesses. In this portion, we'll go through the main concepts and strategies for finding pertaining pathways in a 300-word framework. Integrating multi-omics information collections is a technique to accomplish linked circuit identification [87]. A full image of the molecular landscape of cells and tissues may be obtained by researchers by merging data from the domains of genomes, transcriptomics, proteomics, and metabolomics. By using integrative analytical techniques like as network-based approaches, pathway crosstalk analysis, and route enrichment analysis, scientists may deduce the functional connections across pathways and identify coordinated changes at several molecular levels [88]. Thanks to this science and technology, the underlying cellular mechanisms of pathophysiology develop clearer and it becomes feasible to detect related networks that become imbalanced in disease states. The use of computer simulations and modeling are two more methods for identifying comparable routes. Mathematical models can simulate the dynamic behavior within biological networks and forecast how changes to one way may impact the operation of other routes. These modeling tools include ODEs and agent-based models.

Using the integration of previously collected data and prior knowledge about pathway connections, researchers can construct models that predict for route crosstalk and identify novel features in intricate biological systems [89]. This approach not only simplifies the recognition of crucial links and feedback circuits within linked pathways but also offers a basis for investigating possible processes. Moreover, researchers could systematically interfere with biological processes and track the phenotypic changes that ensue by using high-throughput screening (HTS) and genome-wide CRISPR screens. Researchers may map out the functional linkages amongst circuits and reveal novel links between genes and regulatory mechanisms by combining enormous amounts of data from perturbation research with analysis by computers. This method speeds up the identification of related pathways pertinent to illness and provides an outline for the design of targeted medication and precision healthcare procedures [90].

5.5.3 FINDING BIOMARKERS AND POTENTIAL TREATMENT TARGETS

To aid in early illness diagnosis, prognosis, and the creation of tailored therapies, biomedical researchers must identify biomarkers and therapeutic targets. Targets of treatment are components that, when altered, might offer beneficial consequences; biomarkers are substances that have the potential to be utilized for tracking biological processes or states of disease. This 300-word framework will go over the key concepts and methods for determining treatment targets and biomarkers [91]. Finding biomarkers and possible therapy targets may be accomplished, in part, by combining several multi-omics data sets. By combining data from genomes, transcriptomics, proteomics, metabolomics, and epigenomics, researchers may be able to recognize possible markers for evaluation, prognosis, and therapy response in addition to identifying molecular patterns linked to disease states [92]. Using integrative analytic tools such as machine learning algorithms, network-based approaches, and route enhancement evaluation, experts may find unbalanced networks as well as significant biological actors that underlie the underlying causes of illness. This approach provides insights into potential targets of therapy for intervention and accelerates up the finding of new markers with analytical or prognostic value [93]. Biomarkers and possible therapy targets could be identified via functional genome sequencing and HTS approaches. Two examples of HTS techniques that enable researchers to methodically impede biological processes to identify genes or proteins that are essential for the spread of disorders are CRISPR-based genetic screens and RNAi testing. After sifting through vast datasets about genetic or chemical ingredients, researchers may find molecules that alter disease-relevant properties to be intriguing candidates for more investigation. Moreover, functional genomics methods like as single-celled sequencing and territorial transcriptomics enable researchers to pinpoint pharmacological vulnerabilities, locate biomarkers specific to certain cell types, and describe a range of illness conditions [94]. Furthermore, translational research initiatives such as human trials and preclinical studies of validation are essential for validating therapeutic targets and biomarkers in relevant patient populations and disease models. Preclinical research uses experimental data from patient- and animal-derived materials to validate the efficacy and safety of potential therapeutic approaches [95]. Clinical trials investigate medicinal benefits and possible biomarkers of drugs in large patient cohorts. They also assess the clinical utility of these drugs in predicting disease outcomes and guiding treatment decisions (see Table 5.3).

TABLE 5.3 Alzheimer's Disease, Integrating Neurology, Genetics, and Systems Biology Approaches, as Well as Potential Biomarkers and Treatment Targets

Interdisciplinary Methods	Potential Biomarkers	Potential Treatment Targets	References
Neurology	Amyloid beta (Aβ) plaques	Aβ aggregation inhibitors	[96]
	Tau protein pathology	Tau protein stabilizers	[96]
	Neuroinflammation markers	Anti-inflammatory agents	[96]
	Brain atrophy patterns	Neuroprotective compounds	[96]
Genetics	*APOE* ε4 allele	*APOE*-targeted therapies	[97]
	Variants in *PSEN1*, *PSEN2*, and *APP* genes	Genetic modification strategies	[97]
	Genome-wide association studies	Precision medicine approaches	[97]
Systems Biology	Dysregulated metabolic pathways	Metabolic modulators and network-based therapies	[98]
	Aberrant protein-protein interactions	Protein-protein interaction inhibitors	[98]
	Altered gene expression profiles	Gene expression regulators	[98]

5.6 FUTURE DIRECTIONS AND CHALLENGES

Updating personalized medicine, using the promise of new technologies, and tackling significant obstacles in data integration and interpretation are anticipated to be the main areas of emphasis for future biomedical research [99]. There are still several issues that need to be resolved despite tremendous advancements, such as the need for better computing techniques, data consistency, and ethical issues. We will examine prospects and problems in scientific study using a 300-word essay format. Further study in the biomedical field might focus on constructing personalized treatment plans that take into account a patient's genetic composition, lifestyle, and surroundings [100]. By fusing multi-omics collections of data with automated learning approaches, researchers may create predictive algorithms for early health recognition, treatment response prediction, and risk assessment. Furthermore, new avenues for precision gene editing and targeted medication development are made possible by the introduction of genome editing tools like CRISPR-Cas9. Furthermore, the combination of AI, neural networks, and big data analytics presents opportunities for speeding up drug development and biomedical research [101]. Large-scale omics data sets may be evaluated, drug-target interactions can be predicted, and new therapeutic opportunities can be found using AI-powered algorithms. Furthermore, academics have been able to map intricate biological events with formerly unreachable precision and identify cellular heterogeneity thanks to innovations in single-cell sequencing, high-throughput imaging, and spatial transcriptomics. To fully profit from these developments, a few roadblocks need to be cleared. While more advanced computer techniques and standardized data formats are required, data integration and interpretation remain very difficult tasks. Furthermore, in the big data and medical genomics age, it must be done to ensure data privacy, security, and proper utilization of sensitive health information. Furthermore, it is essential to tackle discrepancies concerning medical care and genetic technology accessibility to ensure an equal distribution of benefits across diverse groups [102].

5.6.1 PERSONALIZED MEDICINE APPROACHES

The process of customizing medical procedures and therapies for each patient based on their unique genetic makeup, environment, lifestyle, and other factors is known as precision medicine, often known as personalized medicine. This strategy has considerable potential for enhancing patient satisfaction, minimizing negative reactions, and increasing treatment efficacy [103]. The 300-word piece that follows will look at customized treatment plans and their effects [104]. A key component of personalized medical care is the use of genetic data to assist therapeutic decisions. Improvements in genome sequencing technology, especially NGS, have helped make it possible to quickly and inexpensively investigate the patient's genetic characteristics. Utilizing GWAS and WGS, genetic variants related to drug metabolism, treatment response, and risk for diseases may be identified. Pharmacogenomics, the study of how a person's genes impact the response to treatment, has enabled therapists to make predictions about success in therapy and hazards based on the patient's genetic history [105].

5.6.2 ETHICAL CONSIDERATIONS AND PRIVACY CONCERNS

Concerns about ethics and confidentiality have grown increasingly pressing as personalized healthcare develops. The combined use of genetic data with client medical records and other private data raises ethical concerns around permission, data ownership, confidentiality, and possible abuse [106]. Here, we will examine the moral dilemmas and issues of privacy surrounding personalized medicine in a 300-word manner. Enlightened consent is one moral element of personalized treatment. Giving patients a full explanation of the advantages and disadvantages of genetic testing, as well as any possible effects on their families, is crucial [107].

5.6.3 OVERCOMING TECHNICAL AND METHODOLOGICAL LIMITATIONS

Methodological and technological obstacles need to be tackled if customized healthcare can advance and fulfill its full promises of benefiting patient outcomes. Resolving technological issues and methodological gaps is crucial to ensuring the quality, dependability, and scalability of personalized medical processes since personalized medicine depends on complicated data integration, computer analysis, and clinical interpretation [108]. In this 300-word framework, the main conceptual and technological obstacles will be investigated jointly with their fixes. The deficiency of omics data requirements and interoperability is one technological barrier to personalized medicine [109]. Data integration and cross-study comparison are challenging due to different platforms, analytical techniques, and data formats. Collaborative research endeavors and data harmonization are helped by the development of ontologies, data-sharing protocols, and common data formats. Organizations like the GA4GH and the ICGC were established to encourage data interchange and sharing in personalized medicine research. Computational issues with data processing, storage, and analysis further impede personalized medicine [110].

5.7 CONCLUSION

Systems biology, genetics, and neurology when combined provide a potent multidisciplinary model for comprehending AD as a whole. Working together, scientists may uncover biomarkers for rapid detection and prediction, explore the multifaceted nature of AD the causes, and pinpoint prospective targets of therapy for therapy. Clinicians may be able to explain the cellular, molecular, and physiological abnormalities that underlie AD effectively by integrating data from these various areas. Ultimately, this might lead to the development of improved therapies and diagnostic approaches. Thanks to neuroscience, medical professionals can now better understand how AD develops and identify the neurodegeneration, memory loss, and cognitive decline that are connected to the illness. Practitioners used cutting-edge neuroimaging methods, like as MRI and PET scans, to determine alterations in the anatomy and functioning of the nervous system in AD patients. The data is necessary for illness surveillance and detection. Furthermore, neurological evaluations make it possible to monitor the progression of a disease over time and to identify significant clinical characteristics, both of which are important for the creation of customized treatment regimens and the outcomes of clinical studies. Determining the susceptibility genes linked to an increased risk for contracting AD and understanding its hereditary basis need an understanding of genetics. Several genetic variations, notably the *APOE* gene, are being linked via GWAS to AD therapy. The intricate relationship between inherited variables, brain anatomy, and cognitive impairment in patients with Alzheimer's may be enhanced by researchers by combining genetic, neuroimaging, and clinical data. In addition, uncommon mutations in genes including *APP*, *PSEN1*, and *PSEN2* have been found via genetic analysis on AD patients in families. These findings provide light on the underlying molecular pathways responsible for the development and course of the disease. The systems biology strategies provide an original viewpoint on physiology and heredity by elucidating a complicated web of molecular interactions that underpin AD development. By simulating the dynamics of gene expression, protein signaling, and metabolic processes linked to AD, scientists could potentially be able to recognize significant mechanisms of regulation and abnormal nodes in the disease. By combining multi-omics data, involving as transcriptomics, proteomics, and genomes, systems biology techniques provide an in-depth knowledge of AD pathogenesis and provide novel opportunities for the development of the illness. The integration of these several approaches has significantly improved our understanding of AD. Researchers have found biomarkers in cerebral and bloodstream samples, such as the proteins tau and amyloid-beta, which might be utilized for rapid diagnosis and tracking the course of a disease. Biological system simulations were additionally utilized to describe the physiology of tau and amyloid-beta, which has helped discover potential targets for the development of new drugs. Although there have been breakthroughs, some obstacles still need to be addressed before research results may be used in therapeutic settings. AD's complicated etiology and heterogeneous physical manifestations provide significant obstacles to the development of innovative therapeutics. Ethics related to personalized medicine, including patient privacy and data security, should also be taken into account. In summary, the fields of neurology, genetics, and systems biology working together may greatly advance our knowledge of AD and lead to the development of efficient diagnostic and therapeutic approaches. Through the integration of evidence from multiple places, scientists could identify new biomarkers, pinpoint potential targets for assistance, and gain deeper insights into the pathophysiology of AD. Future studies' results could modify the way AD sufferers receive healthcare. Multidisciplinary collaboration and continuous innovation are required for inclusion into care.

ACKNOWLEDGMENTS

The authors are thankful to the Principal, Department of Pharmacy, Teerthanker Mahaveer University, Moradabad, for providing the necessary facilities.

COMPETING INTEREST STATEMENT

The authors declare no conflict of interest.

DISCLAIMER

None

KEYWORDS

- **Alzheimer's disease**
- **biomarkers**
- **genetics**
- **interdisciplinary research**
- **molecular pathways**
- **neurology**
- **precision medicine**
- **systems biology**

REFERENCES

1. Chiesa, P. A., Cavedo, E., Lista, S., Thompson, P. M., & Hampel, H. (2017). Revolution of resting-state functional neuroimaging genetics in Alzheimer's Disease. *Trends in Neuroscience*, *40*, 469–480. https://doi.org/10.1016/j.tins.2017.06.002.
2. Hampel, H., Vergallo, A., Caraci, F., Cuello, A. C., Lemercier, P., Vellas, B., Giudici, K. V., Baldacci, F., Hänisch, B., Haberkamp, M., Broich, K., Nisticò, R., Emanuele, E., Llavero, F., Zugaza, J. L., Lucía, A., Giacobini, E., & Lista, S. (2021). Future avenues for Alzheimer's disease detection and therapy: Liquid biopsy, intracellular signaling modulation, systems pharmacology drug discovery. *Neuropharmacology*, *185*, 108081. https://doi.org/10.1016/j.neuropharm.2020.108081.
3. Chao, H.-T., Liu, L., & Bellen, H. J. (2017). Building dialogues between clinical and biomedical research through cross-species collaborations, Semin. *Cell & Developmental Biology*, *70,* 49–57. https://doi.org/10.1016/j.semcdb.2017.05.022.
4. Mitra, D., Mitra, D., Sabri Bensaad, M., Sinha, S., Pant, K., Pant, M., Priyadarshini, A., Singh, P., Dassamiour, S., Hambaba, L., Panneerselvam, P., & Das Mohapatra,

P. K. (2022). Evolution of bioinformatics and its impact on modern bio-science in the twenty-first century: Special attention to pharmacology, plant science and drug discovery. *Computational Toxicology, 24,* 100248. https://doi.org/10.1016/j.comtox.2022.100248.

5. Awasthi, A., Bhaskar, S., Panda, S., & Roy, S. (2024). A review of brain injury at multiple time scales and its clinicopathological correlation through in silico modeling. *Brain Multiphysics, 6,* 100090. https://doi.org/10.1016/j.brain.2024.100090.
6. Mutalik, S. P., Pandey, A., & Mutalik, S. (2020). Nanoarchitectronics: A versatile tool for deciphering nanoparticle interaction with cellular proteins, nucleic acids and phospholipids at biological interfaces. *International Journal of Biological Macromolecules, 151,* 136–158. https://doi.org/10.1016/j.ijbiomac.2020.02.150.
7. Li, H., Zhang, B., Shi, Z., Wu, H., Mei, S., Ai, T., Shi, A., Wang, L., Zhu, M., & Tang, J. (2023). Derivation of induced pluripotent stem cell SJTUi003-A from a 69-year-old Chinese Han Sporadic Alzheimer's disease patient with APOEε3/ε4 genetic background. *Stem Cell Research, 71,* 103142. https://doi.org/10.1016/j.scr.2023.103142.
8. Abe-Dohmae, S., & Yokoyama, S. (2021). ABCA7 links sterol metabolism to the host defense system: Molecular background for potential management measure of Alzheimer's disease. *Gene, 768,* 145316. https://doi.org/10.1016/j.gene.2020.145316.
9. Iwata, A., Iwatsubo, T., Ihara, R., Suzuki, K., Matsuyama, Y., Tomita, N., Arai, H., Ishii, K., Senda, M., Ito, K., Ikeuchi, T., Kuwano, R., & Matsuda, H. (2018). Effects of sex, educational background, and chronic kidney disease grading on longitudinal cognitive and functional decline in patients in the Japanese Alzheimer's Disease Neuroimaging Initiative study. *Alzheimer's & Dementia: Translational Research Clinical Interventions, 4,* 765–774. https://doi.org/10.1016/j.trci.2018.06.008.
10. Dai, Q., Zhang, W., Guo, J., Di, W., Zhao, J., Zhang, B., & Wang, Y. (2020). Generation of an induced pluripotent stem cell line (SIAISi003-A) from a 79-year-old patient with Alzheimer's disease having APOE3/4 genetic background. *Stem Cell Research, 48,* 101949. https://doi.org/10.1016/j.scr.2020.101949.
11. Lenoble, Q., Corveleyn, X., Szaffarczyk, S., Pasquier, F., & Boucart, M. (2018). Attentional capture by incongruent object/background scenes in patients with Alzheimer disease. *Cortex, 107,* 4–12. https://doi.org/10.1016/j.cortex.2018.06.002.
12. Gómez, C., Hornero, R., Abásolo, D., Fernández, A., & Escudero, J. (2007). Analysis of the magnetoencephalogram background activity in Alzheimer's disease patients with auto-mutual information. *Computer Methods and Programs Biomedicine, 87,* 239–247. https://doi.org/10.1016/j.cmpb.2007.07.001.
13. Lok, K. Zhao, H. Shen, H., Wang, Z., Gao, X., Zhao, W., & Yin, M. (2013). Characterization of the APP/PS1 mouse model of Alzheimer's disease in senescence accelerated background. *Neuroscience Letters, 557,* 84–89. https://doi.org/10.1016/j.neulet.2013.10.051.
14. Katzav, A. Faust-Socher, A., Kvapil, F., Michaelson, D. M. Blank, M., Pick, C. G., Shoenfeld, Y., Korczyn, A. D., & Chapman, J. (2011). Antiphospholipid syndrome induction exacerbates a transgenic Alzheimer disease model on a female background. *Neurobiology of Aging, 32,* 272–279. https://doi.org/10.1016/j.neurobiolaging.2009.02.007.

15. Rustay, N. R., Cronin, E. A., Curzon, P., Markosyan, S., Bitner, R. S., Ellis, T. A., Waring, J. F., Decker, M. W., Rueter, L. E., & Browman, K. E. (2010). Mice expressing the Swedish APP mutation on a 129 genetic background demonstrate consistent behavioral deficits and pathological markers of Alzheimer's disease. *Brain Research, 1311,* 136–147. https://doi.org/10.1016/j.brainres.2009.11.040.
16. Gómez, C., Hornero, R., Abásolo, D., Fernández, A., & López, M. (2006). Complexity analysis of the magnetoencephalogram background activity in Alzheimer's disease patients. *Medical Engingeering & Physical, 28,* 851–859. https://doi.org/10.1016/j.medengphy.2006.01.003.
17. Wang, Y., Yu, H., Zhang, Y., Zhang, J., Chen, K., Sun, H., Meng, S., Li, G., Lei, Y., & Zhao, J. (2018). Establishment of TUSMi003-A, an induced pluripotent stem cell (iPSC) line from a 62-year old Chinese Han patient with Alzheimer's disease with ApoE3/4 genetic background. *Stem Cell Research, 27,* 57–60. https://doi.org/10.1016/j.scr.2018.01.005.
18. Abásolo, D., Hornero, R., Espino, P., Poza, J., Sánchez, C. I., & de la Rosa, R. (2005). Analysis of regularity in the EEG background activity of Alzheimer's disease patients with approximate entropy. *Clinical Neurophysiology, 116,* 1826–1834. https://doi.org/10.1016/j.clinph.2005.04.001.
19. Romem, A., Zalcman, B. G., Jakubowitz, O., & Pinchas-Mizrachi, R. (2023). Knowledge and attitudes towards patients with Alzheimer's disease across different educational levels of nursing and in different care settings: A cross-sectional study. *Nurse Education in Practice, 71,* 103685. https://doi.org/10.1016/j.nepr.2023.103685.
20. Olin, J. T., Katz, I. R., Meyers, B. S., Schneider, L. S., & Lebowitz, B. D. (2002). Provisional diagnostic criteria for depression of Alzheimer disease: Rationale and background. *American Journal of Geriatric Psychiatry, 10,* 129–141. https://doi.org/10.1097/00019442-200203000-00004.
21. Wang, D., Mangal, R. K., Daniel, A., Gould, M., Stead, T. S., & Ganti, L. (2024). Racial disparities in subjective cognitive decline and its implications among Alzheimer's caretakers. *Journal of the National Medical Association,116*(2 Pt 1), 170–173. https://doi.org/10.1016/j.jnma.2024.01.006.
22. Poza, J., Hornero, R., Abásolo, D., Fernández, A., & García, M. (2007). Extraction of spectral based measures from MEG background oscillations in Alzheimer's disease. *Medical Engingeering & Physics, 29,* 1073–1083. https://doi.org/10.1016/j.medengphy.2006.11.006.
23. Ávila-Jiménez, J. L., Cantón-Habas, V. del M., Carrera-González, P., Rich-Ruiz, M., & Ventura, S. (2024). A deep learning model for Alzheimer's disease diagnosis based on patient clinical records. *Computers in Biology and Medicine, 169,* 107814. https://doi.org/10.1016/j.compbiomed.2023.107814.
24. Shanmugavadivel, K., Sathishkumar, V. E., Cho, J., & Subramanian, M. (2023). Advancements in computer-assisted diagnosis of Alzheimer's disease: A comprehensive survey of neuroimaging methods and AI techniques for early detection. *Ageing Research Reviews, 91,* 102072. https://doi.org/10.1016/j.arr.2023.102072.
25. Ku, H.-L., & Chi, N.-F. (2011). Cerebral lobar microhemorrhages detection by high magnetic field susceptibility weighted image: A potential diagnostic neuroimage technique of Alzheimer's disease. *Medical Hypotheses, 76,* 840–842. https://doi.org/10.1016/j.mehy.2011.02.032.

26. Bazarbekov, I., Razaque, A., Ipalakova, M., Yoo, J., Assipova, Z., & Almisreb, A. (2024). A review of artificial intelligence methods for Alzheimer's disease diagnosis: Insights from neuroimaging to sensor data analysis. *Biomedical Signal Processing and Control, 92,* 106023. https://doi.org/10.1016/j.bspc.2024.106023.
27. Casagrande, C. C., Rempe, M. P., Springer, S. D., & Wilson, T. W. (2023). Comprehensive review of task-based neuroimaging studies of cognitive deficits in Alzheimer's disease using electrophysiological methods. *Ageing Research Reviews, 88,* 101950. https://doi.org/10.1016/j.arr.2023.101950.
28. Prajjwal, P., Marsool, M. D. M., Inban, P., Sharma, B., Asharaf, S., Aleti, S., Gadam, S., Al Sakini, A. S., & Hadi, D. D. (2023). Vascular dementia subtypes, pathophysiology, genetics, neuroimaging, biomarkers, and treatment updates along with its association with Alzheimer's dementia and diabetes mellitus. *Disease-a-Month, 69,* 101557. https://doi.org/10.1016/j.disamonth.2023.101557.
29. Wang, D., Honnorat, N., Fox, P. T., Ritter, K., Eickhoff, S. B., Seshadri, S., & Habes, M. (2023). Deep neural network heatmaps capture Alzheimer's disease patterns reported in a large meta-analysis of neuroimaging studies. *Neuroimage, 269* 119929. https://doi.org/10.1016/j.neuroimage.2023.119929.
30. Gu, X., Zhao, M., Han, X., & Liu, L. (2020). Presenilin-1 mutation is associated with a hippocampus defect in alzheimer's disease: Meta-Analysis for neuroimaging research. *Clinical Neurology and Neurosurgery, 191,* 105679. https://doi.org/10.1016/j.clineuro.2020.105679.
31. Goenka, N., & Tiwari, S. (2022). AlzVNet: A volumetric convolutional neural network for multiclass classification of Alzheimer's disease through multiple neuroimaging computational approaches. *Biomedical Signal Processing and Control, 74,* 103500. https://doi.org/10.1016/j.bspc.2022.103500.
32. Taylor, A., Zhang, F., Niu, X., Heywood, A., Stocks, J., Feng, G., Popuri, K., Beg, M. F., & Wang, L. (2022). Investigating the temporal pattern of neuroimaging-based brain age estimation as a biomarker for Alzheimer's Disease related neurodegeneration. *Neuroimage, 263,* 119621. https://doi.org/10.1016/j.neuroimage.2022.119621.
33. Gregory, S., Pullen, H., Ritchie, C. W., Shannon, O. M., Stevenson, E. J., & Muniz-Terrera, G. (2023). Mediterranean diet and structural neuroimaging biomarkers of Alzheimer's and cerebrovascular disease: A systematic review. *Experimental Gerontology, 172,* 112065. https://doi.org/10.1016/j.exger.2022.112065.
34. Martí-Juan, G., Sanroma-Guell, G., & Piella, G. (2020). A survey on machine and statistical learning for longitudinal analysis of neuroimaging data in Alzheimer's disease. *Computers Methods and Programs Biomedicine, 189,* 105348. https://doi.org/10.1016/j.cmpb.2020.105348.
35. Jomeiri, A., Navin, A. H., & Shamsi, M. (2024). Longitudinal MRI analysis using a hybrid DenseNet-BiLSTM method for Alzheimer's disease prediction. *Behavioural Brain Research, 463,* 114900. https://doi.org/10.1016/j.bbr.2024.114900.
36. Sharma, H., Rani, T., & Khan, S. (2023). An insight into neuropathic pain: A systemic and up-to-date review. *IJPSR, 14,* 607–621.
37. Jiao, C.-N., Gao, Y.-L., Ge, D.-H., Shang, J., & Liu, J.-X. (2024). Multi-modal imaging genetics data fusion by deep auto-encoder and self-representation network for Alzheimer's disease diagnosis and biomarkers extraction. *Engineering*

Applications of Artificial Intelligence, 130, 107782. https://doi.org/10.1016/j.engappai.2023.107782.

38. Timsina, J., Ali, M., Do, A., Wang, L., Western, D., Sung, Y. J., & Cruchaga, C. (2024). Harmonization of CSF and imaging biomarkers in Alzheimer's disease: Need and practical applications for genetics studies and preclinical classification. *Neurobiology of Disease, 190,* 106373. https://doi.org/10.1016/j.nbd.2023.106373.
39. Bao, J., Wen, J., Wen, Z., Yang, S., Cui, Y., Yang, Z., Erus, G., Saykin, A. J., Long, Q., Davatzikos, C., & Shen, L. (2023). Brain-wide genome-wide colocalization study for integrating genetics, transcriptomics and brain morphometry in Alzheimer's disease. *Neuroimage, 280,* 120346. https://doi.org/10.1016/j.neuroimage.2023.120346.
40. Amin, A. M., Mostafa, H., & Khojah, H. M. J. (2023). Insulin resistance in Alzheimer's disease: The genetics and metabolomics links. *Clinica Chimica Acta, 539* 215–236. https://doi.org/10.1016/j.cca.2022.12.016.
41. Ashique, S., Pal, R., Sharma, H., Mishra, N., & Garg, A. (2024).Unraveling the Emerging Niche Role of Extracellular Vesicles (EVs) in Traumatic Brain Injury (TBI). *CNS Neurological Disorders Drug Targets, 23*(11), 1357–1370.
42. Sharma, H., Chandra, P., Verma, A., Pandey, S. N., Kumar, P., & Sigh, A. (2023). Therapeutic approaches of nutraceuticals in the prevention of neurological disorders, *European Chemical Bulletin, 12,* 1575–1596. https://doi.org/10.48047/ecb/2023.12.si5a.038.
43. Sheng, J., Wang, L., Cheng, H., Zhang, Q., Zhou, R., & Shi, Y. (2021). Strategies for multivariate analyses of imaging genetics study in Alzheimer's disease. *Neuroscience Letters, 762,* 136147. https://doi.org/10.1016/j.neulet.2021.136147.
44. Chen, J., Wang, L., De Jager, P. L., Bennett, D. A., Buchman, A. S., & Yang, J. (2022). A scalable Bayesian functional GWAS method accounting for multivariate quantitative functional annotations with applications for studying Alzheimer disease. *Human Genetics and Genomics Advances, 3,* 100143. https://doi.org/10.1016/j.xhgg.2022.100143.
45. Sharma, H., & Chandra, P. (2023). Challenges and future prospects: A benefaction of phytoconstituents on molecular targets pertaining to Alzheimer's disease. *International Journal of Pharmaceutical Investigation, 14,* 117–126. https://doi.org/10.5530/ijpi.14.1.15.
46.. Kim, B.-H., Nho, K., & Lee, J.-M. (2021). Genome-wide association study identifies susceptibility loci of brain atrophy to NFIA and ST18 in Alzheimer's disease. *Neurobiology of Aging, 102,* 200.e1–200.e11. https://doi.org/10.1016/j.neurobiolaging.2021.01.021.
47. Khaire, A. S., Wimberly, C. E., Semmes, E. C., Hurst, J. H., & Walsh, K. M. (2022). An integrated genome and phenome-wide association study approach to understanding Alzheimer's disease predisposition. *Neurobiology of Aging, 118,* 117–123. https://doi.org/10.1016/j.neurobiolaging.2022.05.011.
48. Shi, H., Belbin, O., Medway, C., Brown, K., Kalsheker, N., Carrasquillo, M., Proitsi, P., Powell, J., Lovestone, S., Goate, A., Younkin, S., Passmore, P., & Morgan, K. (2012). Genetic variants influencing human aging from late-onset Alzheimer's disease (LOAD) genome-wide association studies (GWAS). *Neurobiology of Aging, 33,* 1849.e5–1849.e18. https://doi.org/10.1016/j.neurobiolaging.2012.02.014.

49. Ulaganathan, S., & Pitchaimani, A. (2023). Spontaneous and familial models of Alzheimer's disease: Challenges and advances in preclinical research. *Life Sciences*, *328*, 121918. https://doi.org/10.1016/j.lfs.2023.121918.
50. Cash, M. K., Rockwood, K., Fisk, J. D., & Darvesh, S. (2021). Clinicopathological correlations and cholinesterase expression in early-onset familial Alzheimer's disease with the presenilin 1 mutation, Leu235Pro. *Neurobiology of Aging*, *103,* 31–41. https://doi.org/10.1016/j.neurobiolaging.2021.02.025.
51. Guo, X., Jiang, M., Dai, X., Shen, J., & Wang, X. (2023). Presenilin-1, mutated in familial Alzheimer's disease, maintains genome stability via a γ-secretase dependent way. *DNA Repair (Amst),* 131, 103580. https://doi.org/10.1016/j.dnarep.2023.103580.
52. Kourti, M., & Metaxas, A. (2024). A systematic review and meta-analysis of tau phosphorylation in mouse models of familial Alzheimer's disease. *Neurobiology Disease*, *192,* 106427. https://doi.org/10.1016/j.nbd.2024.106427.
53. Dehury, B., Somavarapu, A. K., & Kepp, K. P. (2020). A computer-simulated mechanism of familial Alzheimer's disease: Mutations enhance thermal dynamics and favor looser substrate-binding to γ-secretase. *Journal of Structural Biology*, *212,* 107648. https://doi.org/10.1016/j.jsb.2020.107648.
54. Gurdon, B., & Kaczorowski, C. (2021). Pursuit of precision medicine: Systems biology approaches in Alzheimer's disease mouse models. *Neurobiology of Disease*, *161,* 105558. https://doi.org/10.1016/j.nbd.2021.105558.
55. Miyoshi, E., Morabito, S., & Swarup, V. (2021). Systems biology approaches to unravel the molecular and genetic architecture of Alzheimer's disease and related tauopathies. *Neurobiology Disease*, *160,* 105530. https://doi.org/10.1016/j.nbd.2021.105530.
56. Chowdhury, U. N., Islam, M. B., Ahmad, S., & Moni, M. A. (2020). Systems biology and bioinformatics approach to identify gene signatures, pathways and therapeutic targets of Alzheimer's disease. *Informatics in Medicine Unlocked*, *21,* 100439. https://doi.org/10.1016/j.imu.2020.100439.
57. Bitra, V. R., Challa, S. R., Adiukwu, P. C., & Rapaka, D. (2023). Tau trajectory in Alzheimer's disease: Evidence from the connectome-based computational models. *Brain Research Bulletin*, *203,* 110777. https://doi.org/10.1016/j.brainresbull.2023.110777.
58. Eno, E. A., Louis, H., Ekoja, P., Benjamin, I., Adalikwu, S. A., Orosun, M. M., Unimuke, T. O., Asogwa, F. C., & Agwamba, E. C. (2022). Experimental and computational modeling of the biological activity of benzaldehyde sulphur trioxide as a potential drug for the treatment of Alzheimer disease. *Journal of Indian Chemical Society*, *99*, 100532. https://doi.org/10.1016/j.jics.2022.100532.
59. Ugbaja, S. C., Sanusi, Z. K., Appiah-Kubi, P., Lawal, M. M., & Kumalo, H. M. (2021). Computational modelling of potent β-secretase (BACE1) inhibitors towards Alzheimer's disease treatment. *Biophysical Chemistry*, *270,* 106536. https://doi.org/10.1016/j.bpc.2020.106536.
60. Joshi, A., Wang, D.-H., Watterson, S., McClean, P. L., Behera, C. K., Sharp, T., & Wong-Lin, K. (2020). Opportunities for multiscale computational modelling of serotonergic drug effects in Alzheimer's disease. *Neuropharmacology*, *174,* 108118. https://doi.org/10.1016/j.neuropharm.2020.108118.

61. Demirtaş, M., Falcon, C., Tucholka, A., Gispert, J. D., Molinuevo, J. L., & Deco, G. (2017). A whole-brain computational modeling approach to explain the alterations in resting-state functional connectivity during progression of Alzheimer's disease. *NeuroImage Clinical, 16,* 343–354. https://doi.org/10.1016/j.nicl.2017.08.006.
62. Geerts, H., Hofmann-Apitius, M., & Anastasio, T. J. (2017). Knowledge-driven computational modeling in Alzheimer's disease research: Current state and future trends. *Alzheimer's & Dementia, 13,* 1292–1302. https://doi.org/10.1016/j.jalz.2017.08.011.
63. Yu, M., Nie, Y., Yang, J., Yang, S., Li, R., Rao, V., Hu, X., Fang, C., Li, S., Song, D., Guo, F., Snyder, M. P., Chang, H. Y., Kuo, C. J., Xu, J., & Chang, J. (2023). Integrative multi-omic profiling of adult mouse brain endothelial cells and potential implications in Alzheimer's disease. *Cell Reports, 42,* 113392. https://doi.org/10.1016/j.celrep.2023.113392.
64. Abbas, Z., Tayara, H., & Chong, K. T. (2022). Alzheimer's disease prediction based on continuous feature representation using multi-omics data integration. *Chemometrics and Intelligent Laboratory Systems, 223,* 104536. https://doi.org/10.1016/j.chemolab.2022.104536.
65. Shokhirev, M. N., & Johnson, A. A. (2022). An integrative machine-learning meta-analysis of high-throughput omics data identifies age-specific hallmarks of Alzheimer's disease. *Ageing Research Reviews, 81,* 101721. https://doi.org/10.1016/j.arr.2022.101721.
66. Tan, M. S., Cheah, P.-L., Chin, A.-V., Looi, L.-M., & Chang, S.-W. (2021). A review on omics-based biomarkers discovery for Alzheimer's disease from the bioinformatics perspectives: Statistical approach vs machine learning approach. *Computers in Biology and Medicine, 139,* 104947. https://doi.org/10.1016/j.compbiomed.2021.104947.
67. Hampel, H., Nisticò, R., Seyfried, N. T., Levey, A. I., Modeste, E., Lemercier, P., Baldacci, F., Toschi, N., Garaci, F., Perry, G., Emanuele, E., Valenzuela, P. L., Lucia, A., Urbani, A., Sancesario, G. M., Mapstone, M., Corbo, M., Vergallo, A., & Lista, S. (2021). Omics sciences for systems biology in Alzheimer's disease: State-of-the-art of the evidence. *Ageing Research Reviews, 69,* 101346. https://doi.org/10.1016/j.arr.2021.101346.
68. O'Connor, L. M., O'Connor, B. A., Bin Lim, S., Zeng, J., & Lo, C. H. (2023). Integrative multi-omics and systems bioinformatics in translational neuroscience: A data mining perspective. *Journal of Pharmaceutical Analysis, 13,* 836–850. https://doi.org/10.1016/j.jpha.2023.06.011.
69. Terao, I., & Kodama, W. (2024). Comparative efficacy, tolerability and acceptability of donanemab, lecanemab, aducanumab and lithium on cognitive function in mild cognitive impairment and Alzheimer's disease: A systematic review and network meta-analysis. *Ageing Research Reviews, 94,* 102203. https://doi.org/10.1016/j.arr.2024.102203.
70. Guo, P., Meng, C., Zhang, S., Cai, Y., Huang, J., Shu, J., Wang, J., & Cai, C. (2024). Network-based analysis on the genes and their interactions reveals link between schizophrenia and Alzheimer's disease. *Neuropharmacology, 244,* 109802. https://doi.org/10.1016/j.neuropharm.2023.109802.

71. Cai, H.-Y., Chen, S.-R., Wang, Y., Jiao, J.-J., Qiao, J., Hölscher, C., Wang, Z.-J., Zhang, S.-X., & Wu, M.-N. (2023). Integrated analysis of the lncRNA-associated ceRNA network in Alzheimer's disease. *Gene, 876,* 147484. https://doi.org/10.1016/j.gene.2023.147484.
72. Zhi, J., Yin, L., Zhang, Z., Lv, Y., Wu, F., Yang, Y., Zhang, E., Li, H., Lu, N., Zhou, M., & Hu, Q. (2024). Network pharmacology-based analysis of Jin-Si-Wei on the treatment of Alzheimer's disease. *Journal of Ethnopharmacology*, *319,* 117291. https://doi.org/10.1016/j.jep.2023.117291.
73. Jaramillo-Jimenez, A., Giil, L. M., Borda, M. G., Tovar-Rios, D. A., Kristiansen, K. A., Bruheim, P., Aarsland, D., Barreto, G. E., & Berge, R. K. (2023). Serum TCA cycle metabolites in Lewy bodies dementia and Alzheimer's disease: Network analysis and cognitive prognosis. *Mitochondrion*, *71,* 17–25. https://doi.org/10.1016/j.mito.2023.05.002.
74. Klepl, D., He, F., Wu, M., Blackburn, D. J., & Sarrigiannis, P. G. (2023). Cross-frequency multilayer network analysis with bispectrum-based functional connectivity: A study of Alzheimer's disease. *Neuroscience*, *521,* 77–88. https://doi.org/10.1016/j.neuroscience.2023.04.008.
75. Bubier, J. A., & Chesler, E. J. (2012). Accelerating discovery for complex neurological and behavioral disorders through systems genetics and integrative genomics in the laboratory mouse. *Neurotherapeutics*, *9*, 338–348. https://doi.org/10.1007/s13311-012-0111-3.
76. Hampel, H., Gao, P., Cummings, J., Toschi, N., Thompson, P. M., Hu, Y., Cho, M., & Vergallo, A. (2023). The foundation and architecture of precision medicine in neurology and psychiatry. *Trends in Neuroscience*, *46,* 176–198. https://doi.org/10.1016/j.tins.2022.12.004.
77. Fujimura, K., Guise, A. J., Nakayama, T., Schlaffner, C. N., Meziani, A., Kumar, M., Cheng, L., Vaughan, D. J., Kodani, A., Van Haren, S., Parker, K., Levy, O., Durbin, A. F., Bosch, I., Gehrke, L., Steen, H., Mochida, G. H., & Steen, J. A. (2023). Integrative systems biology characterizes immune-mediated neurodevelopmental changes in murine Zika virus microcephaly. *IScience*, *26,* 106909. https://doi.org/10.1016/j.isci.2023.106909.
78. Nabirotchkin, S., Peluffo, A. E., Rinaudo, P., Yu, J., Hajj, R., & Cohen, D. (2020). Next-generation drug repurposing using human genetics and network biology. *Current Opinion in Pharmacology*, *51,* 78–92. https://doi.org/10.1016/j.coph.2019.12.004.
79. Julian, D., Hollingsworth, E. W., Julian, K., & Imitola, J. (2019). Convergence of human cellular models and genetics to study neural stem cell signaling to enhance central nervous system regeneration and repair. *Seminars Cell & Developmental Biology*, *95,* 84–92. https://doi.org/10.1016/j.semcdb.2019.07.002.
80. Islam, T., Rahman, M. R., Khan, A., & Moni, M. A. (2023). Integration of Mendelian randomisation and systems biology models to identify novel blood-based biomarkers for stroke. *Journal of Biomedical Informatics*, *141,* 104345. https://doi.org/10.1016/j.jbi.2023.104345.
81. Liu, J., & Shi, H. (2023). Finger exercise effects on cognitive functions in older adults with MCI/AD/dementia: A systematic review and meta-analysis of randomized controlled trials. *Brain Behavior and Immunity Integrative*, *4*, 100034. https://doi.org/10.1016/j.bbii.2023.100034.

82. Juhász, G., Földi, I., & Penke, B. (2011). Systems biology of Alzheimer's disease: How diverse molecular changes result in memory impairment in AD. *Neurochemistry International*, *58*, 739–750. https://doi.org/10.1016/j.neuint.2011.02.008.
83. Liu, P., Zhou, Y., Shi, J., Wang, F., Yang, X., Zheng, X., Wang, Y., He, Y., Xie, X., & Pang, X. (2023). Myricetin improves pathological changes in 3×Tg-AD mice by regulating the mitochondria-NLRP3 inflammasome-microglia channel by targeting P38 MAPK signaling pathway. *Phytomedicine*, *115*, 154801. https://doi.org/10.1016/j.phymed.2023.154801.
84. Dhami, M., Raj, K., & Singh, S. (2023). Relevance of gut microbiota to Alzheimer's Disease (AD): Potential effects of probiotic in management of AD. *Aging and Health Research*, *3*, 100128. https://doi.org/10.1016/j.ahr.2023.100128.
85. Zhang, J., He, X., Qing, L., Gao, F., & Wang, B. (2022). BPGAN: Brain PET synthesis from MRI using generative adversarial network for multi-modal Alzheimer's disease diagnosis. Computer Methods and Programs in Biomedicine, *217*, 106676. https://doi.org/10.1016/j.cmpb.2022.106676.
86. He, J., Liu, F., Xu, T., Ma, J., Yu, H., Zhao, J., Xie, Y., Luo, L., Yang, Q., Lou, T., He, L., & Sun, D. (2023). The role of hydrogen therapy in Alzheimer's disease management: Insights into mechanisms, administration routes, and future challenges. Biomedicine & Pharmacotherapy, *168*, 115807. https://doi.org/10.1016/j.biopha.2023.115807.
87. Zivari-Ghader, T., Valioglu, F., Eftekhari, A., Aliyeva, N., Beylerli, O., Davran, S., Cho, W. C., Beilerli, A., Khalilov, R., & Javadov, S. (2024). Recent progresses in natural based therapeutic materials for Alzheimer's disease. *Heliyon*, e26351. https://doi.org/10.1016/j.heliyon.2024.e26351.
88. Jayaswamy, P. K., Gollapalli, P., Vijaykrishnaraj, M., Alexander, L. M., Patil, P., & Shetty, P. (2023). Identification of network-based differential gene expression signatures and their transcriptional factors to develop progressive blood biomarkers for Alzheimer's disease. *Human Gene*, *37*, 201202. https://doi.org/10.1016/j.humgen.2023.201202.
89. Alahmady, N. F., Alkhulaifi, F. M., Abdullah Momenah, M., Ali Alharbi, A., Allohibi, A., Alsubhi, N. H., & Ahmed Alhazmi, W. (2024). Biochemical characterization of chamomile essential oil: Antioxidant, antibacterial, anticancer and neuroprotective activity and potential treatment for Alzheimer's disease. *Saudi Journal of Biological Science*, *31*, 103912. https://doi.org/10.1016/j.sjbs.2023.103912.
90. Rather, H. A., Almousa, S., Craft, S., & Deep, G. (2023). Therapeutic efficacy and promise of stem cell-derived extracellular vesicles in Alzheimer's disease and other aging-related disorders. *Ageing Research Reviews*, *92*, 102088. https://doi.org/10.1016/j.arr.2023.102088.
91. Gobom, J., Brinkmalm, A., Brinkmalm, G., Blennow, K., & Zetterberg, H. (2024). Alzheimer's disease biomarker analysis using targeted mass spectrometry. *Molecular Cellular Proteomics*, *23*, 100721. https://doi.org/10.1016/j.mcpro.2024.100721.
92. Bhole, R. P., Chikhale, R. V., & Rathi, K. M. (2024). Current biomarkers and treatment strategies in Alzheimer disease: An overview and future perspectives. *IBRO Neuroscience Reports*, *16*, 8–42. https://doi.org/10.1016/j.ibneur.2023.11.003.
93. Mantellatto Grigoli, M., Pelegrini, L. N. C., Whelan, R., & Cominetti, M. R. (2024). Present and future of blood-based biomarkers of Alzheimer's disease: Beyond the classics. *Brain Research*, 148812. https://doi.org/10.1016/j.brainres.2024.148812.

94. Armenta-Castro, A., Núñez-Soto, M. T., Rodriguez-Aguillón, K. O., Aguayo-Acosta, A., Oyervides-Muñoz, M. A., Snyder, S. A., Barceló, D., Saththasivam, J., Lawler, J., Sosa-Hernández, J. E., & Parra-Saldívar, R. (2024). Urine biomarkers for Alzheimer's disease: A new opportunity for wastewater-based epidemiology? *Environment International*, *184*, 108462. https://doi.org/10.1016/j.envint.2024.108462.
95. Monteiro, A. R., Barbosa, D. J., Remião, F., & Silva, R. (2023). Alzheimer's disease: Insights and new prospects in disease pathophysiology, biomarkers and disease-modifying drugs. *Biochemical Pharmacology*, *211,* 115522. https://doi.org/10.1016/j.bcp.2023.115522.
96. Mirzaie, A., Nasrollahpour, H., Khalilzadeh, B., Jamali, A. A., Spiteri, R. J., Yousefi, H., Isildak, I., & Rahbarghazi, R. (2023). Cerebrospinal fluid: A specific biofluid for the biosensing of Alzheimer's diseases biomarkers. *TrAC Trends in Analytical Chemistry*, *166,* 117174. https://doi.org/10.1016/j.trac.2023.117174.
97. Wojdała, A. L., Bellomo, G., Gaetani, L., Toja, A., Chipi, E., Shan, D., Chiasserini, D., & Parnetti, L. (2023). Trajectories of CSF and plasma biomarkers across Alzheimer's disease continuum: Disease staging by NF-L, p-tau181, and GFAP. *Neurobiology of Disease*, *189,* 106356. https://doi.org/10.1016/j.nbd.2023.106356.
98. Georgakas, J. E., Howe, M. D., Thompson, L. I., Riera, N. M., & Riddle, M. C. (2023). Biomarkers of Alzheimer's disease: Past, present and future clinical use. *Biomarkers in Neuropsychiatry*, *8*, 100063. https://doi.org/10.1016/j.bionps.2023.100063.
99. Singh, B., Day, C. M., Abdella, S., & Garg, S. (2024). Alzheimer's disease current therapies, novel drug delivery systems and future directions for better disease management. *Journal of Controlled Release*, *367*, 402–424. https://doi.org/10.1016/j.jconrel.2024.01.047.
100. Liu, N., Liang, X., Chen, Y.,.& Xie, L. (2024). Recent trends in treatment strategies for Alzheimer's disease and the challenges: A topical advancement. *Ageing Research Reviews*, *94*, 102199. https://doi.org/10.1016/j.arr.2024.102199.
101. Sharma, M., Pal, P., & Gupta, S. K. (2024). The neurotransmitter puzzle of Alzheimer's: Dissecting mechanisms and exploring therapeutic horizons. *Brain Research*, *1829*, 148797. https://doi.org/10.1016/j.brainres.2024.148797.
102. Khojaste-Sarakhsi, M., Haghighi, S. S., Ghomi, S. M. T. F., & Marchiori, E. (2022). Deep learning for Alzheimer's disease diagnosis: A survey. *Artifical Intelligence in Medicine*, *130*, 102332. https://doi.org/10.1016/j.artmed.2022.102332.
103. Ly, M., Yu, G. Z., Mian, A., Cramer, A., Meysami, S., Merrill, D. A., Samara, A., Eisenstein, S. A., Hershey, T., Babulal, G. M., Lenze, E. J., Morris, J. C., Benzinger, T. L. S., & Raji, C. A. (2023). Neuroinflammation: A modifiable pathway linking obesity, Alzheimer's disease, and depression. *American Journal of Geriatric Psychiatry*, *31,* 853–866. https://doi.org/10.1016/j.jagp.2023.06.001.
104. Glavind, I. M. L., & Mogensen, H. O. (2022). Fragile storytelling: Methodological considerations when conducting ethnographic fieldwork among people with Alzheimer's disease. *SSM - Qualitative Research Health*, *2*, 100103. https://doi.org/10.1016/j.ssmqr.2022.100103.
105. Wang, C., Piao, S., Huang, Z., Gao, Q., Zhang, J., Li, Y., & Shan, H. (2024). Joint learning framework of cross-modal synthesis and diagnosis for Alzheimer's disease by mining underlying shared modality information. *Medical Image Analysis*, *91*, 103032. https://doi.org/10.1016/j.media.2023.103032.

106. Wilkenfeld, D. A., Orbell, S. L., & Lingler, J. H. (2021). Ethical considerations in communicating Alzheimer's disease neuroimaging biomarker test results to symptomatic individuals. *Neurotherapeutics, 18,* 673–685. https://doi.org/10.1007/s13311-021-01047-0.
107. McBrien, B. (2007). Exercising restraint: Clinical, legal and ethical considerations for the patient with Alzheimer's disease. *Accident and Emergency Nursing, 15,* 94–100. https://doi.org/10.1016/j.aaen.2006.12.007.
108. Sekhar, J. C., Rajyalakshmi, C., Nagaraj, S., Sankar, S., Saturi, R., & Harshavardhan, A. (2023). Deep generative adversarial networks with marine predators algorithm for classification of Alzheimer's disease using electroencephalogram. *Journal of King Saud University – Computer and Information Sciences, 35,* 101848. https://doi.org/10.1016/j.jksuci.2023.101848.
109. Ayus, I., & Gupta, D. (2024). A novel hybrid ensemble based Alzheimer's identification system using deep learning technique. *Biomedical Signal Processing and Control, 92,* 106079. https://doi.org/10.1016/j.bspc.2024.106079.
110. Zhang, Q., Long, Y., Cai, H., & Chen, Y.-W. (2024). Lightweight neural network for Alzheimer's disease classification using multi-slice sMRI. *Magnetic Resonance Imaging, 107,* 164–170. https://doi.org/10.1016/j.mri.2023.12.010.

CHAPTER 6

Personalized Medicine Approaches Based on Genetic Profiling in the Treatment of Alzheimer's Disease

SAKSHI SONI*, PARVEEN RAWAT, VANDANA SONI, and SUSHIL K. KASHAW

Department of Pharmaceutical Sciences, Dr. Harisingh Gour Vishwavidyalaya, Sagar, Madhya Pradesh, India

*Corresponding author

ABSTRACT

Alzheimer's disease (AD), a chronic neurodegenerative disorder, presents a significant challenge to medical practitioners due to its complex etiology and limited treatment success under the traditional one-size-fits-all paradigm. This chapter explores the potential of personalized medicine strategies in AD treatment, particularly focusing on genetic profiling to tailor interventions to individual patients' unique disease characteristics. Beginning with an examination of the complex genetic landscape of AD, including established risk factors like the APOE ε4 allele and emerging insights into other influential genes, this chapter discusses how genetic profiling can inform various aspects of personalized medicine for AD. This includes facilitating early diagnosis, stratifying patients based on their genetic makeup for more targeted therapies, guiding the development of novel therapeutics by understanding specific genetic pathways, and optimizing drug response by predicting individual reactions to existing AD medications. While acknowledging challenges, such as ethical considerations, data privacy concerns, and the need for robust validation of

Cellular and Molecular Mechanisms in Alzheimer's Disease Therapeutics, Volume 1: Cellular Mechanisms. Prashant Tiwari, Sunil Kumar Kadiri, and Kunjbihari Sulakhiya (Eds.)

DOI: 10.1201/9781779641793-6

genetic markers, this chapter concludes with an optimistic outlook on the transformative potential of genetic profiling in reshaping AD diagnosis, treatment, and prevention.

6.1 INTRODUCTION

Alzheimer's disease (AD) casts a long and devastating shadow, posing a critical healthcare challenge for individuals, families, and societies worldwide. Its prevalence rises with age, making the need for effective interventions more pressing than ever. While decades of research have yielded some progress, current treatment approaches remain limited, primarily managing symptoms rather than modifying the disease course [1, 36, 37, 80]. This is where personalized medicine emerges as a beacon of hope. By tailoring treatments to individual genetic profiles, it offers immense potential for addressing the complexity and heterogeneity of AD [2, 38, 39, 81]. The burden of AD is multifaceted. Beyond the profound cognitive decline, memory loss, and functional impairment experienced by patients, the emotional, social, and economic impacts ripple outward. Families grapple with caregiving challenges, witnessing the heartbreaking decline of loved ones. Societal costs reach staggering heights, encompassing healthcare expenses, long-term care, and lost productivity. Estimates suggest the global cost of dementia will balloon to trillions in the coming decades [3, 40–42]. Current treatment options, primarily acetylcholinesterase inhibitors and N-methyl-D-aspartate receptor antagonists, offer modest symptomatic relief for some patients. However, they fall short of altering the neurodegenerative process or halting progression. Similarly, emerging therapies targeting amyloid-beta and tau pathologies have yielded disappointing results, highlighting the intricate and multifaceted nature of AD. Treatment resistance, disease heterogeneity, and comorbidities further complicate the development of effective interventions [4, 43–45]. Therefore, a paradigm shift is essential. Personalized medicine, leveraging the power of genetic profiling, holds immense promise for revolutionizing AD diagnosis, treatment strategies, and ultimately, patient outcomes. By unlocking the unique characteristics of each patient's disease, we can move beyond a one-size-fits-all approach toward a future of targeted therapies and a brighter outlook for those living with AD. Personalized medicine offers a revolutionary approach to diagnosing, treating, and managing AD. It tailors interventions to each patient's unique characteristics, aiming to maximize benefits, minimize side effects, and improve overall well-being. This approach considers genetic, molecular, environmental, and lifestyle factors to guide treatment decisions and optimize care [5, 46–48]. Genetic profiling plays a key role, identifying variants linked to disease risk, progression, and treatment response. For example, the APOE ε4 gene variant increases AD risk, while ε2 may offer some protection. Knowing this allows for personalized risk assessment and management strategies. Beyond APOE, researchers are exploring other genetic markers and building polygenic risk scores (PRS) to give a more comprehensive picture of an individual's genetic predisposition. This information can be used for early intervention and targeted prevention [6, 49, 50]. By leveraging genetic profiles, personalized medicine offers exciting possibilities for optimizing AD treatment, like precision pharmacotherapy, gene therapy, immunotherapy, and lifestyle and behavioral interventions. This chapter will explore the profound burden of AD, the limitations of current therapies, and the exciting potential of personalized medicine in this domain. We will delve into the role of genetic profiling, a key tool for personalized medicine, and how it can revolutionize AD diagnosis, treatment strategies, and ultimately, patient outcomes.

6.2 GENETICS OF ALZHEIMER'S DISEASE

AD is a complex neurodegenerative disorder with a significant genetic component. While the exact causes of AD are not fully understood, genetics play a major role in an individual's risk of developing the disease. This section will explore the major risk genes and susceptibility alleles associated with AD, delve into the role of the *APOE* gene and other key genes, discuss how genetic variants can influence disease subtypes, and highlight ongoing genetic research focused on identifying new targets for treatment. Several genes have been identified as increasing the risk of developing AD. These genes contain variations called alleles, and some alleles can significantly elevate an individual's susceptibility to the disease. Here are some of the major risk genes also shown in Figure 6.1 [7, 51–53]:

1. **APOE:** This gene is the strongest genetic risk factor for AD. The *apolipoprotein E (APOE)* gene is the most well-established genetic risk factor for AD. It exists in three main allelic forms: *APOE ε2, APOE ε3*, and *APOE ε4*.

a. ***APOE ε4:*** This allele is the strongest genetic risk factor for late-onset AD. Having one copy of the APOE ε4 allele increases risk, and having two copies significantly increases risk.

b. ***APOE ε3:*** This is the most common allele and is considered neutral in terms of AD risk.

c. ***APOE ε2:*** This allele is relatively rare and is associated with a lower risk of AD [8, 9, 54–56].

2. **Presenilin 1 (*PSEN1*) and presenilin 2 (*PSEN2*):** Mutations in these genes cause a rare form of early-onset AD.
3. **Amyloid beta precursor protein (*APP*):** Mutations in this gene are also linked to early-onset AD.
4. **Clusterin (*CLU*):** Variations in this gene may increase the risk of AD.
5. **Microtubule-associated protein tau (*MAPT*):** Variations in this gene are associated with an increased risk of AD, particularly frontotemporal dementia, a subtype of AD [10, 11, 57–60].
6. ***BIN1:*** This gene is involved in endocytosis, a cellular process important for clearing waste products from the brain. Variations in *BIN1* have been associated with an increased risk of AD.
7. ***CR1:*** This gene is part of the complement system, which is involved in the immune response. Variations in *CR1* may increase the risk of AD by affecting the clearance of amyloid-beta plaques.
8. ***TREM2:*** This gene plays a role in the function of microglia, which are immune cells in the brain. Mutations in *TREM2* can impair the ability of microglia to clear amyloid-beta plaques and contribute to AD development [12, 61–63].

Genetic variations can influence the subtypes of AD a person may develop. AD is a heterogeneous disease, meaning it can manifest differently in various individuals. Some genetic variations are associated with specific subtypes, such as early-onset AD or frontotemporal dementia. For example, mutations in the *PSEN1* and *PSEN2* genes are known to cause a rare form of early-onset AD, while variations in the *MAPT* gene are more commonly linked to frontotemporal dementia. Understanding the genetic underpinnings of different AD subtypes can aid in developing more targeted treatment strategies. Genetic research on AD is ongoing, with scientists constantly striving to identify new genes and genetic variants associated with the disease. This research is crucial for developing more effective therapies and potentially even preventive measures for AD [13, 82–84].

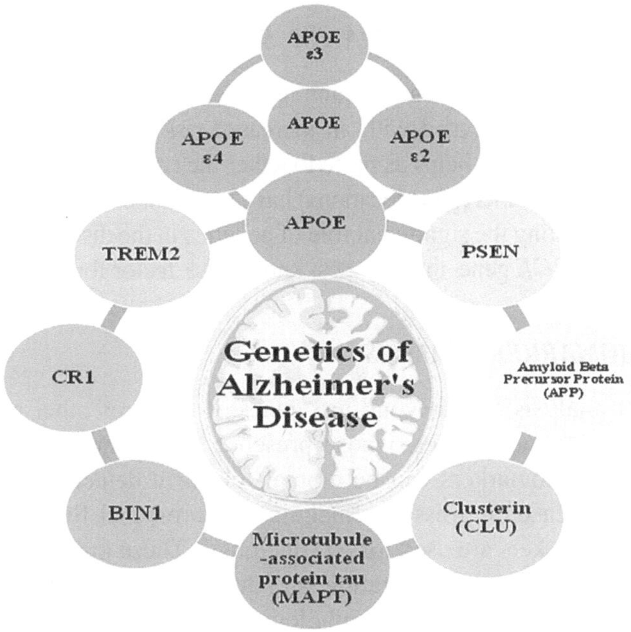

FIGURE 6.1 Some major risk genes in Alzheimer's disease.

6.3 GENETIC PROFILING TECHNIQUES

Understanding the genetic underpinnings of AD is crucial for developing effective treatments and potentially even preventive strategies. The following sections explore a range of powerful techniques used to create detailed genetic profiles in the context of AD as shown in Figure 6.2.

6.3.1 NEXT-GENERATION SEQUENCING (NGS)

NGS revolutionizes genetic analysis, enabling rapid and accurate sequencing of DNA, RNA, and other genetic material. In AD research, NGS can sequence an individual's entire genome or zoom in on specific genes of interest. This comprehensive analysis provides a deeper understanding of the genetic landscape associated with AD. NGS has been instrumental in: Identifying the genetic causes of early-onset AD (EOAD), examining bacterial populations in the brains of AD patients, potentially linking gut health to AD, and discovering rare genetic variations in genes, like *APP*, *TREM2*, and *PLD3* that may increase susceptibility to AD [14, 64, 65, 85].

6.3.2 GENOME-WIDE ASSOCIATION STUDIES (GWAS)

GWAS takes a broad approach, scanning the entire genome for common genetic variations associated with an increased risk of AD. By identifying these variations, GWAS helps us piece together the complex genetic architecture of AD. Over 70 loci (gene locations) have been linked to AD risk through GWAS, highlighting the significant role of genetics in the disease. GWAS has identified the *APOE* gene, the strongest known risk factor for AD [15, 66].

6.3.3 BIOMARKER ANALYSIS

Biomarker analysis goes beyond just genes. It explores other molecules, like proteins or metabolites, whose presence or absence can indicate a disease state. Biomarkers can be crucial for early detection of AD and monitoring disease progression. Imaging, cerebrospinal fluid (CSF), and blood-based markers are used for AD diagnosis. These markers can detect amyloid plaques, neurodegeneration, and the presence of proteins like Aβ and tau, key components of AD pathology [16, 67].

6.3.4 TRANSCRIPTOMICS

Transcriptomics analyzes messenger RNA (mRNA) produced by genes, providing insights into gene expression and cellular activity. Altered gene expression patterns, such as changes in synaptic genes, have been linked to cognitive decline in AD. Transcriptomics helps identify genes associated with specific vulnerabilities in the brain, like the hippocampus, that are affected in AD [17, 68].

6.3.5 PROTEOMICS

Proteomics analyzes proteins, offering a deeper understanding of cellular processes related to AD. Proteomic studies have identified protein changes in brain tissue and biofluids, potentially revealing novel biomarkers and therapeutic targets. Studies have identified differentially expressed proteins related to immunity, metabolism, and cytoskeletal structure in AD. Proteomics can also reveal various protein modifications (PTMs) that influence disease progression [18, 69].

By employing this diverse arsenal of genetic profiling techniques, researchers are gaining valuable insights into the intricate genetic web of AD. This knowledge will pave the way for the development of more targeted therapies, personalized medicine approaches, and potentially even preventive strategies for this devastating disease.

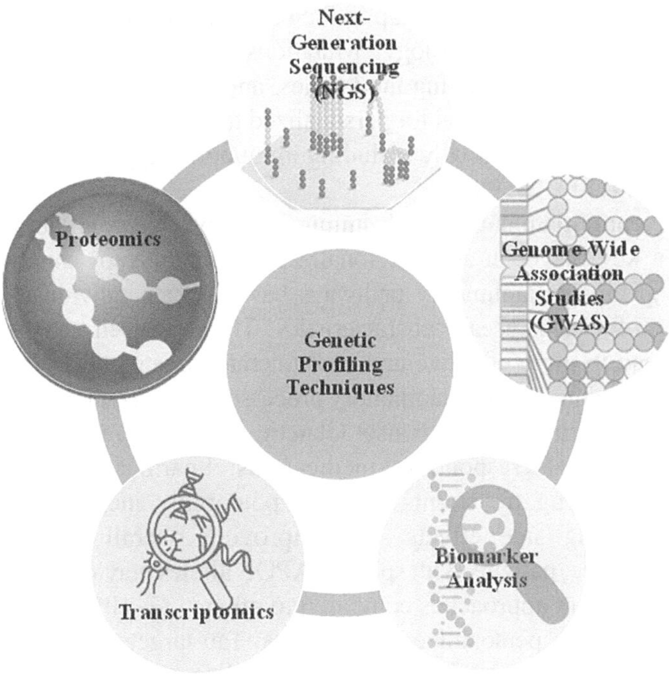

FIGURE 6.2 Powerful techniques used in genetic profiling of Alzheimer's disease.

6.4 PERSONALIZED TREATMENT STRATEGIES BASED ON GENETIC PROFILING

The fight against AD is getting a powerful weapon—personalized medicine. This approach leverages an individual's unique genetic makeup to develop targeted treatment strategies. Here's how it's transforming AD treatment:

1. Early intervention and prevention: By analyzing genes, environment, and lifestyle, precision medicine can identify individuals at high risk for late-onset AD. This allows for early intervention and potentially even prevention strategies. Genes like *APOE* and *MTHFR* play a role in AD susceptibility. By understanding an individual's genetic profile, clinicians can develop preventive strategies tailored to their specific risk factors [19, 20, 100, 101].
2. Targeting amyloid pathology: Mutations in genes like *PSEN1*, *PSEN2*, and *APP* increase the risk of amyloid plaque formation, a hallmark of AD. This genetic information can be used to develop personalized therapies, such as gene therapy or immunotherapy, to target amyloid buildup in the brain [21, 88, 89].
3. Targeting tau pathology: Mutations in the *MAPT* gene elevate the risk of developing tau tangles, another hallmark of AD. Early identification allows for personalized treatment plans that target tau pathology, potentially including immunotherapy or other emerging drugs [22, 23, 90].
4. Neuroinflammation and immunotherapy: Neuroinflammation is a key player in AD. Personalized immunotherapy aims to target specific inflammatory pathways based on an individual's genetic profile and disease characteristics. This approach can potentially lead to more effective and well-tolerated treatments by addressing the unique neuroinflammatory processes in each patient [24, 86, 87].
5. Optimizing drug response: Genetic variations can influence how individuals respond to medications. Pharmacogenomics helps tailor drug treatments for AD, potentially increasing efficacy, reducing side effects, and improving overall outcomes. For instance, patients with specific APOE alleles may require different treatment approaches compared to others [25, 105, 106].
6. Emerging personalized approaches: Tau-targeted immunotherapy is a new approach that uses antibodies to target and clear tau protein buildup in the brain. Blood-based biomarkers are being developed to identify individuals at risk of developing AD in its early stages, allowing for earlier intervention. Combining various biomarkers, including genetic, cognitive, and imaging data, can provide a comprehensive picture of the disease and guide the most effective treatment strategies for each patient. Combining various drugs or therapies targeting different aspects of AD pathology holds promise for increased efficacy and reduced side effects [102–104].

Personalized medicine offers a new era of hope in the fight against AD. By harnessing the power of individual genetics, researchers and clinicians are developing more effective and targeted treatment strategies, paving the way for a future where AD can be prevented, managed, or even potentially reversed [26, 70, 71].

6.5 PHARMACOGENOMICS IN ALZHEIMER'S TREATMENT

Pharmacogenomics is a field that studies how genes affect an individual's response to medications. This is particularly important in AD, as there is no cure and treatments can have variable effectiveness and side effects. By analyzing a person's genes, pharmacogenomics can help predict how they will respond to a particular Alzheimer's drug. This can help doctors tailor treatment plans to the individual patient, potentially improving outcomes and reducing side effects. Here are some of the ways pharmacogenomics is being used in Alzheimer's treatment:

1. Identifying patients who are more likely to respond to a particular drug [91–94]. For example, the *APOE4* gene is a risk factor for AD. People who carry the *APOE4* allele may be less likely to respond to acetylcholinesterase inhibitors, which are a common type of Alzheimer's drug.
2. Identifying patients who are at risk for side effects from a particular drug [95–97]. For example, some people have a genetic variation that makes them metabolize drugs more slowly. This could lead to them taking too much of a drug and experiencing side effects. Pharmacogenomics is still a relatively new field, but it has the potential to revolutionize the way AD is treated. By tailoring treatments to the individual patient, pharmacogenomics could help to improve the effectiveness of treatment and reduce side effects [27, 28, 72].

6.6 PRECISION DIAGNOSIS AND PROGNOSIS

Genetic variations can indicate specific subtypes of AD. This is crucial because different subtypes may have distinct underlying causes and progression patterns. Identifying the subtype allows for a more accurate diagnosis and paves the way for targeted therapies. Certain genes can predict an increased risk of developing AD. Early detection allows for preventative measures and intervention strategies before symptoms become severe [29, 73, 98, 99].

Genetic analysis can reveal how a patient might respond to specific medications. This helps doctors choose drugs with the highest chance of effectiveness and minimize the risk of side effects. For example, the *APOE4* gene can influence the response to acetylcholinesterase inhibitors, a common AD treatment. Genetic profiling can categorize patients based on their risk of progression. This allows doctors to personalize treatment plans, focusing on slowing disease progression for high-risk individuals and managing symptoms in others. Genetic analysis can help select suitable participants for clinical trials of new AD therapies. This ensures trials target the most relevant patient populations, leading to more meaningful results [30, 74–76].

6.7 ETHICAL CONSIDERATIONS AND CHALLENGES

Personalized medicine for AD using genetic profiling offers immense potential, but it also raises significant ethical considerations and challenges.

1. Genetic information is highly sensitive. Robust data security measures are crucial to prevent breaches and misuse of genetic data by insurers, employers, or even third parties.
2. Patients undergoing genetic testing for AD need clear and comprehensive information about the test's implications, potential limitations, and how their data will be used.
3. Learning about a genetic predisposition to AD can cause anxiety and distress. Proper counseling and support systems are vital to address these concerns.
4. Genetic information could be misused for discrimination in areas like employment or insurance. Clear regulations and ethical frameworks are needed to prevent such misuse.
5. The cost of genetic testing and personalized treatments might create disparities in access. Strategies are needed to ensure all patients have fair access to these advancements [31–34].

Our understanding of the complex interplay between genes and the environment in AD is still evolving. More research is needed to determine the true predictive power of specific genes. Currently, there is no cure for AD. Even with personalized medicine, treatment options may be limited to managing symptoms and slowing progression. While genetics play a role, AD is a complex disease influenced by other factors like lifestyle and environment. Focusing solely on genetics could lead to neglecting other aspects of disease management [77–79].

6.8 CASE STUDIES AND CLINICAL TRIALS

Table 6.1 shows the case studies employing personalized medicine in AD treatment that have demonstrated promising outcomes, showcasing the potential of tailoring therapies to individual genetic profiles.

6.9 FUTURE DIRECTIONS AND CONCLUSION

Looking ahead, the future of AD research and treatment holds tremendous promise, with innovative directions poised to transform our approach to this complex condition. Refinement of genetic risk scores offers the potential to enhance our ability to predict individual susceptibility to AD, guiding personalized preventive interventions. Integrating lifestyle and environmental factors alongside genetics will provide a more holistic understanding of AD risk, informing comprehensive strategies for disease management. Exploring the therapeutic potential of gene editing technologies like CRISPR presents an exciting avenue for correcting disease-causing mutations at the genetic level. Additionally, the utilization of digital medicine, including wearables and other digital tools, holds promise for real-time monitoring of cognitive function and disease progression. Furthermore, leveraging artificial intelligence to analyze vast datasets may uncover novel drug targets and treatment strategies, driving forward the development of effective therapies.

In conclusion, personalized medicine offers a transformative approach to tackling AD. By leveraging individual genetic profiles and integrating cutting-edge technologies, researchers and clinicians are poised to develop more targeted and effective therapies. This holds the potential for early intervention, improved treatment efficacy, and the possibility of slowing disease progression, ultimately enhancing the quality of life for patients

TABLE 6.1 Case Studies Employing Personalized Medicine in Alzheimer's Disease Treatment [35].

NCT Number	Study Title	Study Status	Conditions	Interventions
NCT01585272	Tolerability of rivastigmine before and after switching from oral formulation to transdermal patch in Alzheimer's dementia.	Completed	Alzheimer's dementia	Drug: ENA713
NCT01409564	Cilostazol augmentation study in dementia.	Completed	Alzheimer's dementia	Drug: Cilostazol Drug: Placebo
NCT00322153	A study of the safety and efficacy of memantine in moderate to severe Alzheimer's disease.	Completed	Dementia of the Alzheimer's type	Drug: Memantine ER Drug: Placebo
NCT03100617	Alzheimer's family caregiver intervention in Vietnam.	Completed	Dementia Alzheimer's	Behavioral: Psychosocial caregiver intervention
NCT04623242	Dominantly inherited Alzheimer network trial: An opportunity to prevent dementia. A study of potential disease modifying treatments in individuals at risk for or with a type of early onset Alzheimer's disease caused by a genetic mutation.	Completed	Alzheimer's disease/ dementia/Alzheimer's disease, familial	Drug: Gantenerumab Drug: Solanezumab Drug: Matching placebo (Gantenerumab) Drug: Matching placebo (Solanezumab)
NCT04023994	A single ascending dose study to investigate the safety, tolerability, immunogenicity, and pharmacokinetics of intravenously administered RO7126209 in healthy participants.	Completed	Alzheimer's disease	Drug: RO7126209 Drug: Placebo
NCT01946243	The feasibility of florbetapir quantitation.	Completed	Alzheimer's disease	Drug: Florbetapir F18
NCT02553928	Comparison of once daily and twice daily dosing on safety and tolerability of memantine in patients with Alzheimer.	Completed	Alzheimer dementia (AD)	Drug: Memantine (once daily) Drug: Memantine (twice daily)
NCT03811847	A pilot of methylphenidate in mild cognitive impairment and dementia participants.	Completed	Alzheimer dementia	Drug: Methylphenidate extended-release oral capsule Other: Placebo
NCT02185053	A phase II study of CPC-201 to treat Alzheimer's disease type dementia.	Completed	Alzheimer's disease	Drug: CPC-201

and their families. While challenges undoubtedly remain, the ongoing pursuit of these innovative directions in AD research brings us closer to a future where AD is no longer a looming threat, but a conquered challenge.

ACKNOWLEDGMENT

Authors want to acknowledge UGC Non-NET Fellowship (Reg No. Y20454003), Department of Pharmaceutical Sciences, Dr. Harisingh Gour University, Sagar, Madhya Pradesh, India.

KEYWORDS

- **Alzheimer's disease**
- **apolipoprotein E**
- **clinical trials**
- **dementia**
- **genetic profiling**
- **personalized medicine**
- **pharmacogenomics**

REFERENCES

1. Zhong, M. Z., Peng, T., Duarte, M. L., Wang, M., & Cai, D. (2024). Updates on mouse models of Alzheimer's disease. *Molecular Neurodegeneration*, *19*(1), 23.
2. Suwinski, P., Ong, C., Ling, M. H., Poh, Y. M., Khan, A. M., & Ong, H. S. (2019). Advancing personalized medicine through the application of whole exome sequencing and big data analytics. *Frontiers in Genetics*, *10*, 422886.
3. Skaria, A. P. (2022). The economic and societal burden of Alzheimer disease: Managed care considerations. *The American Journal of Managed Care*, *28*(10 Suppl), S188–S196.
4. Kumar Vats, D., Biswas, N., Pandian, P., & Kamalakannan, A. (2023). Review current and emerging treatment options for Alzheimer's disease. *Neurology Letters*, *2*(1), 1–5.

5. Sagud, M., Tudor, L., & Pivac, N. (2021). Personalized treatment interventions: Nonpharmacological and natural treatment strategies in Alzheimer's disease. *Expert Review of Neurotherapeutics*, *21*(5), 571–589.
6. Cary, G. A., Wiley, J. C., Gockley, J., Keegan, S., Amirtha Ganesh, S. S., Heath, L., Butler, R. R., Mangravite, L. M., Logsdon, B. A., Longo, F. M., & Levey, A. (2024). Genetic and multi-omic risk assessment of Alzheimer's disease implicates core associated biological domains. *Alzheimer's & Dementia: Translational Research & Clinical Interventions*, *10*(2), e12461.
7. Arora, S., Sharma, D., Layek, B., & Singh, J. (2021). A review of brain-targeted nonviral gene-based therapies for the treatment of Alzheimer's disease. *Molecular Pharmaceutics*, *18*(12), 4237–4255.
8. Roses, A. D. (1996). Apolipoprotein E alleles as risk factors in Alzheimer's disease. *Annual Review of Medicine*, *47*(1), 387–400.
9. Serrano-Pozo, A., Das, S., & Hyman, B. T. (2021). APOE and Alzheimer's disease: Advances in genetics, pathophysiology, and therapeutic approaches. *The Lancet Neurology*, *20*(1), 68–80.
10. D'Argenio, V., & Sarnataro, D. (2020). New insights into the molecular bases of familial Alzheimer's disease. *Journal of Personalized Medicine*, *10*(2), 26.
11. Giau, V. V., Bagyinszky, E., Yang, Y. S., Youn, Y. C., An, S. S., & Kim, S. Y. (2019). Genetic analyses of early-onset Alzheimer's disease using next generation sequencing. *Scientific Reports*, *9*(1), 8368.
12. Raulin, A. C., Doss, S. V., Trottier, Z. A., Ikezu, T. C., Bu, G., & Liu, C. C. (2022). ApoE in Alzheimer's disease: Pathophysiology and therapeutic strategies. *Molecular Neurodegeneration*, *17*(1), 72.
13. Abanish, S., Konstantin, G., Anatoliy, Y., & Akushevich, I. (2022). Genetic architecture of Alzheimer's disease risks. *Innovation in Aging*, *6*(Suppl 1), 2.
14. Emery, D. C., Shoemark, D. K., Batstone, T. E., Waterfall, C. M., Coghill, J. A., Cerajewska, T. L., Davies, M., West, N. X., & Allen, S. J. (2017). 16S rRNA next generation sequencing analysis shows bacteria in Alzheimer's post-mortem brain. *Frontiers in Aging Neuroscience*, *9*, 195.
15. Bertram, L., & Tanzi, R. E. (2009). Genome-wide association studies in Alzheimer's disease. *Human Molecular Genetics*, *18*(R2), R137–R145.
16. Márquez, F., & Yassa, M. A. (2019). Neuroimaging biomarkers for Alzheimer's disease. *Molecular Neurodegeneration*, *14*(1), 21.
17. Williams, J. B., Cao, Q., & Yan, Z. (2021). Transcriptomic analysis of human brains with Alzheimer's disease reveals the altered expression of synaptic genes linked to cognitive deficits. *Brain Communications*, *3*(3), fcab123.
18. Bai, B., Vanderwall, D., Li, Y., Wang, X., Poudel, S., Wang, H., Dey, K. K., Chen, P. C., Yang, K., & Peng, J. (2021). Proteomic landscape of Alzheimer's Disease: novel insights into pathogenesis and biomarker discovery. *Molecular Neurodegeneration*, *16*(1), 55.
19. Sagud, M., Tudor, L., & Pivac, N. (2021). Personalized treatment interventions: nonpharmacological and natural treatment strategies in Alzheimer's disease. *Expert Review of Neurotherapeutics*, *21*(5), 571–589.

20. Di Meco, A., & Vassar, R. (2021). Early detection and personalized medicine: Future strategies against Alzheimer's disease. *Progress in Molecular Biology and Translational Science*, *177*, 157–173.
21. Loera-Valencia, R., Piras, A., Ismail, M. A., Manchanda, S., Eyjolfsdottir, H., Saido, T. C., Johansson, J., Eriksdotter, M., Winblad, B., & Nilsson, P. (2018). Targeting Alzheimer's disease with gene and cell therapies. *Journal of Internal Medicine*, *284*(1), 2–36.
22. Granholm, A. C., & Hamlett, E. D. (2024). The role of tau pathology in Alzheimer's disease and down syndrome. *Journal of Clinical Medicine*, *13*(5), 1338.
23. Congdon, E. E., & Sigurdsson, E. M. (2018). Tau-targeting therapies for Alzheimer disease. *Nature Reviews Neurology*, *14*(7), 399–415.
24. Li, S., Lu, C., Zhao, Z., Lu, D., & Zheng, G. (2023). Uncovering neuroinflammation-related modules and potential repurposing drugs for Alzheimer's disease through multi-omics data integrative analysis. *Frontiers in Aging Neuroscience*, *15*, 1161405.
25. Cacabelos, R. (2005). Pharmacogenomics, nutrigenomics and therapeutic optimization in Alzheimer's disease. *Aging Health*, *1*(2), 303–48.
26. Granholm, A. C., & Hamlett, E. D. (2024). The role of tau pathology in Alzheimer's disease and down syndrome. *Journal of Clinical Medicine*, *13*(5), 1338.
27. Cacabelos, R., Naidoo, V., Martínez-Iglesias, O., Corzo, L., Cacabelos, N., Pego, R., & Carril, J. C. (2022). Pharmacogenomics of Alzheimer's Disease: Novel strategies for drug utilization and development. In Pharmacogenomics in drug discovery and development (pp. 275–387). New York, United States: Springer.
28. Cacabelos, R., Torrellas, C., & Carrera, I. (2015). Opportunities in pharmacogenomics for the treatment of Alzheimer's disease. *Future Neurology*, *10*(3), 229–252.
29. Lee, B. N., Wang, J., Nho, K., Saykin, A. J., & Shen, L. (2023). Discovering precision AD biomarkers with varying prognosis effects in genetics driven subpopulations. *AMIA Summits on Translational Science Proceedings*, *2023*, 340–349.
30. Azar, J., Salama, M., Chidambaram, S. B., Al-Balushi, B., Essa, M. M., & Qoronfleh, M. W. (2021). Precision health in Alzheimer disease: Risk assessment-based strategies. *Precision Medical Sciences*, *10*(2), 54–70.
31. Cacabelos, R., Torrellas, C., & Carrera, I. (2015). Opportunities in pharmacogenomics for the treatment of Alzheimer's disease. *Future Neurology*, *10*(3), 229–252.
32. Kazakov, A. S., & Zyryanov, S. K. (2023). Ethical issues of pharmacotherapy and clinical trials in patients with dementia. *Medical Ethics*, (2), 24–26.
33. Gustavsson, E., Raaschou, P., Lärfars, G., Sandman, L., & Juth, N. (2022). The ethics of disease-modifying drugs targeting Alzheimer disease: Response to our commentators. *Journal of Medical Ethics*, *48*(3), 193.
34. Eriksdotter, M. (2021). Commentary to "Novel drug candidates targeting Alzheimer's disease: ethical challenges with identifying the relevant patient population". *Journal of Medical Ethics*, *47*(9), 617.
35. *Clinicaltrials.gov*. https://clinicaltrials.gov/search?cond=Alzheimer%27s+Disease. (accessed April 28, 2024).
36. Zhenyi, Z. (2023). Emerging treatments for Alzheimer's Disease. *Highlights in Science Engineering and Technology*, *54*, 280–286.

37. Kumar Vats, D., Biswas, N., Pandian, P., & Kamalakannan, A. (2023). Review current and emerging treatment options for Alzheimer's disease. *Neurology Letters*, *2*(1), 1–5.
38. Galluzzi, S., Pievani, M., Zanetti, O., Benussi, L., The Italian-DIAfN Working Group, Frisoni, G. B., & Di Maria, E. (2022). Disclosure of genetic risk factors for Alzheimer's disease to cognitively healthy individuals—From current practice towards a personalised medicine scenario. *Biomedicines*, *10*(12), 3177.
39. Cacabelos, R., Naidoo, V., Martínez-Iglesias, O., Corzo, L., Cacabelos, N., Pego, R., & Carril, J. C. (2022). Personalized management and treatment of Alzheimer's disease. *Life (Basel)*, *12*(3), 460.
40. Skaria, A. P. (2022). The economic and societal burden of Alzheimer disease: Managed care considerations. *The American Journal of Managed Care*, *28*(10 Suppl), S188–S196.
41. Li, X., Feng, X., Sun, X., Hou, N., Han, F., & Liu, Y. (2022). Global, regional, and national burden of Alzheimer's disease and other dementias, 1990–2019. *Frontiers in Aging Neuroscience*, *14*, 937486.
42. Hahn-Pedersen, J. H. (2022). The cost burden of Alzheimer's disease on care partners. *Alzheimers & Dementia*, *18*(S11), e062278.
43. Miculas, D. C., Negru, P. A., Bungau, S. G., Behl, T., Hassan, S. S. U., & Tit, D. M. (2022). Pharmacotherapy evolution in Alzheimer's disease: Current framework and relevant directions. *Cells*, *12*(1), 131.
44. Lee, D., Clark, E. D., Antonsdottir, I. M., & Porsteinsson, A. P. (2023). A 2023 update on the advancements in the treatment of agitation in Alzheimer's disease. *Expert Opinion on Pharmacotherapy*, *24*(6), 691–703.
45. Nagata, T., Shinagawa, S., Nakajima, S., Noda, Y., & Mimura, M. (2022). Pharmacotherapeutic combinations for the treatment of Alzheimer's disease. *Expert Opinion on Pharmacotherapy*, *23*(6), 727–737.
46. Pathak, L. (2023). Personalized treatment for multiple sclerosis: The role of precision medicine. *Neurology Letters*, *2*(1), 30–34.
47. Arafah, A., Khatoon, S., Rasool, I., Khan, A., Rather, M. A., Abujabal, K. A., Faqih, Y. A. H., Rashid, H., Rashid, S. M., Bilal Ahmad, S., Alexiou, A., & Rehman, M. U. (2023). The future of precision medicine in the cure of Alzheimer's disease. *Biomedicines*, *11*(2), 335.
48. Pinto, M. V., & Neves, J. F. (2022). Precision medicine: The use of tailored therapy in primary immunodeficiencies. *Frontiers in Immunology*, *13*, 1029560.
49. Ayyubova, G. (2023). APOE4 is a risk factor and potential therapeutic target for Alzheimer's disease. *CNS Neurological Disorders-Drug Targets*, *23*(3), 342–352.
50. Kafyra, M., & Dedoussis, G. V. (2023). Polygenic risk scores and personalized approaches to cardiometabolic disease prevention and treatment: A short review. *Journal of Atherosclerosis Prevention and Treatment*, *14*(1), 35–43.
51. Rangan, A. V., Schork, N. J., & Elman, J. A. (2023). Exploring the genetic heterogeneity of Alzheimer's disease: Evidence for disease-relevant and disease-specific clustering. medRxiv. https://doi.org/10.1101/2023.05.02.23289347.
52. Singh, A., Arbeev, K., Yashin, A., & Akushevich, I. (2022). Genetic architecture of Alzheimer's disease risks. *Innovation in Aging*, *6*(Suppl 1), 2.

53. Huma, T., Nawaz, R., Li, X., & Willden, A. (2022). Alzheimer's disease (AD): Risks, treatments, prevention, and future implementations. *Advances in Alzheimer's Disease*, *11*, 11–21.
54. Jiang, J., Zhou, X., Wong, H. Y., Ouyang, L., Ip, F. C. F., Chau, V. M. N., Lau, S. F., Wu, W., Wong, D. Y. K., Seo, H., Fu, W. Y., Lai, N. C. H., Chen, Y., Chen, Y., Tong, E. P. S., Alzheimer's Disease Neuroimaging Initiative, Mok, V. C. T., Kwok, T. C. Y., Mok, K. Y., Shoai, M., Lehallier, B., Losada, P. M., O'Brien, E., Porter, T., Laws, S. M., Hardy, J., Wyss-Coray, T., Masters, C. L., Fu, A. K. Y., & Ip, N. Y. (2022). An *IL1RL1* genetic variant lowers soluble ST2 levels and the risk effects of *APOE*-ε4 in Alzheimer's disease. *Nature Aging, 2*, 616–634.
55. Celis, K., Moreno, M. D. M. M., Rajabli, F., Whitehead, P., Hamilton-Nelson, K., Dykxhoorn, D. M., Nuytemans, K., Wang, L., Flanagan, M., Weintraub, S., Geula, C., Gearing, M., Dalgard, C. L., Jin, F., Bennett, D. A., Schuck, T., Pericak-Vance, M. A., Griswold, A. J., Young, J. I., & Vance, J. M. (2023). Ancestry-related differences in chromatin accessibility and gene expression of APOE ε4 are associated with Alzheimer's disease risk. *Alzheimers & Dementis*, *19*(9), 3902–3915.
56. Ayyubova, G. (2023). APOE4 is a risk factor and potential therapeutic target for Alzheimer's disease. *CNS Neurological Disorders-Drug Targets*, *23*(3), 342–352.
57. Liang, Z., Wu, Y., Li, C., & Liu, Z. (2023). Clinical and genetic characteristics in a central-southern Chinese cohort of early-onset Alzheimer's disease. *Frontiers in Neurology*, *14*, 1119326.
58. Liu, C., Cong, L., Zhu, M., Wang, Y., Tang, S., Han, X., Zhang, Q., Tian, N., Liu, K., Liang, X., Fa, W., Wang, N., Hou, T., & Du, Y. (2022). Screening for genetic mutations associated with early-onset Alzheimer's disease in Han Chinese. *Current Alzheimer Research*, *19*(10), 724–733.
59. Neupane, A., da Fonseca, E. L., Kurup, J. T., Pericak-Vance, M. A., Martin, E. R., Schellenberg, G. D., Reitz, C., Beecham, G. W., Cruchaga, C., & Fernandez, V. (2022). A large-scale genome-wide association study of early-onset Alzheimer disease. *Alzheimers & Dementis*, *18*(S4), e064268.
60. Pagnon de la Vega, M., Näslund, C., Brundin, R., Lannfelt, L., Löwenmark, M., Kilander, L., Ingelsson, M., & Giedraitis, V. (2022). Mutation analysis of disease causing genes in patients with early onset or familial forms of Alzheimer's disease and frontotemporal dementia. *BMC Genomics*, *23*(1), 99.
61. Dean, H. B., Greer, R. A., Kober, D. L., Roberson, E. D., Brett, T. J., & Song, Y. (2023). AD-associated variants directly impair surface interactions between the hydrophobic site of TREM2 and the hinge region of ApoE. *Alzheimer's & Dementia, 19*, e067180.
62. Winfree, R. L., Seto, M., Dumitrescu, L., Menon, V., De Jager, P., Wang, Y., Schneider, J., Bennett, D. A., Jefferson, A. L., & Hohman, TJ. (2023). TREM2 gene expression associations with Alzheimer's disease neuropathology are region-specific: Implications for cortical versus subcortical microglia. *Acta Neuropathologica*, *145*(6), 733–747.
63. Tran, K., Milinkeviciute, G., Tran, K. M., Phan, J., Da Cunha, C., Javonillo, D. I., Kawauchi, S., Tenner, A. J., LaFerla, F., MacGregor, G. R., & Green, K. N. (2022). BIN1 K358R variant in 5xFAD mice ameliorates AB pathology and microgliosis. *Alzheimer's & Dementia*, *18*(S4), e066218

64. Desai, K., Iqbal, S., Baralo, B., Pereira, K., Jain, A., Deshpande, S., Renzu, M., Mehta, V., Rather, M., & Thirumaran, R. (2023). Utility of next generation sequencing (NGS) in advanced malignancies in a community hospital in an underserved area: A game changer? *Journal of Clinical Oncology, 41*(16_suppl), e15037.
65. Menyhárt, O., Győrffy, B., & Szabó, A. (2022). Gyermekkori genetikai rendellenességek diagnosztikája újgenerációs szekvenálással [Diagnosis of genetic disorders in childhood with next-generation sequencing]. *Orvosi Hetilap, 163*(51), 2027–2040.
66. Wang, Y. H., Luo, P. P., Geng, A. Y., Li, X., Liu, T. H., He, Y. J., Huang, L., & Tang, Y. Q. (2023). Identification of highly reliable risk genes for Alzheimer's disease through joint-tissue integrative analysis. *Frontiers in Aging Neuroscience, 15*, 1183119.
67. Xiao, W., Priller, J., Schneider, A., Fliessbach, K., Wiltfang, J., Jessen, F., Duzel, E., Buerger, K., Perneczky, R., Teipel, S., Laske, C., Spottke, A., Heneka, M. T., & Peters, O. (2023). Neurogranin and BACE1 as biomarkers for early detection of Alzheimer's disease. *Alzheimers & Dementia, 19*(S2), e065558.
68. Wu, H., Wang, J., Hu, X., Zhuang, C., Zhou, J., Wu, P., Li, S., & Zhao, R. C. (2023). Comprehensive transcript-level analysis reveals transcriptional reprogramming during the progression of Alzheimer's disease. *Frontiers in Aging Neuroscience, 15*, 1191680.
69. Sung, Y. J., Yang, C., Norton, J., Johnson, M., Fagan, A., Bateman, R. J., Perrin, R. J., Morris, J. C., Farlow, M. R., Chhatwal, J. P., Schofield, P. R., Chui, H., Wang, F., Novotny, B., Eteleeb, A., Karch, C., Schindler, S. E., Rhinn, H., Johnson, E. C. B., Oh, H. S., Rutledge, J. E., Dammer, E. B., Seyfried, N. T., Wyss-Coray, T., Harari, O., & Cruchaga, C. (2023). Proteomics of brain, CSF, and plasma identifies molecular signatures for distinguishing sporadic and genetic Alzheimer's disease. *Science Translational Medicine, 15*(703), eabq5923.
70. Panza, F., Dibello, V., Sardone, R., Castellana, F., Zupo, R., Lampignano, L., Bortone, I., Stallone, R., Cirillo, N., Damiani, C., Altamura, M., Bellomo, A., Daniele, A., Solfrizzi, V., & Lozupone, M. (2023). Clinical development of passive tau-based immunotherapeutics for treating primary and secondary tauopathies. *Expert Opinion on Investigational Drugs, 32*(7), 625–634.
71. Boyd, J. (2022). Characterization of Tau vaccines identifies diverse antibody binding profiles and efficacy, demonstrating that specific forms of Tau can be targeted by vaccines with subsequent functional impact. *Alzheimers & Dementia, 18*(S10), e063138.
72. Gupta, M., Kaur, H., Grover, S., & Kukreti, R. (2008). Pharmacogenomics and treatment for dementia induced by Alzheimer's disease. *Pharmacogenomics, 9*(7), 895–903.
73. Elman, J. A., Schork, N. J., Rangan, A. V.; Alzheimer's Disease Neuroimaging Initiative. (2023). Exploring the genetic heterogeneity of Alzheimer's disease: Evidence for disease-relevant and disease-specific clustering. *medRxiv*. https://doi.org/10.1101/2023.05.02.23289347.
74. Su, W. M., Gu, X. J., Dou, M., Duan, Q. Q., Jiang, Z., Yin, K. F., Cai, W. C., Cao, B., Wang, Y., & Chen, Y. P. (2023). Systematic druggable genome-wide Mendelian randomisation identifies therapeutic targets for Alzheimer's disease. *Journal of Neurology, Neurosurgery, and Psychiatry, 94*(11), 954–961.

75. Li, X., Fernandes, B. S., Liu, A., Chen, J., Chen, X., Zhao, Z., & Dai, Y. (2023). Genetically-regulated pathway-polygenic risk score (GRPa-PRS): A risk stratification method to identify genetically regulated pathways in polygenic diseases. *medRxiv*. https://doi.org/10.1101/2023.06.19.23291621.
76. Galluzzi, S., Pievani, M., Zanetti, O., Benussi, L., The Italian-DIAfN Working Group, Frisoni, G. B., & Di Maria, E. (2022). Disclosure of genetic risk factors for Alzheimer's disease to cognitively healthy individuals—From current practice towards a personalised medicine scenario. *Biomedicines*, *10*(12), 3177.
77. Hao, Z., Liu, K., Zhou, L., & Chen, P. (2023). Precious but convenient means of prevention and treatment: Physiological molecular mechanisms of interaction between exercise and motor factors and Alzheimer's disease. *Frontiers in Physiology*, *14*, 1193031.
78. Galluzzi, S., Pievani, M., Zanetti, O., Benussi, L., The Italian-DIAfN Working Group, Frisoni, G. B., & Di Maria, E. (2022). Disclosure of genetic risk factors for Alzheimer's disease to cognitively healthy individuals—From current practice towards a personalised medicine scenario. *Biomedicines*, *10*(12), 3177.
79. Dunn, P. J., Lea, R. A., Maksemous, N., Smith, R. A., Sutherland, H. G., Haupt, L. M., & Griffiths, L. R. (2022). Investigating a genetic link between Alzheimer's disease and CADASIL-related cerebral small vessel disease. *Molecular Neurobiology*, *59*(12), 7293–7302.
80. Bagga, S., & Kumar, M. (2022). Current status of Alzheimer's disease and pathological mechanisms investigating the therapeutic molecular targets. *Current Molecular Medicine*, *23*(6), 492–508.
81. Oh, S. L., Zhou, M., Chin, E. W. M., Amarnath, G., Cheah, C. H., Ng, K. P., Kandiah, N., Goh, E. L.K, & Chiam, K. H. (2022). Alzheimer's disease blood biomarkers associated with neuroinflammation as therapeutic targets for early personalized intervention. *Frontiers in Digital Health*, *4*, 875895.
82. Elman, J. A., Schork, N. J., Rangan, A. V.; Alzheimer's Disease Neuroimaging Initiative. (2023). Exploring the genetic heterogeneity of Alzheimer's disease: Evidence for disease-relevant and disease-specific clustering. *medRxiv*. https://doi.org/10.1101/2023.05.02.23289347.
83. Xiong, W., Cai, J., Sun, B., Lin, H., Wei, C., Huang, C., Zhu, X., & Tan, H.; Alzheimer' Disease Neuroimaging Initiative. (2023). The association between genetic variations and morphology-based brain networks changes in Alzheimer's disease. *Journal of Neurochemistry*, *168*(8), 1490–1502.
84. Shigemizu, D., Asanomi, Y., Akiyama, S., Mitsumori, R., Niida, S., & Ozaki, K. (2022). Whole-genome sequencing reveals novel ethnicity-specific rare variants associated with Alzheimer's disease. *Molecular Psychiatry*, *27*(5), 2554–2562.
85. Brase, L., You, S. F., del Aguila, J., Dai, Y, Novotny, B. C., Soriano-Tarraga, C., Dykstra, T, Fernandez, M. V., & et al. (2021). A landscape of the genetic and cellular heterogeneity in Alzheimer disease. *medRxiv*. https://doi.org/10.1101/2021.11.30.21267072.
86. Jain, S., Singh, R., Paliwal, S., Swapnil, P., & Sharma, S. (2023). Targeting neuroinflammation as disease modifying approach to Alzheimer's disease: Potential and challenges. *Mini-Reviews in Medicinal Chemistry*, *23*(22), 2097–2116.

87. Xue, Z. (2023). Neuroinflammation as a core pathogenic mechanism of Alzheimer's disease: Recent research advances and related therapeutical approaches. *International Conference on Biological Engineering and Medical Science (ICBioMed2022)*. United Kingdom: Oxford.
88. Bae, H., Shim, K. H., Yoo, J., Yang, Y. S., An, S. S. A., & Kang, M. J. (2023). Double mutations in a patient with early-onset Alzheimer's disease in Korea: An APP Val551Met and a PSEN2 His169Asn. *International Journal of Molecular Sciences*, *24*(8), 7446.
89. Dunn, P. J., Lea, R. A., Maksemous, N., Smith, R. A., Sutherland, H. G., Haupt, L. M., & Griffiths, L. R. (2022). Investigating a genetic link between Alzheimer's disease and CADASIL-related cerebral small vessel disease. *Molecular Neurobiology*, *59*(12), 7293–7302.
90. Easton, A., Jensen, M. L., Wang, C., Hagedorn, P. H., Li, Y., Weed, M., Meredith, J. E., Guss, V., Jones, K., Gill, M., Krause, C., Brown, J. M., Hunihan, L., Natale, J., Fernandes, A., Lu, Y., Polino, J., Bookbinder, M., Cadelina, G., Benitex, Y., Sane, R., Morrison, J., Drexler, D., Mercer, S. E., Bon, C., Pandya, N. J., Jagasia, R., Ou Yang, T. H., Distler, T., Grüninger, F., Meldgaard, M., Terrigno, M., Macor, J. E., Albright, C. F., Loy, J., Hoeg, A. M., Olson, R. E., & Cacace, A. M. (2022). Identification and characterization of a MAPT-targeting locked nucleic acid antisense oligonucleotide therapeutic for tauopathies. *Molecular Therapy Nucleic Acids*, *29*, 625–642.
91. Islam, F., Gorbovskaya, I., & Müller, D. J. (2021). Pharmacogenetic/pharmacogenomic tests for treatment prediction in depression. *Advances in Experimental Medicine and Biology*, *1305*, 231–255.
92. Shukla, R. (2020). Pharmacogenomics: Overview, applications, and recent developments. In A. A. Parikesit, (Ed.), *Drug design-novel advances in the omics field and applications.* IntechOpen.
93. Hockings, J. K., Pasternak, A. L., Erwin, A. L., Mason, N. T., Eng, C., & Hicks, J. K. (2020). Pharmacogenomics: An evolving clinical tool for precision medicine. *Cleveland Clinic Journal of Medicine*, *87*(2), 91–99.
94. Kisor, D. F., Hoefer, C., & Decker, B. S. (2019). Pharmacogenomics and precision medicine. In *Clinical pharmacy education, practice and research*, (pp. 437–451). Elsevier.
95. Islam, F., Gorbovskaya, I., & Müller, D. J. (2021). Pharmacogenetic/pharmacogenomic tests for treatment prediction in depression. *Advances in Experimental Medicine and Biology*, *1305*, 231–255.
96. Xiang, X., & Yuan, Y. (2020). Application of pharmacogenomics in drug discovery and development. In W. Cai, Z. Liu, L. Miao, & X. Xiang, (Eds.), *Pharmacogenomics in precision medicine*. Singapore: Springer.
97. Cacabelos, R., Torrellas, C., Teijido, O., & Carril, J. C. (2016). Pharmacogenetic considerations in the treatment of Alzheimer's disease. *Pharmacogenomics*, *17*(9), 1041–1074.
98. Gao, X. R., Chiariglione, M., & Arch, A. J. (2022). Whole-exome sequencing study identifies genes associated with Alzheimer's disease and related dementias. *medRxiv*. https://doi.org/10.1101/2022.12.22.22283864.
99. An, Q., Zhang, M., Hu, X., Su, C., Wang, F., & Zhu, Y. (2022). Integrative analyses of neuroimaging, clinical, and multi-omics data identified meaningful Alzheimer's

disease progression subtypes. *medRxiv*. https://doi.org/10.1101/2022.11.01.22281820.

100. Caldwell, A. B., Anantharaman, B. G., Ramachandran, S., Nguyen, P., Liu, Q., Trinh, I., Galasko, D. R., Desplats, P. A., & Wagner, S. L., & Subramaniam, S. (2022). Transcriptomic profiling of Early-onset sporadic Alzheimer's disease patients. *Molecular Brain, 15*, 83.
101. Hampel, H., O'Bryant, S. E., Durrleman, S., Younesi, E., Rojkova, K., Escott-Price, V., Corvol, J. C., Broich, K., Dubois, B., & Lista, S. (2017). A precision medicine initiative for Alzheimer's disease: The road ahead to biomarker-guided integrative disease modeling. *Climacteric: The Journal of the International Menopause Society*, *20*(2), 107–118.
102. Boyd, J. (2022). Characterization of Tau vaccines identifies diverse antibody binding profiles and efficacy, demonstrating that specific forms of Tau can be targeted by vaccines with subsequent functional impact. *Alzheimers & Dementia, 18*(S10), e063138.
103. Mashal, Y., Abdelhady, H., & Iyer, A. K. (2022). Comparison of tau and amyloid-β targeted immunotherapy nanoparticles for Alzheimer's disease. *Biomolecules*, *12*(7), 1001.
104. Tashima, T. (2022). Delivery of intravenously administered antibodies targeting Alzheimer's disease-relevant tau species into the brain based on receptor-mediated transcytosis. *Pharmaceutics*, *14*(2), 411.
105. Sheikh, S., Rana, S., Wani, H. A., Ali, A., Danish, Q., Majid, S., Naikoo, N. A., Sehar, N., Binlebdah, A. F., Arafah, A., & Rehman, M. U. (2022). Genetic polymorphisms, role in drug metabolism and effect on clinical pharmacy. In *Genetic polymorphism and disease* (p. 15). CRC Press.
106. Francis Lam, Y. W. (2022). Applying Pharmacogenomics in Drug Therapy of Neurologic and Psychiatric Disorders. *Comprehensive Pharmacology, 2*, 277–301.

CHAPTER 7

Integration of Biomarker Data into Clinical Trials and Personalized Treatment Plans

AHANA HAZRA[1], RIDEB CHAKRABORTY[1], NAUREEN AFROSE[1], PRATIBHA BHOWMICK[1], CHOWDHURY MOBASWAR HOSSAIN[2], and MITHUN BHOWMICK[1*]

[1]*Bengal College of Pharmaceutical Sciences and Research, Basu Sarani, Bidhannagar, Durgapur, West Bengal, India*

[2]*Department of Pharmaceutical Technology, Maulana Abul Kalam Azad University of Technology, Haringata, West Bengal, India*

*Corresponding author

ABSTRACT

A growing number of clinical trials are being prompted by biomarkers in this era of chronic disease management, where the ultimate goal is to optimize and personalize illness therapy. This is due to the increased accessibility and affordability of genetic profiling at the individual patient level. Molecular biomarkers especially are getting more and more significant in clinical practice and research as the targeted therapy era progresses. Recent advances in identifying and confirming biomarkers for the elements of Alzheimer's disease (AD) are extremely precise detection of AD-associated pathophysiological and neuropathological alterations in cerebrospinal fluid and positron emission tomography markers of tau and amyloid-β proteins. Nevertheless, the high expense, limited access, and intrusive nature of these tests prevent them from being widely used as effective first-line methods for identifying pathophysiological patterns.

Cellular and Molecular Mechanisms in Alzheimer's Disease Therapeutics, Volume 1: Cellular Mechanisms. Prashant Tiwari, Sunil Kumar Kadiri, and Kunjbihari Sulakhiya (Eds.)

DOI: 10.1201/9781779641793-7

Hence, the use of multi-agent platform trials has currently represented another significant conceptual change in biomarker-driven medicine. These experiments confront particular operational and behavioral difficulties even if they might have significant economic and infrastructure benefits. This chapter will address biomarkers for inflammatory disorders, synapse disarray, and the degeneration of neurons. It will also emphasize the need for additional research through prevalent pathologies and offer suggestions for how various biomarkers might be used probably in conjugation to speed up the development and clinical application of novel therapy for AD.

7.1 AN OVERVIEW OF ALZHEIMER'S DISEASE

Alzheimer's disorder (AD) is the most common cause of cognition in the elderly, a progressive neurological illness that progressively impairs cognitive function and ultimately results in death. We go over the pathophysiology, management, epidemiology, and clinical characteristics of AD [1].

Alzheimer's syndrome affects 60–70% of elderly individuals with progressive memory impairment. It is estimated that 2.3 million people in the United States have AD overall (range: 1.09–4.8 million). Every 5 years beyond the age of 60, the prevalence of AD doubles, going from 1% in individuals aged between 60–64 to up to 40% in those 85 years of age and older [2]. Reports of new instances are expected to total 360,000 per year, or 980 new cases per day, or 40 new cases every hour. In the next 50 years, the number of AD patients will nearly triple if the current trend continues [3].

Examining the development of AD usage in the post-1968 contemporary generation is the reason behind this evaluation of the antemortem biomarkers' current state of technology. A primary estimate intended for hypothesizing a sickness that was previously poorly understood was a clinical-pathological mannequin of a scientific condition that was once inextricably associated with distinct detrimental dyad [4]. The diagnostic time "likely AD" was first chosen in 1984 by the committee of the National Institute of Neurosurgery and Communicating Diseases and Stroke—Alzheimer's Conditions and Related Conditions Organization [5]. At that time, it was almost generally accepted in the scientific community.

There was formerly a very similar definition in the DSM-III-R; it was called "degenerative dementia of the Alzheimer type." A diagnosis of credible AD of the Alzheimer's type requires a significant impairment in daily functioning and at least two cognitive domains. In the past, cognitive illness was associated with a decrease in previous levels of functioning and a gradually worsening state. Although the uncertainty implied by the term "probable" was often disregarded, the clinicopathologic model expressed with the help of potential AD was originally tailored in an abridged structure for population-based studies and standard scientific practice [6]. The term "AD" originated from the clinicopathologic mannequin, which represented all dementia that was not caused by another clinically evident cause. This term characterized the increasing community consciousness of cognitive impairment in the Middle Ages. On the other hand, neuropathologists determined in the 1990s that AD was defined by an adequate load of postmortem examination, plaques containing neuritis, and tangles of neurons were observed, independent regarding the medical state of the patient before death [7]. The diagnostic term "Alzheimer's disease" has several special meanings, which poses a nosologic challenge for the field of AD research. AD was most likely added in 1984 to denote a newly discovered, clinically diagnosed form of amnestic dementia for which no other explanation could be supported. The term "probable AD" refers to a clinicopathologic entity that postulated a precise and delicate correlation between the neuropathology of brain tangles carrying tau, neuritic deposits carrying beta-amyloid, and amnestic neurodegeneration [8].

7.2 NEUROTRANSMITTERS OF ALZHEIMER'S DISEASE

7.2.1 ACETYLCHOLINE

A common characteristic of AD is the loss of basal forebrain cholinergic cells. Amyloid beta (Aβ) pathology is exacerbated by impaired cortical cholinergic neurotransmission, which also raises tau protein phosphorylation. Discerning beginning of the inflammatory acetylcholine receptor M1-M3-but not M2-M4-mAChR promotes secretion of sAPPα and reduces the production of total Aβ. The mitochondrial ace influence *APP* processing through the phosphate-containing ethanolamine signal pathway and possibly the tyrosine kinase MAP kinase pathway [9]. On the other hand, *BACE1* expression was downregulated as a result of protein kinase A and M2-mAChR pathways being activated. Nicotine additionally mentioned demonstrated to affect the handling of *APP* by supporting the non-amyloidogenic route through its impact on the nicotinic acetylcholine receptor, or nicotinic nAChR [10]. Nicotine also disrupts preformed Aβ fibrils and inhibits the creation of new Aβ fibrils. Aβ damages M1 receptors and lowers the amount of acetylcholine within cells. Aβ may interfere with the function of nAChR by directly and very definitively binding to the α7 subtype of the receptor. Tau phosphorylation is significantly increased upon activation of nAChR, whereas tau phosphorylation may be prevented upon activation of mAChR [11].

7.2.2 GLUTAMATE

Glutamatergic excitotoxicity and a self-reinforcing neuronal death cascade may arise from abnormalities in glutamate-glutamine cycle components. N-methyl D aspartate (NMDA) located extrasynaptically may become activated as a result of persistent brain damage. Discs large homolog 4 and GNB2L1 are two scaffolding proteins that this interacts with Ref. [12]. In AD patients, persistent NMDAR activation is influenced by pathways involving the dissociation of GNB2L1 from the Fyn-DLG4 relationship. Ca^{+2} continues to enter the cytoplasm as a result of Fyn's ongoing activation of extra-synaptic NMDA (NR2B subunit). Mitochondrial dysfunction is caused by high intracellular Ca^{+2}. An indication for CREB shutdown is sent by persistent over-activation of extra-synaptic NMDAR, which causes phospho-CREB levels to drop [13]. As a result, pro-survival signals, such as brain-derived neurotrophic factors are produced at a lower rate. Over time, all of these occurrences result in cellular malfunction and neuronal death.

Through nitric oxide activation, Aβ increases the release of glutamate. Aβ also contributes to glutamatergic excitotoxicity by preventing glial absorption of glutamate. The AMPA's GluR2 subunit participates in the neuronal death cascade, and overproduction of this subunit offers protection. The sensitivity of neurons to Ca^{2+}-mediated neurotoxicity, which can result in neurodegeneration, can be influenced by the loss of GluR2. Additionally, there is proof that AD patients have reduced glutamate absorption as a result of glial transporter deficit [13]. In AD, there is a relationship between oxidative radicals, glutamate transporters, and Aβ.

7.2.3 SEROTONIN

Evidence of serotoninergic involvement in AD can be shown in the following observations: (i) depletion of 5HT sensors and cells that synthesize 5HT in AD; (ii) cerebrospinal fluid (CSF) modification of 5HT and 5HIAA for AD [14]. Five HT polymorphisms are associated with AD; and (iii) serotonergic drugs have been shown to alleviate behavioral symptoms of AD, particularly agitation [15].

7.2.4 NOREPINEPHRINE

In AD, locus coeruleus (LC) neurons are lost. Patients with more severe intellectual decline and those who are younger tend to have lower levels of norepinephrine in various brain regions. In AD, α-2 receptor activity is lost [16]. The consistency of a minimum number of peripheral acetylcholine activities is dependent upon the noradrenergic system remaining intact, according to the findings. Norepinephrine is therefore implicated in AD by several lines of evidence [17].

7.2.5 DOPAMINE

There is proof that micromolar quantities of L-dopa or dopamine are adequate to considerably prevent the creation of new Aβ fibrils or break up those that are already present [17]. D1 agonists are being tested to improve cognition in AD because the D1 receptor appears to be more crucial in regulating development and some aspects of mental functioning, like recollection and spatial comprehension [18].

7.3 MOLECULAR ALZHEIMER'S ILLNESS

AD, one of the more prevalent forms of memory loss, is an illness of the brain that affects the elderly. It is distinguished by the development of two major protein gathers: senile deposits and neurofibrillary knots, both of which play a role in the process of progressive degeneration of neurons and death. Neurodegeneration in dementia is a cellular pathology rather than an increased aging process. Senile lesions are caused by the accumulation of β-amyloid peptide strands in the brain, which are formed from the proteolytic division of the precursor protein for amyloid. The protein tau is the primary aspect of paired circular filaments, which develop a

The underlying cause of these two distinct aphasic presentations remains unknown, even though they both share the trait of being language-predominant disorders [29].

7.4.4 RIGHT VARIANT

In AD, visual impairment is not usually the first or main symptom to manifest. When it appears, it might herald the arrival of PCA. In other instances, visuospatial dysfunction points to a more subtle, non-memory variety that is linked to more atrophy and pathology in the right hemisphere relative to the left, a clear pattern that persists over time and as the disease progresses [30]. The relationship between visuoperceptive AD and PCA is similarly poorly understood as the one between linguistic variation of AD and LPA. Furthermore, it is unclear if those who have visuoperceptive dysfunction represent a true subtype of AD rather than a prodromal phase of widespread Lewy body disease, despite occasional anecdotal reports to the contrary [31]. Even though Lewy bodies can co-occur in posterior cortical atrophy (PCA), perpetuating clinically obvious hallucinations and parkinsonism associated with Alzheimer's disease (AD) pathology, AD-related alterations can also be observed in isolation, with the underlying molecular landscape appearing to be the sole factor influencing phenotypic characteristics and disease progression. Research on pathology and biomarkers in imaging could provide answers to some of these queries.

7.4.5 FRONTAL VARIANT

The incredibly rare EOAD subtype known as frontal variant AD was initially identified through Johnson et al. [32], who discovered four belongings out of 60 people by diagnostically proven AD. This subtype is linked to notable frontal cognitive and behavioral symptoms. In their 100 case series, Alladi and colleagues also found just two examples, of which only one had a real dysexecutive syndrome and the other only had behavioral symptoms [32]. Pathologically, NFTs are more prevalent in the frontal regions (10 times more than in normal AD), and their loading is comparable in other regions and the entorhinal lobe. Frontal variant AD is distinguished from FTD by amyloid plaques as well as the absence of in layers II and III of tumors, microvacuolarization, and cellular death. Based on biochemical study, frontal version AD shows reduced calcium-independent phospholipase A2 production within the frontal areas [33]. These patients had frontal variant AD, an extremely rare subtype of EOAD associated with significant brain intellectual and behavioral manifestations. Only two cases were found by Alladi and colleagues out of their 100 case series; only one of these cases had a real dysexecutive syndrome, while the other just showed behavioral characteristics. The entorhinal cortex and other regions show equal loading in pathological conditions, whereas the frontal regions show a ten-fold increase in NFTs compared to normal AD. Amyloid plaques and the lack of cell loss, microvacuolarization, and tumors in layers II and III are what set frontal variant AD apart from FTD. When compared to typical AD, frontal variant AD shows primarily smaller calcium-independent phospholipase A2 action within the frontal area in the post-mortem protein assay. This isoform of phospholipase is specific to neurons and serves as an indirect measure of neuronal health [34]. Aβ levels in CSF were shown to be lower in one case report, despite borderline tau levels. Because frontal variant AD is so uncommon, diagnosing it clinically has proven to be difficult. In one study, the cognitive behavioral inventory classified an AD group

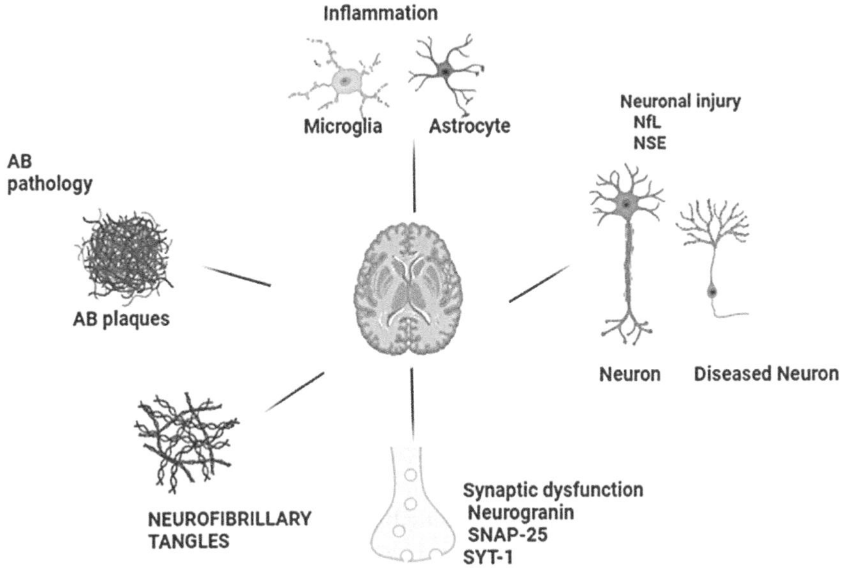

FIGURE 7.1 Alzheimer's fluid biomarkers as of right now.

as "high frontality," if frontal symptoms, such as disinhibition, utilization behavior, foreign limb sensations, and indifference were present [34]. A comparison of Woodward's "excessive frontal perspective" group with FTD using global cognition screening tests did not find any differences, with the exception of three items on the anterior cognitive inventory: hyperorality, persistence, and aspontaneity. Additionally, Woodward et al. [34] discovered in a different investigation that frontal variant AD is linked to co-occurring FTD pathology. Given this common disease and diagnostic uncertainty, CSF biomarker profiles and amyloid imaging will likely be necessary to definitively show prefrontal variance AD in a group sufficient in size to do the necessary neuropsychological scans and genetic studies to characterize this type of AD [35].

7.5 ALZHEIMER'S CONDITION: LOOKING FOR INDICATORS

The older neurological disorder known as AD impairs recollection and other aspects of thinking. It is widely accepted as the most frequent senile cause. The major increase in life expectancy is likely to lead to a steep rise in the prevalence of AD patients over time, which will have a huge influence on healthcare worldwide since aging is the main risk indicator for the development of the illness [36]. The pathogenesis of the spontaneous form of AD is complicated and involves both ambient and inherited variables. The major effect that AD's rising incidence would have on world health is mostly due to our insufficient and incomplete understanding of its pathophysiology. Even though tau neurons knots up and beta-amyloid lesions are commonly acknowledged as the characteristics that characterize the neural pathology of AD, it is still unknown how exactly they contribute to the progression of neuronal degeneration because they could just be the outcome of continuous damage. This insight has given rise to the theory that the pathophysiology of AD may also be significantly influenced by other disease processes [37]. Disruption of brain equilibrium, inflammation of the brain, degeneration of neurons and synapses, irregular metabolism, vascular malfunction, and oxidative damage are a few examples. The presence of tau nerve tangles and Aβ deposits are commonly acknowledged as the characteristics that characterize the neuropathology of AD; it is still unknown how exactly they contribute to the progression of neuronal degeneration because they could just be the outcome of continuous damage. Alteration of brain homeostasis, neuroinflammation, loss of neurons and synapses, irregular metabolism, vascular malfunction, and oxidative stress are a few examples [38]. The presence of tau nerve tangles despite all of the advances in modern medicine, the final identification of AD still depends on the discovery of specific diseased lesions in the brain, including plaques of amyloid and tangles of neurofibrillary fibers, that mimic the ones that Alzheimer himself first documented in 1907. Furthermore, based mostly on somatic and psychological assessments performed *in vivo*, mainstream healthcare can identify AD as either possible or likely [39]. The presence of tau nerve tangles and the most recent recommendation is to select a comprehensive strategy that incorporates both the clinical assessment and the *in vivo* neuropathological evaluation. In the past few decades, attempts have been made to figure out the disease's *in vivo* neuropathological assessment irrespective of clinical symptoms. As a matter of fact, the therapies administered to patients with more serious conditions do not completely stop the biological events leading to AD; they can only manage its symptoms. However, the so-called "beneficial methods," which disrupt the pathogenesis of the disease and influence its cause and course, are more effective when brain injury is still minimal. With all of this, it is evident that tools for sensitive and precise diagnosis and prognosis are required for the effective treatment of this neurodegenerative disease (NDD). These tools will be used for early diagnosis, monitoring the progression of the illness, and assessing the effectiveness of treatment [38].

7.6 DIFFERENT TYPES OF BIOMARKER PROFILES FOR TYPICAL AND ATYPICAL PHENOTYPES OF AD

In contrast to healthcare symptoms, which are just a patient's expression and assessment of their state of wellness, biomarkers, also known as "biological markers," are different. The National Institutes of Research Biomarkers terminology committee defined biomarkers as "proof of any natural, infectious, or pharmacogenomic reaction to any therapy adaptation" in 1998 [41]. Any substances, configurations, or bodily processes that are measurable both internally and externally and that have the capacity to influence alterations in the body or the probability of neurological illness prevalence are considered biological indicators. Finally, a progressive loss of cognitive function is the hallmark of the AD clinical condition, which is a manifestation of neurotoxicity [42].

7.6.1 BLOOD-BASED BIOMARKERS: BRAIN–TO–PERIPHERY COMMUNICATION

The CSF, closely related to the central nervous system, is a plausible source of AD biomarkers. In actuality, the CSF is a reflection of the state of the central nervous system, hence pathological changes within the mind that alter the quantities of different substances might impact the CSF's composition [43]. Over the years, biomarker detection studies in the CSF have greatly benefited both research and clinical practices. Although these investigations have significantly improved our comprehension of the natural history of AD, they have also added a new dimension to the diagnosis of the disease *in vivo* by including the measurement of AD-specific biomarkers in the CSF [44]. However, the high intrusion of CSF extraction by lumbar puncture—as well as the risks and expenses connected with the procedure—has severely restricted its adoption in the standard therapeutic daily life, given patients' reluctance to endure invasive and costly testing repeatedly. Because of its drawbacks, the primary focus of biomarker research has switched from CSF to biological sources that are more readily available. Right now, blood collection is the primary emphasis, which is a routine procedure that is certainly simpler than lumbar puncture [45]. It is also less costly, easy to implement in big populations, and reproducible at regular intervals. Blood-based biomarkers appear to be a simple instrument to use in clinical settings as a result to monitor the course of a disease or the efficacy of a treatment over time.

Studies indicate that patients with AD have a functionally compromised blood–brain barrier and suffer an increased leakage of CSF molecules; blood is a good source of diagnostics for this neurologic illness. These molecules then serve as CNS informative blood biomarkers by reflecting pathological changes in the peripheral AD brain [46]. But because of the complexity of AD disease and the nature of the sample, blood biomarker analysis presents a number of detection issues, and it can frequently be challenging to correlate changes in the central nervous system with blood biomarker alterations [44]. Firstly, given the pathological heterogeneity of AD, the biomarkers' concentrations in the bloodstream are probably much lower than those of the CSF due to significant dilutions; additionally, the BBB integrity may be affected differently, which could have an impact on the degree of analyte crossing. Furthermore, peripheral fluids may undergo proteolysis of brain proteins that penetrate the blood–brain barrier [47]. Secondly, keep in mind that AD can be linked to an overall inflammatory reaction that might alter blood levels of proteins that indicate the course of the illness; as a result, the fraction of biomarkers linked to the brain might be obscured by the fraction produced in the periphery [48]. Last but not least, a complex matrix consisting of several antibodies and other compounds found in blood may interfere with biomarker readings and lower the accuracy and specificity of monitoring procedures.

7.6.2 AB PATHOLOGY BIOMARKERS

The external accumulation of Aβ, which results from *BACE1* and γ-secretase cleaving the precursor protein for amyloid to form deposits, is a significant pathogenic characteristic of AD. It is believed to be the disease's primary pathogenic event. Aβ 42, a breakdown product of *APP*, is generally transported by the lymphatic system through the brain interstitial fluid to the blood and CSF [49]. Amyloid PET has undergone significant standardization in terms of identifying abnormality cut-points and quantifying pathology, and it has been validated in comparison with neuropathology. Additionally, it possesses sufficient usage requirements [50].

7.6.3 TAU PATHOLOGY BIOMARKERS

One of the fundamental pathogenic features of AD is the accumulation of elevated tau in the neuronal soma, which results in the creation of neurofibrillary tangles. However, tau aggregates in cells or glial cells in general have also been connected to other neurodegenerative dementias [51]. In the AD research criteria, the CSF Aβ 42/Aβ 40 ratio and the cornerstone indicators totaled tau (T-tau) and phosphorylation of tau (P-tau) are considered diagnostic, since they have been suggested as biomarkers for physiologically characterizing AD. In all neurodegenerative dementias, the levels of T- and P-tau in the CSF are suggestive of the pathophysiology of AD [51]. The most likely explanation for the enhanced tau levels in the CSF is that neurons are responding to Aβ exposure by phosphorylating and releasing tau more often.

7.6.3.1 ENTIRE TAU

In 1993, the first ELISA method for determining tau levels in CSF was published. This ELISA was based on a sandwich configuration of the polyclonal antitau antiserum and the recombinant antitau antibody against the mid-domain. The first sandwich ELISA technique based only on monoclonal antibodies was introduced 2 years later with the publication of the "Innogenetics" or "Innotest" assay [52]. Since the assay is based on mid region monoclonal antibodies that identify all six tau variants regardless of their kinase state, it was named a "all" tau assay. It was discovered that CSF t-tau significantly increased in AD dementia patients; this discovery has since been validated in hundreds of articles, also employing a range of other assay techniques [53].

According to certain theories, CSF T-tau acts as an "environmental marker," signifying the severity of either neurodegeneration or acute neuronal damage. After acute brain trauma, CSF T-tau levels are actually dynamic; they increase days after the injury and stay high for weeks until they revert to normal [54]. When compared to AD, CSF T-tau levels are higher in chronic neurological conditions, especially in Creutzfeldt-Jakob disorder, where they are 10- to 20-fold higher. A quicker progression of symptoms in the AD spectrum is indicated by higher levels of CSF T-tau and P-tau, demonstrating the usefulness of CSF T-tau as a diagnostic tool for the severity of neuronal disorders [55].

7.6.3.2 PHYSIOLOGY OF TAU AND OTHER PATHOGENIC PROTEINS

The brain is susceptible to tau disease due to the misfolded protein's capacity to move across cells and release itself into the extracellular space by a mechanism akin to that of prion-like particles [56]. These ideas have been applied therapeutically to address how tauopathies and other NDDs progress [57]. A limited number of studies have shown crucial *in vivo* data linking tau disease and dysfunction to other pathogenic proteins, including soluble Aβ oligomers and synthetic Aβ 42 fibrils, which together cause synergistic toxicity [58]. The information gathered from these investigations indicated a robust molecular connection between wild type tau, amyloid β proteins, and synaptic disease [59].

Based on the connection between tau inclusions and additional tau pathology-related proteins that worsen the disease, tauopathies are classified as primary or secondary [60]. The primary tauopathies are a subtype of frontotemporal lobar degeneration (FTLD), which is characterized by significant tau accumulation in both neuronal and glial cells.

These conditions cause degeneration of the cortex of the frontal and temporal lobes [61]. On the other hand, AD, the primary neurological illness, is a subordinate tauopathy that is defined by extracellular amyloid plaques together with tau pathology [62].

7.6.3.3 TARGETING TAU AGGREGATION

Tau's ability to bind microtubules is decreased as a result of many post-translational changes, which improve tau's ability to separate from microtubules. This phenomenon progressively raises the concentration of tau in the cytosol, increasing the likelihood of tau-tau contact and eventually, tau polymerization [63]. A wide range of tau multimer types, including PHFs, twisted ribbons, straight filaments, and oligomeric aggregations are detected at higher levels in a number of tauopathies, particularly AD [64]. Furthermore, by impairing vital cellular processes, these tau clumps would gradually overstrain neurons, resulting in the healthy neurons to die. Growing research on tau has demonstrated that tau oligomers are among the most hazardous species, capable of causing neurotoxicity and neurodegeneration in AD patients [65]. Therefore, one important tactic for addressing tau is the prevention of tau aggregations. Currently, the majority of tau aggregation inhibitors (TAIs) are compounds of methylene blue, including Rember TM, LMTX, and methylthioninium chloride. MTC, a medication for methemoglobinemia that has FDA approval, has been repurposed to treat AD and other tauopathies that are similar [66]. Despite numerous study reports demonstrating the effectiveness of methylene derivatives like MTC in the reversal of proteolytic persistence of PHF through the inhibition of tau-tau binding without affecting tau-tubulin interaction in *in vitro* studies, there is an uneven efficacy when contrasting both *in vitro/vivo* and clinical trials on human subjects. Thus, these drugs have not had much of an impact or effectiveness on humans [67]. As a result, there is a critical need for drugs in this tactical application of therapies against tau.

7.6.3.4 FOCUSING ON TAU INTERFACIAL TRANSFER

A plethora of signs indicate that tau pathology in AD can be slowed down by blocking tau's intercellular transmission. The impact of the disease can reasonably decrease if the pathological spread of aggregative tau is stopped, as tau pathology is linked to numerous damaging roles against neurons [68]. Three distinct approaches are used for blocking the inter-neuronal transmission of tau. This entails preventing tau from being released, lowering tau oligomerization and extracellular tau levels, and preventing tau from being absorbed [69]. Targeting the release of tau via any one of these three distinct channels could be like aiming for a single change that has two distinct appearances. The availability and levels of extracellular tau decrease when tau release is inhibited. This decreases the obtainability of outside of cells tau and prevents tau from being taken up by nearby neuronal cells [70].

7.6.3.5 TARGETING MICROTUBULE STABILIZATION

Unwanted dysfunctional actions of microtubule assemblies and axonal transport result from tau's improper function. One tactic that does not emphasize tau is to target microtubule stabilization. The idea behind this approach is to give AD patients a beneficial neuroprotective impact by using microtubule stabilizers [71]. The notable molecule, epothilone D, which maintains MT, is a small one. Spatial memory deficits may be restored by chemical microtubule-stabilizer derived from Taxol by decreasing tau pathology and hippocampal neuronal loss in transgenic rats. It can also easily pass through the blood–brain barrier. Additionally, epothilone D significantly enhanced the number of microtubules and lowered the amount of axons with abnormal shape in genetic mouse strains of tau in both young and old mice [72]. Recent developments in the subject include the use of davunetide and abeotaxane as additional microtubule stabilizers. Nevertheless, a significant disadvantage of both current and emerging developments aimed at microtubule stabilization is their uneven performance when compared to studies conducted on humans and animal models [73]. Therefore, a significant amount of data from successful clinical trials remains to be gathered in order to build this strategy.

7.6.3.6 INVASIVE TREATMENT AND THE TRANSMISSION OF TAU

Accordingly, mounting data points to the possibility that preventing tau from being transferred between cells could stop or significantly decrease the development of tau pathology in AD. Furthermore, stopping the progression of tau pathology may have an effect on disease-related cognitive decline because it is linked to synapse loss and subsequent cell death. Therapeutic techniques to stop or lessen the interneuronal transfer of tau, include inhibiting tau release, lowering the amount of extracellular tau that accumulates or oligomerizes, or stopping tau absorption [74]. By focusing on tau release, extracellular tau levels can be lowered, which would limit the amount of tau that nearby neurons may absorb. As tau is released *in vivo* by neurons, presumably in an activity-dependent manner, more investigation is required to pinpoint the exact mechanism [75]. This will be important in order to discover potential targets for therapeutic intervention. It's interesting to note that extracellular tau in the above-described absence of neurodegeneration raises the possibility that tau has an as-yet-unidentified purpose outside of cells. In fact, studies have demonstrated that exogenously administered recombinant tau has a higher affinity than acetylcholine for binding and activating muscarinic acetylcholine receptors in cultured hippocampal neurons [76–78].

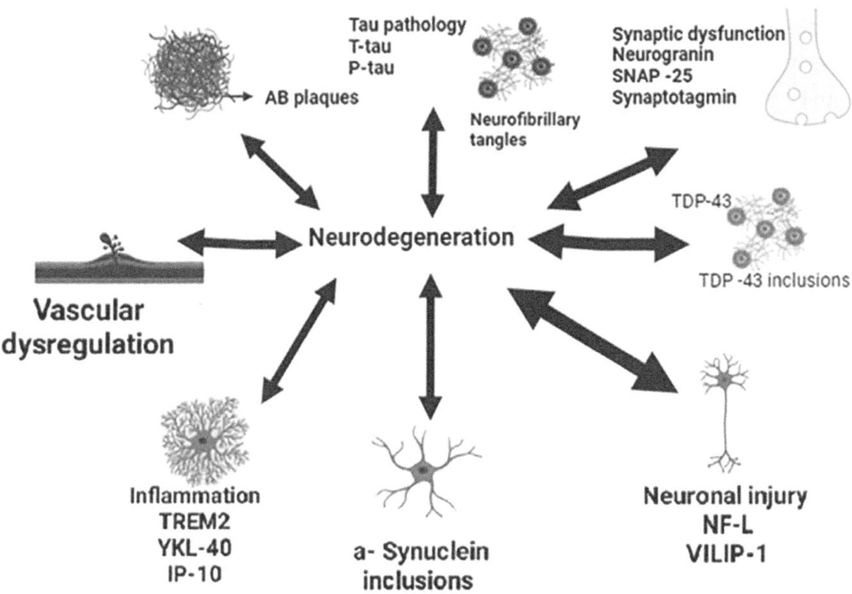

FIGURE 7.2 The primary pathogenic pathways underlying Alzheimer's disease and the fluid biomarkers that are linked to them.

7.7 CONCLUSION

A single proteinopathy may be linked to many symptoms and multiple proteinopathies may coexist to produce a single phenotype, making the clinical syndromes of NDD extremely diverse and overlapping. NDD can have pure or single pathogenic characteristics, but co-pathologies, or mixed diseases, are more common than previously believed. These co-pathologies include various forms of concomitant cerebrovascular disease, particularly in the elderly. Generally speaking, chronic non-destructive disorder (NDD) phenotypes are not exclusive to a single illness, and this should be considered, particularly in clinical practice. Potentially, the now recognized classical phenotypes of NDD may be combined to form more expansive and theoretically adaptable clinical groups known as "macro-phenotypes" (e.g., combining AD and DLB into a single clinical entity). Partially due to its etiology, infection of the nervous system is essential for the onset of AD, alongside Aβ plaques. This is especially important to consider in light of anti-amyloid therapy's partial rejection efforts. Two pathways by which the immune system in the peripheral environment contributes to neuroinflammation are phosphocytic lymphocytes and inflammatory agents that can get across a compromised barrier between the blood and the brain and enter the brain parenchyma, beyond just reflecting the pathogenic alterations occurring within the AD central nervous system. Peripatetic neuroinflammation biomarkers ought to be regarded as authentic key indicators of AD that are available for use in clinical follow-up and identification of patients. Additional research using cutting-edge, delicate methods is needed. It is therefore advisable to compare the central and peripheral compartments, in a manner similar to the recent findings on phosphorylated tau, for example. Lastly, there is still a significant need for effective disease-modifying treatments for AD, and studying the dysregulated peripheral inflammatory responses is an intriguing avenue of inquiry. In actuality, slowing down the production of cytokines and chemokines, monocyte chemotaxis, neuro-invasion, and Aβ phagocytosis may ultimately prevent AD-related neurodegeneration. In theory, we might even consider using medications that don't penetrate the blood–brain barrier to directly target neuroinflammation at the periphery. Last but not least, addressing peripheral inflammatory processes may potentially be of importance for treating behavioral signs of AD particularly, given the clinical and pathologic associations with the neurobehavioral syndrome of delirium. Clinical research is currently needed to determine whether medications that can influence these processes will eventually be able to lessen the medical characteristic and postpone transformation in individuals with preliminary AD.

KEYWORDS

- **Alzheimer's disease**
- **biomarker**
- **clinical trials**
- **diagnosis**
- **neurobiology**

REFERENCES

1. Scheltens, P., Blennow, K., Breteler, M. M., De Strooper, B., Frisoni, G. B., Salloway, S., & Van der Flier, W. M. (2016). Alzheimer's disease. *The Lancet, 388*, 505–517.
2. Goedert, M., & Spillantini, M. G. (2006). A century of Alzheimer's disease. *Science 314*, 777–781. https://doi.org/10.1126/science.1132814.
3. Bush, A. I. (2003). The metallobiology of Alzheimer's disease. *Trends in Neurosciences, 26,* 207–214.
4. Jack, C. R., & Holtzman, D. M. (2013). Biomarker modeling of Alzheimer's disease. *Neuron, 80,* 1347–1358.
5. Dolci, G. A., Damanti, S., Scortichini, V., Galli, A., Rossi, P. D., Abbate, C., Arosio, B., Mari, D., Arighi, A., & Fumagalli, G. G. (2017). Alzheimer's disease diagnosis: Discrepancy between clinical, neuroimaging, and cerebrospinal fluid biomarkers criteria in an Italian cohort of geriatric outpatients: A retrospective cross-sectional study. *Frontiers in Medicine, 4*, 203.
6. Lopez, O. L., Becker, J. T., Klunk, W., Saxton, J., Hamilton, R. L., Kaufer, D. I., Sweet, R. A., Cidis Meltzer, C., Wisniewski, S., Kamboh, M. I., & DeKosky, S. T. (2000). Research evaluation and diagnosis of probable Alzheimer's disease over the last two decades: I, *Neurology, 55,* 1854–1862. https://doi.org/10.1212/WNL.55.12.1854.
7. Frisoni, G. B., Winblad, B., & O'Brien, J. T. (2011). Revised NIA-AA criteria for the diagnosis of Alzheimer's disease: A step forward but not yet ready for widespread clinical use. *International Psychogeriatrics. 23,* 1191–1196.
8. Jellinger, K. A. (2020). Neuropathology of the Alzheimer's continuum: An update. *Free Neuropathology, 1*. https://www.ncbi.nlm.nih.gov/pmc/articles/PMC10209886/ (accessed April 6, 2024).

9. Francis, P. T. (2005). The interplay of neurotransmitters in Alzheimer's disease. *CNS Spectrums, 10,* 6–9.
10. Reddy, P. H. (2017). A critical assessment of research on neurotransmitters in Alzheimer's disease. *Journal of Alzheimer's Disease, 57,* 969–974.
11. Kandimalla, R., & Reddy, P. H. (2017). Therapeutics of neurotransmitters in Alzheimer's disease. *Journal of Alzheimer's Disease, 57*, 1049–1069.
12. Francis, P. T. (2003). Glutamatergic systems in Alzheimer's disease. *International Journal of Geriatr Psychiatry, 18,* S15–S21. https://doi.org/10.1002/gps.934.
13. Cowburn, R. F., Hardy, J. A., & Roberts, P. J. (1990). Glutamatergic neurotransmission in Alzheimer's disease. *Biochemical Society Transactions, 18*, 390–392.
14. Rodríguez, J. J., Noristani, H. N., & Verkhratsky, A. (2012). The serotonergic system in ageing and Alzheimer's disease. *Progress in Neurobiology, 99,* 15–41.
15. Seyedabadi, M., Fakhfouri, G., Ramezani, V., Mehr, S. E., & Rahimian, R. (2014). The role of serotonin in memory: Interactions with neurotransmitters and downstream signaling. *Experimental Brain Research, 232,* 723–738. https://doi.org/10.1007/s00221-013-3818-4.
16. Chalermpalanupap, T., Kinkead, B., Hu, W. T., Kummer, M. P., Hammerschmidt, T., Heneka, M. T., Weinshenker, D., & Levey, A. I. (2013). Targeting norepinephrine in mild cognitive impairment and Alzheimer's disease. *Alzheimers Research Therapy, 5,* 21. https://doi.org/10.1186/alzrt175.
17. Pan, X., Kaminga, A. C., Wen, S. W., Wu, X., Acheampong, K., & Liu, A. (2019). Dopamine and dopamine receptors in Alzheimer's disease: A systematic review and network meta-analysis. *Frontiers in Aging Neuroscience, 11,* 175.
18. Snowden, S. G., Ebshiana, A. A., Hye, A., Pletnikova, O., O'Brien, R., Yang, A., Troncoso, J., Legido-Quigley, C., & Thambisetty, M. (2019). Neurotransmitter imbalance in the brain and Alzheimer's disease pathology. *Journal of Alzheimer's Disease, 72,* 35–43.
19. Goedert, M., & Spillantini, M. G. (1990). Molecular neuropathology of Alzheimer's disease: In situ hybridization studies. *Cellular and Molecular Neurobiology, 10,* 159–174. https://doi.org/10.1007/BF00733642.
20. Trejo-Lopez, J. A., Yachnis, A. T., & Prokop, S. (2023). Neuropathology of Alzheimer's disease. *Neurotherapeutics, 19,* 173–185.
21. Schellenberg, G. D., & Montine, T. J. (2012). The genetics and neuropathology of Alzheimer's disease. *Acta Neuropathology, 124*, 305–323. https://doi.org/10.1007/s00401-012-0996-2.
22. Jicha, G. A., & Carr, S. A. (2010). Conceptual evolution in Alzheimer's disease: Implications for understanding the clinical phenotype of progressive neurodegenerative disease. *Journal of Alzheimer's Disease, 19,* 253–272.
23. Ryan, N. S., Nicholas, J. M., Weston, P. S., Liang, Y., Lashley, T., Guerreiro, R., Adamson, G., Kenny, J., Beck, J., & Chavez-Gutierrez, L. (2016). Clinical phenotype and genetic associations in autosomal dominant familial Alzheimer's disease: A case series. *The Lancet Neurology, 15*, 1326–1335.
24. Van Der Flier, W. M., Schoonenboom, S. N. M., Pijnenburg, Y. A. L., Fox, N. C., & Scheltens, P. (2006). The effect of *APOE* genotype on clinical phenotype in Alzheimer disease. *Neurology, 67,* 526–527. https://doi.org/10.1212/01.wnl.0000228222.17111.2a.

25. Perry, R. J., & Hodges, J. R. (2000). Differentiating frontal and temporal variant frontotemporal dementia from Alzheimer's disease. *Neurology, 54,* 2277–2284. https://doi.org/10.1212/WNL.54.12.2277.
26. Bertoux, M., de Souza, L. C., Corlier, F., Lamari, F., Bottlaender, M., Dubois, B., & Sarazin, M. (2014). Two distinct amnesic profiles in behavioral variant frontotemporal dementia. *Biological Psychiatry, 75,* 582–588.
27. Gorno-Tempini, M. L., Brambati, S. M., Ginex, V., Ogar, J., Dronkers, N. F., Marcone, A., Perani, D., Garibotto, V., Cappa, S. F., & Miller, B. L. (2008). The logopenic/phonological variant of primary progressive aphasia. *Neurology, 71*, 1227–1234. https://doi.org/10.1212/01.wnl.0000320506.79811.da.
28. Henry, M. L., Wilson, S. M., Babiak, M. C., Mandelli, M. L., Beeson, P. M., Miller, Z. A., & Gorno-Tempini, M. L. (2016). Phonological processing in primary progressive aphasia. *Journal of Cognitive Neuroscience, 28,* 210–222.
29. Conca, F., Esposito, V., Giusto, G., Cappa, S. F., & Catricalà, E. (2022). Characterization of the logopenic variant of primary progressive aphasia: A systematic review and meta-analysis. *Ageing Research Reviews, 82,* 101760.
30. Furey-Kurkjian, M. L., Pietrini, P., Graff-Radford, N. R., Alexander, G. E., Freo, U., Szczepanik, J., & Schapiro, M. B. (1996). Visual variant of Alzheimer's disease: Distinctive neuropsychological features. *Neuropsychology, 10*, 294.
31. Mandal, P. K., Joshi, J., & Saharan, S. (2012). Visuospatial perception: An emerging biomarker for Alzheimer's disease. *Journal of Alzheimer's Disease, 31*, S117–S135.
32. Johnson, J. K., Head, E., Kim, R., Starr, A., & Cotman, C. W. (1999). Clinical and pathological evidence for a frontal variant of Alzheimer disease. *Archives of Neurology, 56,* 1233–1239.
33. Kossard, S., Lee, M.-S., & Wilkinson, B. (1997). Postmenopausal frontal fibrosing alopecia: a frontal variant of lichen planopilaris. *Journal of the American Academy of Dermatology, 36,* 59–66.
34. Woodward, M., Jacova, C., Black, S. E., Kertesz, A., Mackenzie, I. R., Feldman, H., & The ACCORD investigator group. (2010). Differentiating the frontal variant of Alzheimer's disease. *International Journal of Geriatric Psychiatry, 25,* 732–738. https://doi.org/10.1002/gps.2415.
35. Blennerhassett, R., Lillo, P., Halliday, G. M., Hodges, J. R., & Kril, J. J. (2014). Distribution of pathology in frontal variant Alzheimer's disease. *Journal of Alzheimer's Disease, 39*, 63–70.
36. Pratico, D. (2013). Alzheimer's disease and the quest for its biological measures. *Journal of Alzheimer's Disease, 33,* S237–S241.
37. Trojanowski, J. (2004). Searching for the biomarkers of Alzheimer's. *Practical Neurology, 3,* 30–34.
38. Blennow, K., & Zetterberg, H. (2018). Biomarkers for Alzheimer's disease: Current status and prospects for the future. *Journal of Internal Medicine, 284*, 643–663. https://doi.org/10.1111/joim.12816.
39. Blennow, K., & Zetterberg, H. (2018). Biomarkers for Alzheimer's disease: Current status and prospects for the future. *Journal of Internal Medicine, 284,* 643–663. https://doi.org/10.1111/joim.12816.
40. Humpel, C. (2011). Identifying and validating biomarkers for Alzheimer's disease. *Trends in Biotechnology, 29,* 26–32.

41. Baldacci, F., Mazzucchi, S., Della Vecchia, A., Giampietri, L., Giannini, N., Koronyo-Hamaoui, M., Ceravolo, R., Siciliano, G., Bonuccelli, U., Elahi, F. M., Vergallo, A., Lista, S., Giorgi, F. S., & Hampel, H. (2020). The path to biomarker-based diagnostic criteria for the spectrum of neurodegenerative diseases. *Expert Review of Molecular Diagnostics, 20,* 421–441. https://doi.org/10.1080/14737159.2020.1731306.
42. Wallon, D., Rousseau, S., Rovelet-Lecrux, A., Quillard-Muraine, M., Guyant-Maréchal, L., Martinaud, O., Pariente, J., Puel, M., Rollin-Sillaire, A., & Pasquier, F. (2012). The French series of autosomal dominant early onset Alzheimer's disease cases: Mutation spectrum and cerebrospinal fluid biomarkers. *Journal of Alzheimer's Disease, 30,* 847–856.
43. Wenk, G. L. (2003). Neuropathologic changes in Alzheimer's disease. *Journal of Clinical Psychiatry, 64,* 7–10.
44. Angiulli, F., Conti, E., Zoia, C. P., Da Re, F., Appollonio, I., Ferrarese, C., & Tremolizzo, L. (2021). Blood-based biomarkers of neuroinflammation in Alzheimer's disease: A central role for periphery?*Diagnostics, 11*, 1525.
45. Janigro, D., Bailey, D. M., Lehmann, S., Badaut, J., O'Flynn, R., Hirtz, C., & Marchi, N. (2021). Peripheral blood and salivary biomarkers of blood–brain barrier permeability and neuronal damage: Clinical and applied concepts. *Frontiers in Neurology, 11,* 577312.
46. Ashton, N. J., Hye, A., Rajkumar, A. P., Leuzy, A., Snowden, S., Suárez-Calvet, M., Karikari, T. K., Schöll, M., La Joie, R., & Rabinovici, G. D. (2020). An update on blood-based biomarkers for non-Alzheimer neurodegenerative disorders. *Nature Reviews Neurology, 16,* 265–284.
47. Di Battista, A. P., Rhind, S. G., & Baker, A. J. (2013). Application of blood-based biomarkers in human mild traumatic brain injury. *Frontiers in Neurology, 4*, 44. https://doi.org/10.3389/fneur.2013.00044.
48. Hardy-Sosa, A., León-Arcia, K., Llibre-Guerra, J. J., Berlanga-Acosta, J., Baez, S. de la C., Guillen-Nieto, G., & Valdes-Sosa, P. A. (2022). Diagnostic accuracy of blood-based biomarker panels: A systematic review. *Frontiers in Aging Neuroscience,* 683689.
49. Ashton, N. J., Schöll, M., Heurling, K., Gkanatsiou, E., Portelius, E., Höglund, K., Brinkmalm, G., Hye, A., Blennow, K., & Zetterberg, H. (2018). Update on biomarkers for amyloid pathology in Alzheimer's disease. *Biomarkers in Medicine, 12,* 799–812. https://doi.org/10.2217/bmm-2017-0433.
50. Park, S. A., Han, S. M., & Kim, C. E. (2020). New fluid biomarkers tracking non-amyloid-β and non-tau pathology in Alzheimer's disease. *Experimental & Molecular Medicine, 52,* 556–568.
51. Rosenmann, H. (2012). CSF biomarkers for amyloid and tau pathology in Alzheimer's disease. *Journal of Molecular Neuroscience, 47*, 1–14. https://doi.org/10.1007/s12031-011-9665-5.
52. Blennow, K., Vanmechelen, E., & Hampel, H. (2001). CSF total tau, Aβ42 and phosphorylated tau protein as biomarkers for Alzheimer's disease. *Molecular Neurobiology 24*, 087–098. https://doi.org/10.1385/MN:24:1-3:087.
53. Rubenstein, R., Chang, B., Yue, J. K., Chiu, A., Winkler, E. A., Puccio, A. M., Diaz-Arrastia, R., Yuh, E. L., Mukherjee, P., & Valadka, A. B. (2017). Comparing plasma

phospho tau, total tau, and phospho tau–total tau ratio as acute and chronic traumatic brain injury biomarkers. *JAMA Neurology, 74,* 1063–1072.
54. Avila, J., Lucas, J. J., Pérez, M., & Hernández, F. (2004). Role of tau protein in both physiological and pathological conditions. *Physiological Reviews, 84,* 361–384. https://doi.org/10.1152/physrev.00024.2003.
55. Spillantini, M. G., & Goedert, M. (1998). Tau protein pathology in neurodegenerative diseases. *Trends in Neurosciences, 21,* 428–433.
56. Iqbal, K., del C. Alonso, A., Chen, S., Chohan, M. O., El-Akkad, E., Gong, C.-X., Khatoon, S., Li, B., Liu, F., & Rahman, A. (2005). Tau pathology in Alzheimer disease and other tauopathies. *Biochimica et Biophysica Acta (BBA)-Molecular Basis of Disease, 1739,* 198–210.
57. Hampel, H., Blennow, K., Shaw, L. M., Hoessler, Y. C., Zetterberg, H., & Trojanowski, J. Q. (2010). Total and phosphorylated tau protein as biological markers of Alzheimer's disease. *Experimental Gerontology, 45,* 30–40.
58. Castellani, R. J., Nunomura, A., Lee, H., Perry, G., & Smith, M. A. (2008). Phosphorylated tau: toxic, protective, or none of the above. *Journal of Alzheimer's Disease, 14,* 377–383.
59. Sjögren, M., Davidsson, P., Tullberg, M., Minthon, L., Wallin, A., Wikkelso, C., Granérus, A.-K., Vanderstichele, H., Vanmechelen, E., & Blennow, K. (2001). Both total and phosphorylated tau are increased in Alzheimer's disease. *Journal of Neurology, Neurosurgery & Psychiatry, 70,* 624–630.
60. Kosik, K. S., & Shimura, H. (2005). Phosphorylated tau and the neurodegenerative foldopathies. *Biochimica et Biophysica Acta (BBA)-Molecular Basis of Disease, 1739,* 298–310.
61. Friston, K. J., Frith, C. D., Liddle, P. F., Dolan, R. J., Lammertsma, A. A., & Frackowiak, R. S. J. (1990). The relationship between global and local changes in PET scans. *Journal of Cerebral Blood Flow & Metabolism, 10,* 458–466. https://doi.org/10.1038/jcbfm.1990.88.
62. Rao, D. C. (1998). CAT scans, PET scans, and genomic scans. *Genetic Epidemiology, 15,* 1–18. https://doi.org/10.1002/(SICI)1098-2272(1998)15:1<1::AID-GEPI1>3.0.CO;2-B.
63. Seam, P., Juweid, M. E., & Cheson, B. D. (2007). The role of FDG-PET scans in patients with lymphoma, Blood. *The Journal of the American Society of Hematology, 110,* 3507–3516.
64. Friedberg, J. W., & Chengazi, V. (2003). PET scans in the staging of lymphoma: current status. *The Oncologist, 8,* 438–447.
65. Rosenmann, H., Blum, D., Kayed, R., & Ittner, L. M. (2012). Tau protein: Function and pathology. *International Journal of Alzheimer's Disease, 2012,* 707482. https://www.hindawi.com/journals/ijad/2012/707482/abs/ (accessed April 11, 2024).
66. Spillantini, M. G., & Goedert, M. (2013). Tau pathology and neurodegeneration. *The Lancet Neurology, 12,* 609–622.
67. Wong, P. T., & Choi, S. K. (2015). Mechanisms of drug release in nanotherapeutic delivery systems. *Chemical Review, 115,* 3388–3432. https://doi.org/10.1021/cr5004634.

68. Clavaguera, F., Grueninger, F., & Tolnay, M. (2014). Intercellular transfer of tau aggregates and spreading of tau pathology: Implications for therapeutic strategies. *Neuropharmacology, 76,* 9–15.
69. Zhang, K., Sun, Z., Chen, X., Zhang, Y., Guo, A., & Zhang, Y. (2021). Intercellular transport of Tau protein and β-amyloid mediated by tunneling nanotubes. *American Journal of Translational Research, 13,* 12509.
70. Pooler, A. M., Polydoro, M., Wegmann, S., Nicholls, S. B., Spires-Jones, T.L., & Hyman, B. T. (2013). Propagation of tau pathology in Alzheimer's disease: Identification of novel therapeutic targets. *Alzheimers Research & Therapy, 5,* 49. https://doi.org/10.1186/alzrt214.
71. Devi Tangutur, A., Kumar, D., Vamsi Krishna, K., & Kantevari, S. (2017). Microtubule targeting agents as cancer chemotherapeutics: An overview of molecular hybrids as stabilizing and destabilizing agents. *Current Topics in Medicinal Chemistry, 17,* 2523–2537.
72. Steinmetz, M. O., & Prota, A. E. (2018). Microtubule-targeting agents: strategies to hijack the cytoskeleton. *Trends in Cell Biology, 28,* 776–792.
73. Kaverina, I., Rottner, K., & Small, J. V. (1998). Targeting, capture, and stabilization of microtubules at early focal adhesions. *The Journal of Cell Biology, 142,* 181–190.
74. Nisbet, R. M., & Götz, J. (2018). Amyloid-β and Tau in Alzheimer's disease: Novel pathomechanisms and non-pharmacological treatment strategies. *Journal of Alzheimer's Disease, 64,* S517–S527.
75. Wu, X.-L., Piña-Crespo, J., Zhang, Y.-W., Chen, X.-C., & Xu, H.-X. (2017). Tau-mediated neurodegeneration and potential implications in diagnosis and treatment of Alzheimer's disease. *Chinese Medical Journal, 130,* 2978–2990. https://doi.org/10.4103/0366-6999.220313.
76. Shaimardanova, A. A., Solovyeva, V. V., Chulpanova, D. S., James, V., Kitaeva, K. V., & Rizvanov, A. A. (2020). Extracellular vesicles in the diagnosis and treatment of central nervous system diseases. *Neural Regeneration Research, 15,* 586–596.
77. Fan, L., Mao, C., Hu, X., Zhang, S., Yang, Z., Hu, Z., Sun, H., Fan, Y., Dong, Y., & Yang, J. (2020). New insights into the pathogenesis of Alzheimer's disease. *Frontiers in Neurology, 10,* 1312.
78. Demaegd, K., Schymkowitz, J., & Rousseau, F. (2018). Transcellular spreading of Tau in tauopathies. *ChemBioChem, 19,* 2424–2432. https://doi.org/10.1002/cbic.201800288.

CHAPTER 8

Role of Oxidative Stress and Mitochondrial Dysfunction in Progression of Alzheimer's Disease and its Emerging Phyto-therapeutic and Nano Medicinal Interventions

SNIGDHA BHARDWAJ[1], RADHA GOEL[2*], ROSALINE MISHRA[3], and MOHD YASIR[4]

[1]*Department of Pharmaceutics, KIET School of Pharmacy, Ghaziabad, Delhi-NCR, Uttar Pradesh, India*

[2]*Lloyd Institute of Management & Technology, Greater Noida, Uttar Pradesh, India*

[3]*Department of Pharmacy, Metro College of Health Sciences and Research, Uttar Pradesh, India*

[4]*Department of Pharmacy, College of Health Science, Arsi University, Asella, Oromia Region, Ethiopia*

*Corresponding author

ABSTRACT

Alzheimer's disease (AD), marked by age-related cognitive decline and progressive fatal brain destruction, is usually associated with an abnormal breakdown of amyloid-β (Aβ) precursor protein, which leads to the generation and accumulation of harmful Aβ. Oxidative stress (OS) is the leading cause of age-related neurodegeneration and cognitive impairment. OS promotes Aβ synthesis, as well as β-secretase expression and activity.

Cellular and Molecular Mechanisms in Alzheimer's Disease Therapeutics, Volume 1: Cellular Mechanisms. Prashant Tiwari, Sunil Kumar Kadiri, and Kunjbihari Sulakhiya (Eds.)

DOI: 10.1201/9781779641793-8

Furthermore, OS has numerous consequences that may lead to a decrease in ATP synthesis, mitochondrial dysfunction, synaptic dysfunction, membrane damage, cytoskeleton changes, and cell death in AD. The present treatment regimen improves cognitive function and provides momentary symptom relief; however, the disease's progression is not halted or delayed by it. Plant-derived bioactive chemicals have been shown in numerous studies to have potential benefits for memory recovery and for delaying the start of memory deterioration. Certain bioactive chemicals originating from plants possess antioxidative properties, which led to their application in restoring OS-induced alterations in the aging brain. Furthermore, an approach based on nanotechnology has the impact to enhance the specificity and bioavailability of the medicine in the brain. These qualities are not possible with traditional oral therapies, which lack brain specialization and have unfavorable side effects that lead to low patient compliance. This chapter focuses on the aspects related to memory impairment and the course of AD that is mediated by OS. It also provides a thorough summary of possible therapies based on phytoconstituents, as well as applications of nanomedicine strategies developed based on blood–brain barrier (BBB) evolutions and associated applications in AD.

8.1 INTRODUCTION

Alzheimer's disease (AD) is the most widespread and it has become one of the most expensive and dangerous diseases today. The cause of the disease is poorly defined and is not well known [1]. Two signs of sickness in AD regions of the brain are the abnormal storage of Aβ amyloid (Aβ) as extracellular plaques, leading to synaptic dysfunction and neurodegeneration [2, 3]. The second one is tau protein that leads to the generation of intracellular neurofibrillary tangles and contributes in degeneration. The traditional classifications of AD depended on phenotypical criteria 1 and included early-onset versus late-onset forms. Mutations in three genes, namely *APP*, *presenilin 1*, and *presenilin 2* that belong to catalytic core of γ-secretase that processes Aβ have been shown to be contributing to the aetiology of AD, particularly in familial or early-onset type I and II forms, which account for around 10% of AD cases [4]. The late-onset condition happens by a complicated interaction of inherited and environmental variables.

Among them many factors show relation to pathophysiology of disease but Aβ is the most important factor. Aβ consists of polypeptides, with lengths varying between 40 (majority) to more than 43 amino acids in plaques [3]. Aβ 40, the predominant form of amyloid is detected in 90% plaques. On the other hand, (Aβ 42), the longer form is of great concern in the initial phase, resulting in a faster aggregation than Aβ 40 [5]. AB is generated from *APP* [6, 7]. *APP* is cleaved by *BACE1* to generate sβAPP and C99 [6]. C99 breaks down inside the cell membrane by the enzyme γ-secretase [7]. Aβ particles of different lengths are produced by the γ-cleavage, mainly Aβ 40 and Aβ. Aβ is the main reason in etiology of AD, and supported by a number of writers in the last 30 years as per "amyloid cascade" theory [8, 9]. According to this theory, AD patients have cognitive and memory impairment, as a result of tau aggregation brought on by the deposition and accumulation of Aβ [10].

Approximately 90% of the amyloid present in plaques is of type Aβ 40, which is the most common kind. But in the initial stages, the longer form (Aβ 42) predominates and accumulates more quickly than Aβ 40 [5]. For amyloidogenic processing of *APP*, β- and γ-secretases must cleave the protein twice.

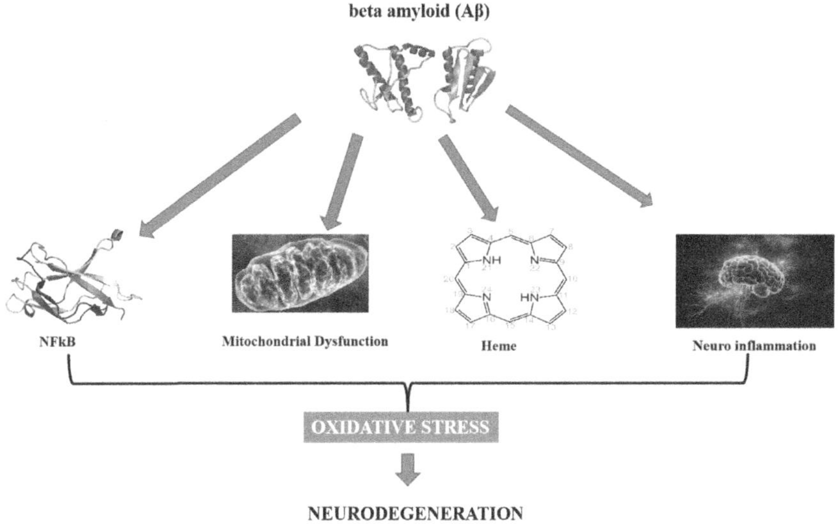

FIGURE 8.1 Pathological mechanisms of neurodegeneration.

Even though there has been considerable success in treating AD, the current approaches only work to reduce symptoms rather than reversing the neurodegenerative illness. The application of nanotechnology opens up new therapeutic options for neurological conditions. Numerous nanocarriers and nanoparticle prodrugs have shown minimal cytotoxicity in animal studies, opening up new avenues for the creation of powerful novel medication formulations for AD [8]. Over the last decade, researchers have developed a variety of nanotechnology-based ways to treat AD, with an emphasis on neuroprotective and neurogenerative techniques. These methods consist of metal-based carriers, carbon nanotubes, solid lipid carriers, lipo-carriers, emulsions, and polymers. Research also shows that nanotechnological methods can improve bioavailability and therapeutic effectiveness, which can help in early AD diagnosis. The ability of traditional AD medications to reach the central nervous system (CNS) is limited by several physiological variables, encompassing the blood-cerebrospinal fluid barrier, the blood–brain barrier (BBB), and P-glycoprotein-mediated drug efflux. To combat this, researchers are using nanotechnology to attach AD medications on nanoparticle surfaces. These obstacles about the CNS are circumvented by this drug delivery strategy [9, 10].

8.2 OXIDATIVE STRESS AND ITS ROLE IN AD

Oxidative stress (OS), a condition that generally intensifies with aging in the brain, is triggered by a cellular redox imbalance, which ends up with overproduction of reactive oxygen species (ROS) or an antioxidant system dysfunction [11].

Alzheimer's patients' brains show significant oxidative damage, including abnormal Aβ production and neurofibrillary tangle development [12]. Overproduction of ROS, which induces a state of OS, can have a damaging effect and plays a crucial part in cell structural damage, and in many disease states and aging. As the nervous system mostly has oxidized lipids, an excessive oxygen consumption rate, and lacks effective antioxidant systems,

OS is recognized to have a key role in evolution of AD [13]. OS leads to AD progression through three primary mechanisms:

1. Macromolecule peroxidation,
2. Aβ metal ion redox potential,
3. Mitochondrial dysfunction.

These variables impact cell homeostasis, produce ROS, and increase Aβ and p-tau production [14].

8.2.1 MACROMOLECULE PEROXIDATION

The start of AD has been associated with lipid peroxidation, breakdown of DNA protein crosslinks, and oxidative death of cells [15]. Peroxidation of double-bonded molecules in polyunsaturated brain lipid products produces biochemically active lipid byproducts that are extremely reactive and capable of stimulating tau phosphorylation and dysfunction, disrupting the intracellular Ca^{2+} signaling pathway, and inducing an apoptotic cascade [16]. This promotes brain tissue destruction, contributing to AD.

8.2.2 METAL ION REDOX POTENTIAL

Metal ion homeostasis is disrupted in AD, with ion levels deviating from healthy ranges [17]. Zn^{2+} and Cu^{2+} cations bind to the hydrophilic N-terminal ends of Aβ peptides, leading to continuous redox reactions that generate high amounts of free radicals. Similarly, metal ions like Zn^{2+} can bind to tau proteins and phosphorylate it [18].

8.2.3 MITOCHONDRIAL DYSFUNCTION

Mitochondrial malfunction and oxidative metabolism are other key factors in development of AD and many neurodegenerative conditions. Autopsied brains of Alzheimer's patients showed reduced cytochrome oxidase activity and messenger RNA levels, decreased adenosine triphosphate formation, abnormal calcium balance, production of ROS, induction of the mitochondrial permeability transition, and excitotoxicity [19].

8.3 REGULATION OF THE ACTIVITY AND EXPRESSION OF ANTIOXIDANT ENZYMES

Antioxidants can mitigate the negative effects of OS. Antioxidant treatment is recommended as a possible way to slow progression of AD by reducing phospholipid peroxidation, protein, and DNA damage.

Antioxidants are primarily divided into two types as per their mode of action:

1. Primary antioxidants: These scavenge ROS and interrupt the chain reaction that generates OS [20].
2. Secondary antioxidants: These are oxidized and degrade hydroperoxides into irreversible forms, creating a beneficial interaction

with primary antioxidants. Secondary antioxidants play a key role in antioxidant renewal, metal deactivation, and singlet oxygen reduction [21].

The antioxidant system is of two types:

1. The endogenous system contains ROS-eliminating enzymes, such as glutathione peroxidase, catalase, glutathione reductase, superoxide dismutase, and others.
2. Exogenous system: Research suggests that food can help treat AD [22]. Nutritional antioxidants include vitamin C and E, as well as carotenoids, flavonoids, and polyphenols [12]. Furthermore, research and clinical data indicate that a healthy diet rich in vitamins, proteins, and minerals will undoubtedly support the drug used to treat AD [23].

8.3.1 ENDOGENOUS SYSTEM

Endogenous antioxidant enzymes serve a significant role in muscle tissue redox state, and under normal conditions, neutralize ROS's pro-oxidant activities. SOD, CAT, and GSH-Px are some of the most active and effective enzyme systems for OS.

8.3.1.1 ANTIOXIDANT ENZYME EXPRESSION

i. **Superoxide dismutase (SOD),** an endogenous protective enzyme, oxidatively degrades the superoxide radical to produce H_2O_2. SOD also prevents dehydratase from becoming inactivated by free radicals like superoxide. SOD exists in three types in human beings.

   a. SOD-1, a copper-zinc-containing SOD (Cu, Zn-SOD), lives in cytoplasm and stimulates transformation in pH-independent conditions. Copper minerals, in particular, play an important role in this enzyme's catalytic function, while zinc is essential for structural stability [24].

   b. SOD-2, a manganese-containing SOD (Mn-SOD), is present in mitochondrial matrix and lowers superoxide radical anions [25]. SOD-2 is essential for aerobic cell survival and for cellular resistance to OS.

c. The SOD-3 enzyme is not triggered via the target or any free radicals, and it is designed to function with cytokines instead of free radicals. Increased levels of SOD-1 defend nerves from Aβ neurotoxicity, whereas lack of SOD-1 results in accelerated aging and reduced lifetime in mice. Impaired SOD-2 is a prevalent etiology associated with OS in Parkinson's and AD [26].

ii. **Catalase (CAT)** is a widely distributed antioxidant enzyme that the body produces when in contact with oxygen. In aerobic species, catalase degrades H_2O_2 produced within the cell. Furthermore, blocking CAT activity increases cytotoxicity and ROS levels, indicating that CAT is essential for maintaining oxidative balance [27].

iii. **Glutathione peroxidase (GPx)** transforms hydroperoxides (ROOH and H_2O_2) into water or comparable alcohols, shielding human cells from OS [18]. GPx1 and GPx4 are present in the majority of tissues, with brain tissue having the highest concentration. GPx1 inhibits H_2O_2 and organic hydroperoxides and can be found in neurons and astrocytes [28].

Because GPx activity is linked to proper growth and cellular metabolism via OS management, GPx4 function may be required for cell survival.

iv. The Thioredoxin (Trx) system is composed of two kinds of proteins (Trx and TrxR) which employ NADPH as an electron source. Trx is a critical cell-cycle controller which reacts to the equilibrium of redox by changing communication channels, transcription factors, and immunological responses, allowing cells to survive in a variety of situations. Trx family contains three isoforms: Trx1 in the cytosol, Trx2 in the mitochondria, and a testis-specific Trx3 in mammalian cells. TrxR also plays a role in the regeneration of additional antioxidants, like dehydroascorbate, lipoic acid, and ubiquinone [29]. Subsequently it is being proposed that TrxR activation promotes a mechanism to compensate while increasing OS, which is hindered by the substrate Trx and contributes to AD.

8.3.1.2 NON-ENZYMATIC ANTIOXIDANTS

Antioxidant enzyme cofactors (selenium, coenzyme Q10, and zinc)

i. **Selenium (Se)** is an enzymatic cofactor involved in a wide range of biological processes, including redox signaling, antioxidant defense systems, thyroid hormone metabolism, and humoral and cell-mediated immunological responses. Se's biological action varies with concentration: Low concentrations promote appropriate developmental processes, medium amounts promote homeostatic function, and high quantities cause induced toxicity. Se is necessary for brain function and has a vital role in etiology of brain diseases, such as AD, Parkinson's, and epilepsy [30].

ii. **Coenzyme Q10 (CoQ)** is a naturally occurring ROS scavenger and catalytic factor found in most organisms. It is found in the brain and gut, which are essential for oxidative phosphorylation. CoQ is an essential antioxidant that shields cell membranes and lipoproteins from harmful impacts of ROS produced during normal metabolic processes [31].

iii. **Zinc and essential metals:** Zinc enhances the function of antioxidant enzymes, such as SOD, GPx, and CAT. Zinc suppresses critical pro-oxidant functions, which regulate oxidant generation and metal-induced oxidative damage. Zinc affects glutathione metabolism and total protein thiol redox state.

   In an AD rat model, zinc and selenium combination therapy significantly reduced mitochondrial dysfunction and lipid peroxidation, while significantly enhancing mental function, SOD, GPx, and CAT activity in brain [32].

   Copper, iron, and magnesium are among the other essential metals required for cell homeostasis and survival.

iv. **Nrf2 antioxidant system:** Nrf2, a transcription factor, regulates redox homeostasis and antioxidant defenses. This protective reaction also activates a detox network that includes oxidation/reduction factors, metabolizing enzymes, efflux transporters, and ROS scavengers [33].

8.3.2 EXOGENOUS SYSTEM

Exogenous antioxidants are antioxidants obtained from our food by eating foods that contain antioxidants and by taking supplements. These include vitamins A, C, and E. Even while exogenous antioxidants can be derived from dietary sources, in today's world; it is nearly difficult to absorb adequate artificial antioxidants to completely neutralize all free radicals. This is why antioxidant drugs are so beneficial. The list of exogenous antioxidants is listed in Table 8.1.

TABLE 8.1 List of Exogenous Antioxidants.

Antioxidant	Functions	References
Vitamin E	Reduces free radical-induced neuronal damage. In mammalian culture cells, inhibits the genesis of dementia.	[34]
Vitamin C	Suppresses the degradation of lipids as a significant defense against ROS in plasma cells.	[35]
β carotene	Limits ROS. Assists in increasing antioxidant availability.	[36]
Vitamin B12	Enhances choline acetyltransferase activity Improves cognitive abilities in AD.	[37]
α-lipoic acid	Promotes reutilization of other antioxidants, such as vitamin C and E. Enhances acetylcholine synthesis to counteract ROS.	[38]
CoQ10	Maintains mitochondrial membrane potential under OS conditions. Prevents neurons from Aβ plaque formation.	[39]
Caffeine	Prevents amyloidosis and Aβ buildup. Decreases Aβ plaque amount in AD-related mice models.	[40]
Curcumin	Inhibits lipoxygenase and cyclooxygenase 2, enzymes that produce inflammatory mediators. Lowers Aβ plaque concentration in AD-related mice models.	[41]
Berberine	Decreases Aβ plaque concentration. Limits production of ROS and RNS.	[42]
Palmatine	Lowers Aβ plaque amount. Prevents the production of ROS and RNS	[43]
Silibinin	Reduce Aβ-induced oxidative damage in mouse models. Inhibits cognitive impairment.	[44]
Quercetin	Blocks the Aβ aggregation *in vitro.*	[45]
Melatonin	Promotes production of GPx, SOD, and NO synthase, which aids in scavenging RNS produced in mitochondria. Reduces the OS in mammalian cells.	[46]
Estrogen	Protects neurons against the toxicity of Aβ.	[47]
Selegiline	Protects the vascular endothelial cells against Aβ toxicity. Helps protect nigral neurons from oxidative deamination.	[48]
Resveratrol	Reduces synthesis of Aβ peptides *in vitro*. Improves memory and cognitive deficits in animal models. Decreases MDA levels in rodent model.	[49]

8.4 AB AND OXIDATIVE STRESS

8.4.1 MECHANISMS BY WHICH AB INDUCES OS

Research suggests that beta amyloid (Aβ) can cause oxidative damage through many methods. Some of which are described in the following sections.

8.4.1.1 TRANSITION METALS

Metals such as copper (Cu(II)), zinc (Zn(II)), and iron (Fe(III)) might exacerbate Aβ's neurotoxic effects. This occurs when these metals are reduced, resulting in the formation of damaging hydrogen peroxide (H_2O_2) [50]. Tirosine-10 (Tyr-10): A culprit: Tyr-10, an amino acid found in Aβ, is essential for the metal-driven production of H_2O_2. Tyr-10's reaction with ROS produces a phenoxy radical, leading to neurotoxicity and Aβ aggregation [51]. Methionine-35 (Met-35) is another key player: Met-35, another amino acid, plays a critical role in Aβ's harmful effects. In a worm model (C. elegans), replacing Met-35 with cysteine reduced protein oxidation. Some studies suggest that Aβ, especially the Aβ 42 type, may interfere with cytochrome c oxidase function by producing reactive methionine [52]. Aβ peptides can harm cell membranes via lipid peroxidation. This impairs the operation of a variety of cellular components, including ion pumps, transporters, and proteins, causing neuronal damage. Toxic Aβ aggregates can lead to increased Ca^{2+} influx into neurons. The calcium overload generated by OS in the membrane renders neurons more vulnerable to excitotoxicity and cell death (apoptosis) [53].

8.4.1.2 NEUROINFLAMMATION

Neuroinflammation is a critical mediator in Aβ-induced OS [54]. Neuroinflammation is a term indicating stimulation of glial cells followed by release of inflammatory mediators [55]. Multiple research have found a clear link between neuroinflammation and milder AD pathology syndromes [56], and inflammatory cytokines rise as moderate cognitive impairment progresses to clinically diagnosed AD.

One study revealed evidence of increased expression of genes in the aging brain and detected small increases in these proteins, as well as suggesting inflammation is also occurring at preclinical phases of AD [55]. Evidently, microglia has an important role in neuroinflammation under present circumstances [54]. When Aβ binds to microglial receptors, it produces oxygen free radicals (OH and O_2), nitric oxide (NO) [19], TNF α [57], cytokines that are associated with inflammation, and chemokines. An emerging cytosolic complex of proteins linked to innate immune response and neuroinflammation is the NLR family pyrin domain containing 3 (*NLRP3*) inflammasome [58]. According to recent studies, *NLRP3* in microglia and astrocytes is activated by Aβ. Damage that cannot be repaired is caused by the generation of *caspase 1* and the release of cytokines. In a transgenic AD animal model, it has been said that inhibiting the *NLRP3* inflammasome reduces Aβ accumulation and provides neuroprotection [59].

8.4.1.3 NF-KB PATHWAY

The NF-kB family is essential for controlling OS. *In vitro* studies indicate that neurons exposed to Aβ experience increased OS via an NF-kB-mediated mechanism. This increase in OS causes the accumulation of lipid peroxides, which ultimately leads to neurodegeneration [60].

8.4.2 MECHANISMS BY WHICH OXIDATIVE STRESS MEDIATES AB PRODUCTION

OS is a state characterized by increased quantities of ROS in cells and/or insufficient cellular defenses against the potentially detrimental effects of ROS [61]. This idea suggests that elevated ROS or RNS generation typically does not give enough use to induce disease, if the antioxidant defense system is able to scavenge the generated reactive species. Nonetheless, there are situations where the aging-related rise in free radical species production is correlated with decrease in endogenous antioxidant defense system's capacity [62]. While OS is linked to an increased risk of many diseases, such as attention deficit hyperactivity disorder, cancer, and so forth [51], natural aging does not. OS growth is seen as an early occurrence in AD, as it causes cell death by leading to cellular damage, cytoskeleton alterations, and so forth [63, 64].

8.4.2.1 APP PROCESSING AND SECRETASES

OS has been linked to Aβ buildup and toxicity, according to multiple studies [65]. OS may contribute to the buildup of Aβ. Oxidizing drugs boosts APP expression and Aβ levels, both intracellular and secreted [65]. We have shown that *BACE1* expression and activity increased in response to OS [66]. It was also found that, there is a substantial link between *BACE1* activity and OS markers in brain tissues of sporadic AD [66], and *BACE1* levels have been reported to be significantly higher than in normal aging [66].

Numerous researchers have showed a sequence of events which include the observation of OS, *BACE1* hyper-expression, and triggering of apoptosis in relation to the accumulation of Aβ [67]. These occurrences cause apoptosis and an excess of Aβ. Moreover, it has been discovered that OS raises gamma secretase activity both *in vivo* and *in vitro*, and that gamma secretase controls the elevated expression of *BACE1* brought on by OS [64]. Understanding the pathophysiology of sporadic AD depends on these results. OS syndrome may be the only recognized condition that can increase the γ-secretase cut on *APP*, in turn leading to enhanced activity of *BACE1* [68, 69].

8.4.2.2 MITOCHONDRIA

Aβ disrupts mitochondrial function, leading to OS [70]. Thus, neurons are high-energy cells that execute a variety of activities, including action potential production, nerve transmission, and axonal transport [71]. Reduced electron transportation effectiveness leads to increased ROS production during the modulation of phosphate oxidation (OXPHOS), mostly at the complex I and complex III levels. These ROS produced at the chain level cause most of the harm to mitochondria molecules [70].

The peptide Aβ enhances ROS production in mitochondria and limits ROS clearance. Aβ has been demonstrated to impair mitochondrial superoxide dismutase (MnSOD), which detoxifies superoxide and protects against peroxidative damage [71]. Aβ can bind and inhibit ABAD, a mitochondrial alcohol dehydrogenase. Through its ability to detoxify aldehydes like 4-hydroxinonenal, ABAD provides protection. Lipid peroxidation, the generation of ROS, and mitochondrial dysfunction are caused by the disruption of the enzyme's detoxification pathway caused by the Aβ peptide's interaction with ABAD.

Moreover, Aβ directly affects mitochondria. Additionally, some scientists have shown that older individuals with AD have changed mitochondrial DNA due to the direct influence of Aβ on mitochondria.

8.5 IMPACT OF BIOACTIVE COMPOUNDS ON PATHWAYS OF NEURODEGENERATION

8.5.1 PHENOLIC AND FLAVONOID COMPOUNDS

Polyphenols are secondary metabolites present in plants, distinguished by aromatic rings containing hydroxyl (OH) groups [72]. Numerous phenolic chemicals have been found, either separately or in precise ratios in botanical extracts. Investigating the health impacts of these natural chemicals remains a major issue in modern medicine [72].

Flavonoids are known for their high antioxidant activity [73]. Polyphenol consumption lowers risk of cardiovascular disease, some cancers, and chronic noncommunicable diseases. A polyphenol-rich diet has been found to be neuroprotective in studies conducted on cell and animal models [74].

Protective effects against oxysterols: Several phenolic compounds have cell-protective characteristics, potentially reducing or easing oxysterol-induced damage to cells and mitochondria (e.g., 7-ketocholesterol) [74]. Antioxidant substances, such as quercetin, resveratrol, and apigenin may influence the generation of ROS caused by oxysterols.

8.5.2 PHENOLIC COMPOUNDS AND ALZHEIMER'S DISEASE

There is a developing hypothesis of common enzymatic targets in AD and type II diabetes mellitus (T2DM). Acetylcholinesterase (AChE) and butyrylcholinesterase (BChE), involved in AD pathogenesis, with IC50 values (half maximum inhibitory concentration) ranging from 347 to 379 μg/mL [75]. Interestingly, enzymes associated with T2DM, α-amylase, and α-glucosidase, have lower IC50 values (127–140 μg/mL). These findings point to potential enzymatic similarities between the two diseases [76]. The *in vitro* investigation suggests that *Sesamia cretica* extract can inhibit major enzymes linked to AD (AChE, BChE) and T2DM (α-glucosidase, α-amylase). This strengthens the case for shared treatment objectives. Furthermore, the identification of active principles in the extract, such as chlorogenic acid and kaempferol, opens the door to future inquiry into their specific inhibitory processes.

Adiantum capillus-veneris L. extract inhibits AChE, α-glucosidase, and α-amylase, supporting the concept of common enzyme targets. The extract also showed action against tyrosinase, indicating a broader therapeutic potential.

These results collectively support the hypothesis that phenolic compounds have shown promise for the development of multi-targeted therapies capable of alleviating symptoms of AD, T2DM, and may be other neurodegenerative disorders.

8.5.3 CAROTENOIDS AND ALZHEIMER'S DISEASE

A study assessing the effect of blood carotenoids in patients with AD discovered a correlation between a lower risk of AD mortality and high baseline serum levels of lutein along with zeaxanthin and lycopene. Gradually decreasing the risk of death was possible by increasing serum levels of lutein, zeaxanthin, and lycopene. The correlation between AD mortality and other serum carotenoids, like α/β-carotenes and β-cryptoxanthin, has not been established. Red blood cells (RBCs) from AD patients have high levels of peroxidized phospholipids (PLOOH) and are highly oxidized. Through regulating their RBC viability, carotenoids may benefit AD patients. Studies on humans and *in vitro* have demonstrated the efficacious suppression of RBC-PLOOH accumulation pathways by carotenoids. RBC carotenoids, particularly lutein, were found to be in low concentrations in the experimental group, indicating an inverse association between RBC [77]. Lycopene has sparked widespread interest due to its antioxidant, anticancer, and neuroprotectant properties. In a rat study, lycopene significantly improved functional deficiencies in memory and learning compared to the control group. Serum levels of TNF-α, IL-1β, and IL-6β increased significantly at the choroid plexus, as did the expression of toll-like receptors, such as TLR4 and NF-κB p65 mRNA and protein levels increased after Aβ1-42 injection, suggesting an inflammatory response. After intragastric injection of lycopene, inflammatory cytokines were significantly reduced [78].

8.5.4 ESSENTIAL OILS AND ALZHEIMER'S DISEASE

Using essential oils (EOs) in aromatherapy may be a successful natural treatment for Alzheimer's. Research on *Pinus halepensis* EOs in a rat model of acute Aβ1-42 poisoning discovered that they reversed the effects of Aβ1-42 on the rise in working and reference memory errors and the decrease in spontaneous alternation. It has been demonstrated that after exposure to Aβ1-42, EOs enhance the oxidant-antioxidant balance and AChE activity in the rat hippocampal tissue. These results suggest that the EOs of *Pinus halepensis* possess nootropic and neuroprotective qualities, making them a viable therapeutic option for lowering Aβ-induced neuronal dysfunctions. Using the AD dementia paradigm in mice, EOs from *Rosmarinus officinalis* were found to significantly improve spontaneous alternation behavior. Additionally, 1,8-cineole was the main ingredient in the inhaled EOs whereas 1,8-pinene has enhanced the cognitive functions.

This supports its use as a traditional medicine for AD therapy. Inhaling EOs from *Tetraclinis articulata* impacted brain OS and memory in an Aβ1-42-induced AD amyloidosis mouse [79].

8.5.5 PHYTOSTEROLS AND ALZHEIMER'S DISEASE

β-sitosterol and lupeol: Potential phytosterols for AD:

The hunt for viable therapies in AD has encouraged researchers to investigate natural compounds with neuroprotective properties. This chapter explores the potential of β-sitosterol from *Polygonum hydropiper* and lupeol from *Pueraria lobata* to slow the progression of AD [80].

β-sitosterol: β-sitosterol effectively blocks BChE and AChE. This finding is consistent with the documented decrease in BChE and AChE activity within the hippocampus and frontal cortex of transgenic AD animal models after β-sitosterol administration. Molecular docking simulations confirmed that β-sitosterol binds directly to cholinesterase enzymes. This improves our understanding of its inhibitory mechanism [81].

8.6 CONVENTIONAL THERAPEUTIC APPROACHES AND THEIR LIMITATIONS IN THE TREATMENT OF AD

Despite escalating socioeconomic costs and patient numbers, there is currently no treatment that effectively stops the progress of AD. Aducanumab, donepezil, galantamine, rivastigmine, memantine, and a synthetic mixture of the two are the only six medications to date to have FDA approval. Aducanumab is the only medication used to remove Aβ plaques; the other five are meant to treat symptoms and have two different functions: They act as agonisms of the cholinergic system or as antagonists of the N-methyl-D-aspartate receptor.

1. ***Aducanumab:*** It is a monoclonal antibody that targets soluble β-amyloid protein aggregates and fibrils with the goal of eradicating Aβ plaques. At least 21 days should elapse between doses of aducanumab 10 mg/kg given intravenously once every four weeks. It is believed that this medicine can work on SP, reducing the course of AD, and the FDA authorized it in April 2021. Three clinical trials demonstrating the treatment's efficacy provided the data necessary for its approval [82].

2. ***Acetylcholinesterase Inhibitors (AChE)***

   a. *Tacrine*: In September 1993, tacrine was the first medication in this class to receive FDA approval, key inhibitor of AChE, this cholinergic analogue produced from acridine raises acetylcholine levels in different parts of the brain [83]. When compared to AChE, its ability to block pseudocholinesterase is much stronger. Dr. Williams Summer conducted the initial AD testing of tacrine in 1989. On the other hand, inconsistent findings in clinical trials raised doubts about its efficacy as its use expanded [84–87]. Thus, researchers found that tacrine helped alleviate symptoms in people with mild to moderate dementia, but it didn't slow down the underlying neurodegeneration's course [88]. Many side effects, including gastrointestinal distress, vomiting, hepatic cytotoxicity, lack of appetite, diarrhoea, and clumsiness caused its withdrawal in 2013 [89, 90].

   b. *Donepezil:* The cholinergic medication donepezil is synthesized from piperidine and acts as a central, non-competitive, reversible inhibitor of AChE enzyme. This enzyme is responsible for hydrolysing acetylcholine [91]. Donepezil can be injected transdermally or taken orally as a pill, liquid, or jelly. The usual course of treatment for mild to severe dementia is 5 mg once a day, with a four- to six-week increase to 10 mg. Following three months of treatment with 10 mg daily, individuals with mild to severe dementia may be eligible for a dosage increase to 23 mg daily [92, 93]. A daily administration of 10 mg of donepezil resulted in improvements in cognitive ability, basic activities of daily living, but did not alter behavior or quality of life [94, 95]. Furthermore, studies with comparisons of the effectiveness of daily doses as low as 10 mg to those as high as 23 mg have returned no statistically significant differences [96, 97]. Nonetheless, no dose has been found to halt the advancement of AD [98]. However, this medicine has few side effects and excellent patient tolerance, particularly in the GI and neurological systems [99].

c. *Galantamine*: It is a tertiary isoquinoline alkaloid that inhibits AChE in a reversible, selective, and competitive manner. Above and beyond that, it enhances acetylcholine's intrinsic activity on nicotinic receptors [100]. In 2001, this medication was given the go light by the FDA. The dosage is 4, 8, 12, 16, and 24 mg, and it can be taken orally in the form of an extended-release capsule once a day or a quick-release solution twice a day. It is recommended that patients begin treatment with an 8 mg daily dose, increasing to a maintenance dose of 16 mg twice a day after 4 to 8 weeks [101, 102]. What makes galantamine so intriguing as a potential treatment for AD is that it mostly affects the CNS, barely affecting the peripheral nervous system. Various compounds that facilitate galantamine's brain distribution have been linked to it along this route, including chitosan, solid lipid nanoparticles, and hydroxyapatite particles containing ceria [103–105]. According to clinical trials on this illness, this drug can help people with mild to severe AD with their cognitive and behavioral symptoms, such as agitation, anxiety, disinhibition, and aberrant movements [106–108]. The drug is the gold standard for treating AD, according to a recent meta-analysis [109]. Despite its excellent safety and tolerability, it can cause seizures, extreme nausea, vomiting, cramping in the stomach, erratic breathing, dizziness, weak muscles, and watery eyes.

It also improves cognitive function, basic ADL performance, and physicians' perspectives on the worldwide situation [110].

d. *Rivastigmine*: Rivastigmine was first licensed in 1997 by the Swiss government and in 2000 by the FDA. It is used to treat mild to moderate AD and mild to moderate Parkinson's disease. It binds to the anionic and stearic sites of AChE to inhibit the enzyme BChE, making it a pseudo-irreversible inhibitor [111]. The suggested initial dosage is 1.5 mg/12 h; if tolerable, this can be increased to 6 mg/12 h, with each dose spaced at least 2 weeks apart; the optimal dose ranges from 3 to 6 mg/12 h. It is available in tablet or drop form. The transdermal patches of this medicine come with a recommended starting dose of 4.6 mg daily, with the possibility of increasing it to 13.3 mg daily, if the patient has no adverse effects [112]. Apart from preventing adverse reactions associated with gastrointestinal and hepatic metabolism, this method permits continuous release for 24 h. Similarly, people with AD who have trouble swallowing or with memory loss may find transdermal patches to be an appealing alternative to the oral route [113]. Furthermore, this administration mechanism has been proven to be the most cost-effective treatment option according to cost-effectiveness studies [114]. The efficacy of rivastigmine in treating AD has been demonstrated in studies. Patients demonstrated improvements in cognitive function as assessed by the ADAS-Cog and MMSE, ADLs, and clinician-rated general impression of change after 26 weeks of treatment with dosages of 6–12 mg orally or 9.5 mg transdermally [115–117]. It should be mentioned that rivastigmine has numerous unpleasant side effects, including nausea, vomiting, diarrhea, weight loss, appetite loss, and stomach discomfort, thus people using it aren't particularly good about sticking to their treatment plans. In addition, chest pain, irregular or slow heartbeat, and irregular or fast breathing are among the many symptoms that may accompany this drug's overdose [118].

3. ***N-Methyl D-Aspartate (NMDA) Antagonists***

Memantine: It is an antagonist of the NMDA receptor that acts by binding to the receptor with moderate affinity and without competing with other substances. Its binding is dependent on the voltage of the receptor. It inhibits the impact of abnormally high levels of glutamate, which can cause malfunction in neurons. The disruption of neurotransmission mediated by glutamate, particularly at NMDA receptors, has a role in both the manifestation of symptoms and the development of neurodegenerative dementia from AD. In 2003, memantine received initial approval from the US FDA as a treatment for mild to severe AD [119]. The drug is available in doses of 7, 14, 21, and 28 mg in slow-release capsule form. Additionally, it comes in tablets with a 10 mg dosage and a solution with a 2 mg/mL concentration. For the first week, a starting dose of 5 mg per day is advised. The dosage is then increased to 5 mg twice daily for the second week and to 15 mg in total for the third week, taking half of a pill in the afternoon and one 10 mg tablet in the morning. Furthermore, memantine and donepezil (Namzaric®) have been licensed by the FDA for the symptomatic treatment of moderate to severe AD. Nevertheless, the European Medicines Agency has not approved Acrescent®, a medicine that contains memantine hydrochloride and donepezil hydrochloride. This decision is based on the absence of reliable evidence demonstrating its efficacy in treating this particular condition [120]. This medicine combination inhibits the detrimental consequences linked to excessive glutamate levels and also prevents the degradation of acetylcholine in the brain. Research indicates that when compared to a placebo, the combination of memantine and donepezil is more effective in improving cognition as assessed by the ADAS-Cog and SIB scale, global evaluation, daily activities, and neuropsychiatric symptoms. It is less acceptable, nevertheless, than taking the medication by itself or a placebo. The scientific literature does not provide definitive conclusions about the treatment's economic viability [121].

8.7 EMERGING ROLE OF NANOTHERAPEUTICS AND ITS EFFECTIVE TARGETS IN AD

There is an immediate need for novel strategies in medication development to circumvent the BBB, a barrier that prevents CNS therapies from reaching patients. Several methods based on nanotechnology can fulfill this need by increasing the drug's efficacy and allowing for its continuous release. A few of the nanotechnology-based techniques include polymers, solid lipid carriers, magnetic nanoparticles, inorganic nanoparticles, antibody-tethered nanoparticles, dendrimers, emulsions, nanoparticle conjugates, cubosomes, intranasal delivery of nanoparticles, and solid lipid carriers. [122, 123]. An abundance of platforms based on nanotechnology have been created to enhance the efficacy of brain therapies. Utilizing innovative design concepts, these assisted drug administration strategies offer numerous benefits over conventional methods. One example is a great deal of leeway in formulation design when it comes to improving drug loading, targeting, and release efficacy. Another is the widespread use of nanoparticles (NPs), a frequent and inexpensive technology which can be used for non-invasive administration. The pharmacokinetics, pharmacodynamics, and bioavailability of medications are improved due to these nanoscale structures, which facilitate their passage across the BBB. Nanocarriers derived from lipids, dendrimers, and polymers are some examples [124, 125]. Table 8.2 presents a few research investigations to develop a new nanoformulation effective against different targets to treat AD.

8.8 APPLICATIONS OF NANOTHERAPEUTICS IN AD MANAGEMENT

Researchers are looking into the use of nanomaterials in precision medicine, namely in treatment of AD as it cannot cross BBB. Numerous nanocarriers have received FDA approval to carry drugs to treat different illnesses, indicating that this constraint may someday be surmounted [135]. Researchers are looking at the potential use of nanocarriers in treatment of neurological conditions. The current methods make use of delivery systems based on nanostructure, which are divided into four categories: lipid vesicular, organic, metallic/non-metallic, and emulsion-based systems [136]. The utilization of NP-functionalized nanomedicine is the optimal way to deliver anti-AD medications to the particular part of the brain afflicted by AD. This is primarily owing to the distinctive characteristics of NP-functionalized nanomedicine, such as its reduced size, enhanced biocompatibility, and to facilitate the efficient delivery of drug across the brain [137–139]. Nanomedicines, roughly 100–10,000 times smaller, can interact with a cell's internal and external proteins and molecules with ease.

TABLE 8.2 Few Investigations Based on Nanotherapeutics and Effective Targets in AD.

Nanodrug-Delivery Systems	Carrier Material	API	Experiment Model	Key Targets	Description	References
Carbon nanotubes	Pristine multi-walled carbon nanotubes (MWCNT); phospholipids and polysorbates	Berberine	β-amyloid induced AD in wistar rats	BBB transcytosis, cholinergic systems	BRB-loaded MWCNT coated with phospholipids and polysorbates showed a notable improvement in memory impairment and a reduction in AChE activity in animals (rats) with AD when compared with uncoated form of the nanotubes.	[126]
Dendrimers	Hydrophobic pyridylphenylene dendrimers	o-phenylene diamine	Amyloid protein aggregates are represented by the inclusion bodies of the ovine prion protein (PrP).	amyloid, cascades	Dendrimers create persistent interactions with proteins, effectively disrupting PrP amyloid clumps at a pH of 7.4, which are useful to reduce Aβ load.	[127]
Gold (Au) (NPs)	AuNPs with PEG	Anthocyanin	Mouse brain endothelial cells	Amyloid cascades and tau hyperphosphorylation	In a model of AD, PEG-AuNPs loaded with anthocyanin showed more neuroprotective effects than the free form of the drug by modulating the p-PI3K/p-Akt/p-GSK3β pathway, preventing tau hyperphosphorylation and amyloid cascades.	[128]
Liposome NPs	Cholesterol, soybean phosphatidylcholine-functionalized with surface wheat germ agglutinin (WGA) and cardiolipin (CL)	Curcumin and nerve growth factor	Human neuroblastoma cell line/AD rat model.	BBB transcytosis, amyloid cascades, and tau hyperphosphorylation	The phosphorylation of tau, JNK, and p38 proteins in Aβ-damaged neurons was reduced by curcumin-CL NPs. AD rat hippocampal Aβ plaque formation and AChE activity were significantly decreased by WGA-curcumin-CL NPs.	[129]

TABLE 8.2 *(Continued)*

Nanodrug-Delivery Systems	Carrier Material	API	Experiment Model	Key Targets	Description	References
Mesoporous silica NPs (MSN)	N-cetyltrimethylammonium bromide, tetraethoxysilane–functionalized with gold nanoparticle	Metal chelator CQ (5-chloro-4-hydroxy-7-iodoquinoline)	PC12 rat adrenal medulla cells/ endothelial cell line.	BBB transcytosis, amyloid cascades	MSN-AuNPs have been found to successfully penetrate an artificial model of BBB in laboratory settings. Additionally, MSN-AuNPs loaded with the metal chelator CQ have been shown to greatly prevent the aggregation of Aβ 40 produced by Cu^{2+}.	[130]
Metallic NPs	Iron crystal structure; functionalized with PEG	Iron oxide	Amyloid fibrillation experiments–*in vitro*	Amyloid cascades	Aβ fibrillation is accelerated when a higher concentration of NPs is subjected to a magnetic field, whereas it is inhibited when the concentration is minimal. Notably, uncharged or negatively charged NPs inhibit fibrillation more effectively.	[131]
Polymeric NPs	Chitosan	Piperine	Colchicine-induced AD rats	Cholinergic and OS systems	Piperine NPs, in contrast to piperine in its free form, significantly reduced behavioral impairment in an AD rat model via reducing OS and AChE levels.	[132]
Polymeric NPs	N-isopropyl acrylamide and N-t-butyl acrylamide – monomers; acrylic acid	Epigallocatechin-3-gallate (EGCG)	Amyloid fibrillation experiments–*in vitro* and neuroblastoma SH-SY5Y cell line viability.	Amyloid cascades	Aβ (Aβ 42 and Aβ 40) fibrillation was simultaneously inhibited by EGCG-loaded negative-charged polymeric NPs. Nonetheless, additional animal research is necessary to validate the same.	[133]
Solid lipid NPs (SLNs)	Lipids and chitosan	RVG-9R/*BACE1* siRNA	Human epithelial adenocarcinoma (Caco-2) cells.	Amyloid cascades	The SLNs formulation improves system's charge and mucoadhesiveness, as well as its drug permeability to a significant degree. Nevertheless, additional verification via *in vivo* experiments is critical.	[134]

The fundamental core structures of NP-functionalized nanomedicines enable the encapsulation of pharmaceuticals and prolonged circulation in bloodstream [140, 141]. Nanomedicines possess the ability to specifically target cells or even specific compartments inside cells, such as Aβ. As a result, they can deliver drugs precisely to the intended diseased site at a predetermined dosage. Nanomedicines can reduce both the amount and frequency of doses, hence enhancing patient adherence. Nanomedicines provide several benefits compared to conventional approaches for treating AD, including brain compatibility, extended half-life, increased bioavailability, regulated release, stability, biocompatibility, biodegradability, and protection against degradation [142].

8.8.1 NANOTHERAPEUTICS POTENTIAL IN MANAGING AD

Diagnostic and therapeutic applications of nanostructure-based delivery systems are expanding the nanomaterial's potential in managing AD pathologies. Functionalized NPs are being used to deliver therapeutic moieties into the brain [143].

1. **Organic Nanostructures**

   *i. **Polymeric-based NPs***

   The efficacy of PEGylated NPs, which are biodegradable and functionalized with antibodies, has been found in transgenic AD mice. These NPs have been shown to effectively address memory impairments and decrease the levels of Aβ soluble peptides, which are related to the onset of AD [143]. Memantine-loaded NPs and vitamin D-binding protein have demonstrated efficacy in reducing Aβ plaques and inflammation in AD, and improving symptoms in transgenic AD mouse models. The results indicated that these compounds may have a therapeutic opportunity for treating AD [144, 145]. SpBMP-9 may be able to treat AD since it inhibits GSK3β and contributes to the formation of cholinergic neurons [132, 146]. Furthermore, the administration of perlerin, a medication targeting AD, has demonstrated notable efficacy in AD Wistar rats. The use of CS-NPs (chitosan NPs) has resulted in increased cognitive abilities without any harmful effects on the brain [147]. CoQ is efficiently delivered to the brains of transgenic mice by trimethyl chitosan-PLGA NPs, preventing AD pathogenesis [148]. Curcumin is a phytoconstituent that functions as an antioxidant, effectively neutralizing harmful free radicals [149–151]. It is derived from natural sources and has a low toxicity level. The administration of curcumin to the brain has been identified as a highly appealing method for treating AD. Curcumin binds to amyloid proteins based on tau protein and exhibits anti-amyloid effects in small concentration (μm) range [152–154]. Inadequate stability and bioavailability issues restrict curcumin from reaching the brain, which lowers the amount of the substance that the brain absorbs [155]. To solve these issues, employing nanocarriers is favored because of their safety and ability to provide increased and sustained brain exposure [156].

ii. ***Nanomicellar***

An aqueous nanomicellar formulation of CoQ (ubisolQ10) is infused into the drinking water of the double transgenic AD mice. Findings show that it lowers the quantity of circulating Aβ plaques and improves long-term memory [157]. It was reported that PEG ceramide nanomicelles speed up the breakdown of tau proteins and promote autophagy in particular cells [158]. A recent study found that curcumin-loaded polymeric nanomicelles substantially suppressed amyloid plaque development in AD mice [159].

iii. ***Dendrimers***

Dendrimers show great potential as an AD treatment [160]. A novel finding involves the transport of memantine to the brain of AD mice in combination with lactoferrin and low-generation dendrimers. A study discovered that the influence on the memory of target mice was statistically significant [161]. Dendrimers with an ethylenediamine core are widely used to increase bioavailability and solubility. This leads to better BBB penetration and subsequent targeting of damaged brain regions in CNS diseases, such as Parkinson's disease and AD [162]. G4HisMal, a dendrimer that has been successfully created with a poly (propylene imine) core and a maltose histidine shell, may significantly reduce the symptoms of AD, including memory impairment [163].

iv. Nanogels

Nanogels show great promise as drug delivery vehicles because of their capacity to bind drugs, macromolecules, and active compounds for a variety of illnesses, including AD. A successful therapy for deferoxamine may be possible with nanogels given via chitosan and tripolyphosphate, according to a recent study [164]. A preclinical study in mice revealed that nanogels could enhance nose-to-brain transfer, suggesting insulin as a possible treatment for AD [165].

2. Lipid-based NPs for AD Therapy

Research highlights the significance of lipid-based nanocarriers as a delivery strategy for drugs to treat conditions affecting the CNS, such AD and dementia, especially when administered through the nose [166].

i. Solid lipid NPs

By delivering a-bisabolol in the form of solid lipid NPs, which show strong inhibitory effects on amyloid aggregation is beneficial for treating AD [167]. An innovative method that makes use of transferrin-functionalized nanostructured lipid carriers can be used to target brain endothelial cells to express p-glycoprotein and breast cancer resistance protein transporters [168]. Using the solvent emulsification diffusion method, the study found that donepezil-loaded solid lipid NPs were more effective than other formulations in improving medication administration through intranasal pathways to the brain [169]. To alleviate memory loss and other symptoms of AD, researchers used lipid carriers containing nanostructures of curcumin to reduce OS parameters in the brain. Similarly, another study found that by increasing the curcumin's bioavailability in the brain, both solid lipid NPs loaded with curcumin and nanostructure lipid carriers significantly reduced the neurodegenerative consequences of AD [170].

ii. *Liposomes*

Drug delivery to brain tissues has made full use of the self-assembling and amphiphilic characteristics of the nanosized vesicular liposomes [171]. It is revealed that liposomes coated with polyethylene glycol (PEG) can avoid being opsonized by RES. Furthermore, there is evidence that liposomes conjugated with glutathione and PEG can improve medication uptake all across the endothelium of BBB [172]. Functionalized liposomes have the potential to transport high concentrations of genes to specific tissues during AD treatment, and the results show that they are both safe and compatible [173]. Osthole (Ost) is thought of as a chemical that fights AD because of its anti-Aβ characteristics and its protective effects on neurons in the hippocampus. The aim was to create an Ost-liposomes carrying system that could increase bioavailability and exposure to targeted areas in AD mouse brains [174].

iii. *Niosomes*

A niosome-lipid nanocarrier system loaded with artemisia absinthium was shown to have significant effects on AD pathology when used to inhibit amyloid aggregation. This indicates that by inhibiting the production of amyloid, the formulation may be utilized to treat AD. Niosomes coated with pentamide-loaded chitosan glutamate were created to enhance the intranasal route of drug administration to AD patients. According to research, pentamide has much greater BBB efficacy. A niosome formulation prepared by the film hydration process contains cholesterol and sorbitan esters. The formulation shows notable increases in drug efficacy to target brain tissues [175].

iv. *Nanoemulsion*

According to Nirale et al. (2020), anti-AD medications can be made more effective and targeted by using nanoemulsion formulations. Intranasal administration of loaded memantine via a nanoemulsion that has been homogenized and ultrasonicated has allowed AD treatment to circumvent the BBB [176]. Naringerin nanoemulsion is being developed for improved therapeutic use and efficacy. The study suggests that naringerin nanoemulsion could be an effective method to combat Aβ neurotoxicity and amyloidogenesis [177].

v. ***Cubosomes***

Cubosomes are another kind of lipid-based NPs that show promise in biology as a means of delivering medications into the brain. Evidence from the administration of donepezil-HCL via cubosomal mucoadhesive *in situ* nasal gel, developed and could make it a desirable method of delivering medicine to the affected brain areas [178].

vi. ***Metallic NPs***

One potential avenue for site-targeted administration of drug in the treatment of AD is metallic NPs. Despite their limits caused by chemistry-based approaches, certain NPs possess notable anti-AD capabilities. These NPs include cerium, selenium, gold, and iron [179].

vii. ***Selenium NPs***

Some of the most important micronutrients for cellular health are the trace minerals selenium (II), sodium selenite (VI), and sodium selenite (IV), all of which serve as antioxidants. With their ability to reduce OS and decrease cell cytotoxicity, they show considerable potential in treating neurological disorders like Alzheimer's through nanoformulation. Researchers have discovered that altered selenium NPs with sialic acid can penetrate BBB, and when exposed to them, can hinder the process of Aβ aggregation [180].

viii. ***Cerium NPs***

As per reported studies, cerium oxide NPs are found to be highly effective as they protected essential brain function in an AD patient from elevated ROS levels and showed no side effects. Ceria therapy for AD is effective because it increases absorption over the BBB and prevents undesired buildup at other biological locations [181].

ix. ***Gold NPs***

Drug delivery across BBB is crucial for treating neurodegenerative diseases, and gold NPs (AuNPs) play a significant role to this process. Neurotoxic Aβ 42 aggregation can be strongly inhibited by D-glutathione-stabilized AuNPs when these particles are administered to AD mouse models. Therapy with anthocyanin-loaded PEG-AuNPs in βamyloid mouse model of AD has the potential to prevent age-related neurodegenerative illness [182]. A recent study discovered that employing a new tetrapeptide as a neuroprotective medication to ward against AD is possible since it enhances central cholinergic system function and reduces AChE levels [181].

x. ***Iron NPs***

The combination of iron oxide NPs and near-infrared fluorescent dye based on phenothiazine can yield novel theranostic drugs for AD. This enables NIR fluorescence of Aβ plaques and magnetic resonance imaging [182].

8.9 FUTURE PROSPECTS AND CONCLUSION

Research in the domains of molecular, cellular, and neurology has revealed a lot of crucial information regarding the causes and mechanisms of AD.

OS, in addition to Aβ and p-tau toxicity, contributes significantly to the progression of AD symptoms and death. The present FDA-approved medicines are not effective and have substantial adverse effects, demanding fresh approaches to combating AD pathogenesis. A multi-targeted therapy approach is necessary due to the immense complexity of AD etiology. Current studies on the application of antioxidants that exist naturally, like polyphenolic chemicals sourced from plants, offer a desirable alternative to synthetic substances because of their harmless nature, antioxidant activity, and capacity to readily cross the BBB to improve cognitive function in *in vivo* research. Effective treatment for neurodegenerative illnesses remains a clinically unmet demand. A great deal of research validates the theory that OS is important for the genesis of disease; consequently, antioxidant therapies may provide an effective treatment. However, numerous obstacles, such as insufficient dose, lack of bioavailability, limited retention and transport to the CNS, as well as insufficient antioxidant capacity to totally counteract the negative effects of ROS, may have hampered their clinical application. In this regard, drug delivery systems based on NPs may be able to assist in resolving some of the previously listed issues.

Antioxidant enzymes have great catalytic activity, which makes them promising. As a result, a lot of research has been done recently to create a nanotherapy-based method of delivering antioxidant enzymes. A robust antioxidant system may have several therapeutic implications because OS is a prevalent pathophysiological event in a range of illnesses. According to a recent study, using phyto- and nanomedicines to treat AD may be advantageous. Many phytochemicals have demonstrated potential for treating AD, especially with customized nanomedicines becoming more and more successful, a way to increase efficacy several times by efficiently delivering active components to the target regions. Biogenic production of metal NPs for AD treatment is an unexplored area of study at the nexus of nanotechnology and medicinal plants.

KEYWORDS

- **Alzheimer's disease**
- **ATP synthesis**
- **mitochondrial dysfunction**
- **neurodegeneration**
- **oxidative stress**
- **phytoconstituents**
- **synaptic dysfunction**

REFERENCES

1. Scheltens, P., De Strooper, B., Kivipelto, M., Holstege, H., Chételat, G., Teunissen, C., Cummings, J., & van der Flier, W. M. (2021). Alzheimer's disease. *Lancet, 397*, 1577–1590.
2. Stöhr, J., Watts, J. C., Mensinger, Z. L., Oehler, A., Grillo, S. K., DeArmond, S. J., Prusiner S. B., & Giles K. (2012). Purified and synthetic Alzheimer's amyloid beta (Aβ) prions. *Proceedings of the National Academy of Sciences USA, 109*, 11025–11030.
3. Yu, H., & Wu, J. (2021). Amyloid-β: A double agent in Alzheimer's disease? *Biomedical and Pharmacology, 139*, 111575.

4. Mendez, M. F. (2017). Early-Onset Alzheimer Disease. *Neurologic Clinics,35*, 263–281.
5. Wälti, M. A., Ravotti, F., Arai, H., Glabe, C. G., Wall, J. S., Böckmann, A., Güntert, P., Meier, B. H., & Riek, R. (2016). Atomic-resolution structure of a disease-relevant Aβ(1–42) amyloid fibril. *Proceedings of the National Academy of Sciences USA, 113*, E4976–E4984.
6. Nguyen, K. V. (2015). The human β-amyloid precursor protein: Biomolecular and epigenetic aspects. *Biomolecular Concepts, 6*, 11–32.
7. Wilkins, H. M., & Swerdlow, R. H. (2017). Amyloid precursor protein processing and bioenergetics. *Brain Research Bulletin,133*, 71–79.
8. Harilal, S., Jose, J., Parambi, D. G. T., Kumar, R., Mathew, G. E., Uddin, M. S., Kim, H., & Mathew, B. (2019). Advancements in nanotherapeutics for Alzheimer's disease: Current perspectives. *Journal of Pharmacy and Pharmacology, 71*(9), 1370–1383.
9. Song, N., Sun, S., Chen, K., Wang, Y., Wang, H., Meng, J., Guo, M., Zhang, X. D., & Zhang, R. (2023). Emerging nanotechnology for Alzheimer's disease: From detection to treatment. *Journal of Controlled Release, 360*, 392–417.
10. Mohapatra, P., Gopikrishnan, M., Doss, C., & Chandrasekaran, N. (2024). How precise are nanomedicines in overcoming the blood–brain barrier? A comprehensive review of the literature. *International Journal of Nanomedicine, 19*, 2441–2467.
11. Li, J., O, W., Li, W., Jiang, Z. G., & Ghanbari, H. A. (2013). Oxidative stress and neurodegenerative disorders *International Journal of Molecular Science, 14*, 24438–24475.
12. Christen, Y. (2000). Oxidative stress and Alzheimer disease. *American Journal of Clinical Nutrition, 71*, 621S–629S
13. Chen, X., Guo, C., & Kong, J. (2012). Oxidative stress in neurodegenerative diseases. *Neural Regeneration Research, 7*(5), 376–385.
14. Hawking, Z. L. (2016). Alzheimer's disease: The role of mitochondrial dysfunction and potential new therapies. *Bioscience Horizones, 9*, 1–11.
15. Al-Abd, N. M., Mohamed Nor, Z., & Mansor, M. (2015). Antioxidant, antibacterial activity, and phytochemical characterization of *Melaleuca cajuputi* extract. *BMC Complementary & Alternative Medicine, 15*, 385.
16. Tamagno, E., Robino, G., Obbili, A., Bardini, P., Aragno, M., Parola, M., & Danni, O. (2003). H2O2 and 4-hydroxynonenal mediate amyloid beta-induced neuronal apoptosis by activating JNKs and p38MAPK. *Experimental Neurology, 180*(2), 144–155.
17. Cheignon, C., Tomas, M., Bonnefont-Rousselot, D., Faller, P., Hureau, C., & Collin, F. (2018). Oxidative stress and the amyloid beta peptide in Alzheimer's disease. *Redox Biology, 14*, 450–464.
18. Cristóvão, J. S., Santos, R., & Gomes, C. M. (2016). Metals and neuronal metal binding proteins implicated in alzheimer's disease. *Oxidative Medicine and Cellullar Longevity, 2016*, 1–11.
19. Islam, B. U., Jabir, N. R., & Tabrez, S. (2019). The role of mitochondrial defects and oxidative stress in Alzheimer's disease. *Journal of Drug Targeting, 27*(9), 932–942.

20. Velagapudi, R., El-Bakoush, A., & Olajide, O. A. (2018). Activation of nrf2 pathway contributes to neuroprotection by the dietary flavonoid tiliroside. *Molecular Neurobiology*, *55*, 8103–8123.
21. Irshad, M., & Chaudhuri, P. S. (2002) Oxidant-antioxidant system: Role and significance in human body. *Indian Journal of Experimental Biology*, *40*(11), 1233–1239.
22. Burgener, S. C., Buettner, L., Buckwalter, K. C., Beattie, E., Bossen, A. L., Fick, D. M., Schreiner, A. (2008). Evidence supporting nutritional interventions for persons in early stage Alzheimer's disease (AD). *Journal of Nutrition Health and Aging*, *12*, 18–21.
23. Desai, P., Shete, H., Adnaik, R., Disouza, J., & Patravale, V. (2015). Therapeutic targets and delivery challenges for Alzheimer's disease. *World Journal of Pharmacology*, *4*(3), 236–264
24. Banci, L., Benedetto, M., Bertini, I., Del Conte, R., Piccioli, M., & Viezzoli, M. S. (1998). Solution structure of reduced monomeric Q133M2 copper, zinc superoxide dismutase (SOD). Why is SOD a dimeric enzyme? *Biochemistry*, *37*, 11780–11791.
25. Guan, Y., Hickey, M. J., Borgstahl, G. E., Hallewell, R. A., Lepock, J. R., O'Connor, D., Hsieh, Y., Nick, H. S., Silverman, D. N., & Tainer, J. A. (1998). Crystal structure of Y34F mutant human mitochondrial manganese superoxide dismutase and the functional role of tyrosine 34. *Biochemistry*, *37*, 4722–4730.
26. Adachi, T., & Wang, X. L. (1998). Association of extracellular-superoxide dismutase phenotype with the endothelial constitutive nitric oxide synthase polymorphism. *FEBS Letters*, *433*, 166–168.
27. Terlecky, S. R., Koepke, J. I., & Walton, P. A. (2006). Peroxisomes and aging. *Biochimica et Biophysica Acta*, *1763*, 1749–1754.
28. Kemp, M., Go, Y. M., & Jones, D. P. (2008). Nonequilibrium thermodynamics of thiol/disulfide redox systems: A perspective on redox systems biology. *Free Radical Biology and Medicine*, *44*, 921–937.
29. Burke-Gaffney, A., Callister, M. E., & Nakamura, H. (2005). Thioredoxin: Friend or foe in human disease? *Trends in Pharmacology Science*, *26*, 398–404.
30. Hardy, G., Hardy, I., & Manzanares, W. (2012). Selenium supplementation in the critically ill. *Nutrition in Clinical Practice*, *27*, 21–33.
31. Hargreaves, I. P. (2003). Ubiquinone: Cholesterol's reclusive cousin. *Annals of Clinical Biochemistry*, *40*, 207–218.
32. Farbood, Y., Sarkaki, A., Mahdavinia, M., Ghadiri, A., Teimoori, A. Seif, F., Dehghani, M. A., & Navabi, S. P. (2020). Protective effects of co-administration of zinc and selenium against streptozotocin-induced Alzheimer's disease: Behavioral, mitochondrial oxidative stress, and GPR39 expression alterations in rats. *Neurotoxicity Research*, *38*, 398–407.
33. Zhang, D., & Hannink, D. D. (2003). Distinct Cysteine residues in keap1 are required for keap1-dependent ubiquitination of Nrf2 and for stabilization of Nrf2 by chemopreventive agents and oxidative stress. *Molecular and Cellular Biology*, *23*, 8137–8151.
34. Suárez, S., Mu, T., Sun, H., & Añón, M. C. (2020). Antioxidant activity, nutritional, and phenolic composition of sweet potato leaves as affected by harvesting period. *International Journal of Food Properties*, *23*, 178–188.

35. Lopa, S. S., Al-Amin, M., Hasan, M., Ahammed, M., Islam, K. M., Alam, A. H. M., Tanaka, T., & Sadik, M. (2021). Phytochemical analysis and cholinesterase inhibitory and antioxidant activities of Enhydra fluctuans relevant in the management of Alzheimer's disease. *International Journal of Food Science, 2021*, 8862025.
36. Ouknin, M., Aghraz, A., Chibane, M., Boumezzourh, A., Costa, J., & Majidi, L. (2021). Enzyme inhibitory, antioxidant activity and phytochemical analysis of essential oil from cultivated Rosmarinus officinalis. *Journal of Food Measurement and Characterization, 15*, 3782–3790.
37. Mielech, A., Pu´scion-Jakubik, A., Markiewicz-Zukowska, R., & Socha, K. (2020). Vitamins in Alzheimer's disease—Review of the latest ˙ reports. *Nutritients, 12*, 3458
38. Gavazza, M., Marmunti, M., Leaden, P., Zeinsteger, P., & Palacios, A. (2021). Alpha-lipoic acid: Antioxidant activity against non-enzymatic peroxidation of rat kidney and liver mitochondria. *Journal of Clinical Biomedical Investigation, 1*, 1–4.
39. Wear, D., Vegh, C., Sandhu, J. K., Sikorska, M., Cohen, J., & Pandey, S. (2021). Ubisol-Q10, a nanomicellar and water-dispersible formulation of coenzyme-Q10 as a potential treatment for Alzheimer's and Parkinson's disease. *Antioxidants, 10*, 764.
40. Londzin, P., Zamora, M., K ̨akol, B., Taborek, A., & Folwarczna, J. (2021). Potential of caffeine in Alzheimer's disease—A review of experimental studies. *Nutritients, 13*, 537.
41. Shabbir, U., Rubab, M., Tyagi, A., & Oh, D.-H. (2021). Curcumin and its derivatives as theranostic agents in Alzheimer's disease: The implication of nanotechnology. *International Journal of Molecular Sciences, 22*, 196.
42. Tarabasz, D., & Kukula-Koch, W. (2020). Palmatine: A review of pharmacological properties and pharmacokinetics. *Phytherapy Research, 34*, 33–50.
43. Qajari, N. M., & Khonakdar-Tarsi, A. (2021). Short communication silibinin and neurological diseases. *Journal of Neurobiology & Physiology, 3*, 8–9.
44. Singh, S. K., Srivastav, S., Castellani, R. J., Plascencia-Villa, G., & Perry, G. (2019). Neuroprotective and antioxidant effect of Ginkgo biloba extract against AD and other neurological disorders. *Neurotherapeuthics, 16*, 666–674.
45. Hossain, M. F., Wang, N., Chen, R., Li, S., Roy, J., Uddin, M. G., Li, Z., Lim, L. W., & Song, Y. Q. (2021). Exploring the multifunctional role of melatonin in regulating autophagy and sleep to mitigate Alzheimer's disease neuropathology. *Ageing Research Reviews, 67*, 101304.
46. Tamagno, E., & Guglielmotto, M. (2022). Estrogens still represent an attractive therapeutic approach for Alzheimer's disease. *Neural Regeneration Research, 17*, 93.
47. Cai, M., & Yang, E. J. (2020). Effect of combined electroacupuncture and selegiline treatment in Alzheimer's disease: An animal model. *Frontiers in Pharmacology, 11*, 2048.
48. Cano, A., Ettcheto, M., Chang, J. H., Barroso, E., Espina, M., Kühne, B. A., Barenys, M., Auladell, C., Folch, J., & Souto, E. B. (2019). Dual-drug loaded nanoparticles of epigallocatechin-3-gallate (EGCG)/Ascorbic acid enhance therapeutic efficacy of EGCG in a APPswe/PS1dE9 Alzheimer's disease mice model. *Journal of Controlled Release, 301*, 62–75.
49. Gomes, B. A. Q., Silva, J. P., Romeiro, C. F., Dos Santos, S. M., Rodrigues, C. A., Goncalves, P. R., Sakai, J. T., Mendes, P. F., Varela, E. L., & Monteiro, M. C. (2018).

Neuroprotective mechanisms of resveratrol in Alzheimer's disease: Role of SIRT1. *Oxidative Medicine and Cell Longevity, 2018*, 8152373.
50. Huang, X., Cuajungco, M. P., Atwood, C. S., Hartshorn, M. A., Tyndall, J. D., Hanson, G. R., Stokes, K. C., Leopold, M., Multhaup, G., Goldstein, L. E., Scarpa, R. C., & Saunders, A. J. (1999). Cu(II) potentiation of Alzheimer abeta neurotoxicity. Correlation with cell-free hydrogen peroxide production and metal reduction. *Journal of Biological Chemistry, 274*, 37111–37116.
51. Barnham, K. J., Haeffner, F., Ciccotosto, G. D., Curtain, C. C., Tew, D., Mavros, C., Beyreuther, K., Carrington, D., Masters, C. L., Cherny, R. A., Cappai, R., & Bush, A. I. (2004). Tyrosine gated electron transfer is key to the toxic mechanism of Alzheimer's disease beta-amyloid. *FASEB Journal, 18*, 1427–1429
52. Crouch, P. J., Barnham, K. J., Duce, J. A., Blake, R. E., Masters, C. L., & Trounce, I. A. (2006). Copper-dependent inhibition of cytochrome c oxidase by Abeta(1-42) requires reduced methionine at residue 35 of the Abeta peptide. *Journal of Neurochemistry, 99*, 226–236.
53. Anantharaman, M., Tangpong, J., Keller, J. N., Murphy, M. P., Markesbery, W. R., Kiningham, K. K., & St Clair, D. K. (2016). Beta-amyloid mediated nitration of manganese superoxide dismutase: Implication for OS in a APPNLH/NLH X PS-1P264L/P264L double knock-in mouse model of Alzheimer's disease. *American Journal of Pathology, 168*, 1608–1618.
54. Ganguly, U., Kaur, U., Chakrabarti, S. S., Sharma, P., Agrawal, B. K., Saso, L., & Chakrabarti, S. (2021). OS, neuroinflammation, and NADPH Oxidase: Implications in the pathogenesis and treatment of Alzheimer's disease. *Oxidative Medicine and Cellular Longevity, 2021*, 7086512.
55. Frost, G. R., Jonas, L. A., & Li, Y. M. (2019). Friend, foe or both? Immune activity in Alzheimer's disease. *Frontiers in Aging Neuroscience, 11*, 337.
56. Leng, F., & Edison, P. (2021). Neuroinflammation and microglial activation in Alzheimer disease: Where do we go from here? *Nature Reviews Neurology, 17*, 157–172.
57. Cribbs, D. H., Berchtold, N. C., Perreau, V., Coleman, P. D., Rogers, J., Tenner A. J., & Cotman, C. W. (2012). Extensive innate immune gene activation accompanies brain aging, increasing vulnerability to cognitive decline and neurodegeneration: A microarray study. *Journal of Neuroinflammation, 9*, 179.
58. Xiang, Z., Haroutunian, V., Ho, L., Purohit, D., & Pasinetti, G. M. (2006). Microglia activation in the brain as inflammatory biomarker of Alzheimer's disease neuropathology and clinical dementia. *Disease Markers, 22*, 95–102.
59. Hanslik, K. L., & Ulland, T. K. (2020). The role of microglia and the Nlrp3 inflammasome in Alzheimer's disease. *Frontiers in Neurology, 11*, 570711.
60. Jha, N. K., Jha, S. K., Kar, R., Nand, P., Swati, K., & Goswami, V. K. (2019). Nuclear factor-kappa β as a therapeutic target for Alzheimer's disease. *Journal of Neurochemistry, 150*, 113–137.
61. Mattson, M. P. (2004). Pathways towards and away from Alzheimer's disease. *Nature, 430*, 631–639.
62. Persson, T., Popescu, B. O., & Cedazo-Minguez, A. (2014). OS in Alzheimer's disease: Why did antioxidant therapy fail? *Oxidative Medicine and Cellular Longevity*, 427318.

63. Atamna, H., & Boyle, K. (2006). Amyloid-β peptide binds with heme to form a peroxidase: Relationship to the cytopathologies of Alzheimer's disease. *Proceedings of the National Academy of Sciences USA, 103*, 3381–3386.
64. Thiabaud, G., Pizzocaro, S., Garcia-Serres, R., Latour, J. M., Monzani, E., & Casella, L. (2013). Heme binding induces dimerization and nitration of truncated β-amyloid peptide Aβ16 under OS. *Angewandte Chemie International Edition, 52*, 8041–8044.
65. Saeedi, M., & Rashidy-Pour, A. (2021). Association between chronic stress and Alzheimer's disease: Therapeutic effects of Saffron. *Biomedicine & Pharmacotherapy, 133*, 110995.
66. Zhang, J. S., Zhou, S. F., Wang, Q., Guo, J. N., Liang, H. M., Deng, J. B., & He, W. Y. (2016). Gastrodin suppresses BACE1 expression under OS condition via inhibition of the PKR/eIF2α pathway in Alzheimer's disease. *Neuroscience, 25*, 1–9.
67. Hernández-Rodríguez, M., Arciniega-Martínez, I. M., García-Marín, I. D., Correa-Basurto, J., & Rosales-Hernández, M. C. (2020). Chronic administration of scopolamine increased GSK3βP9, Beta Secretase, Amyloid Beta, and OS in the Hippocampus of Wistar Rats. *Molecular Neurobiology, 57*, 3979–3988.
68. Jo, D. G., Arumugam, T. V., Woo, H. N., Park, J. S., Tang, S. C., Mughal, M., Hyun, D. H., Park, J.-H., Choi, Y. H., & Gwon, A. R. (2010). Evidence that gamma-secretase mediates OS-induced beta-secretase expression in Alzheimer's disease. *Neurobiology Aging, 31*, 917–925.
69. Giliberto, L., Borghi, R., Piccini, A., Mangerini, R., Sorbi, S., Cirmena, G., Garuti, A., Ghetti, B., Tagliavini, F., & Mughal, M. R. (2009). Mutant presenilin 1 increases the expression and activity of BACE1. *Journal of Biological Chemistry, 284*, 9027–9038.
70. Bhatia, V., & Sharma, S. (2021). Role of mitochondrial dysfunction, OS and autophagy in progression of Alzheimer's disease. *Journal of the Neurological Sciece, 421*, 117253.
71. Icone, P., Nuzzo, D., Caruana, L., & Scafidi, Di Carlo M. (2014). Mitochondrial dysfunction: Different routes to Alzheimer's disease therapy. *Oxidative Medicine and Cellular Longevity, 2014*, 780179.
72. Bahrami A., Delshadi R., Assadpour E., Jafari S. M., & Williams L. (2020). Antimicrobial-loaded nanocarriers for food packaging applications. *Advances in Colloid and Interface Science, 278*, 102140.
73. Loizzo, M. R., Menichini, F., Tundis, R., Bonesi, M., Conforti, F., Nadjafi, F., Statti, G. A., Frega, N. G., & Menichini, F. (2009). In vitro biological activity of Salvia Leriifolia benth essential oil relevant to the treatment of Alzheimer's disease. *Journal of Oleo Science, 58*(8), 443–446.
74. Bourre, J. M. (2004). Roles of unsaturated fatty acids (Especially Omega-3 Fatty Acids) in the brain at various ages and during ageing. *Journal of Nutrition Health and Aging, 8*(3), 163–174.
75. Bazan, N. G., Musto, A. E., & Knott, E. J. (2011). Endogenous signaling by Omega-3 docosahexaenoic acid-derived mediators sustains homeostatic synaptic and circuitry integrity. *Molecular Neurobiology, 44*(2), 216–222.
76. Bahadori M. B., & Kirkan B. (2019). Sarikurkcu C. Phenolic ingredients and therapeutic potential of Stachys Cretica Subsp. Smyrnaea for the management of

OS, Alzheimer's Disease, Hyperglycemia, and Melasma. *Industrial Crops and Production, 127*, 82–87.

77. Abdulqadir A., Cakmak Y. S., & Zengin G. (2018). Phenolic compounds, antioxidant properties and enzyme inhibition ability of Adiantum Capillus Veneris L. linked to Alzheimer's Disease, Diabetes Mellitus and Skin Disorders. *Current Organic Chemistry, 22*(17), 1697–1703.
78. Kiko, T., Nakagawa, K., Tsuduki, T., Suzuki, T., Arai, H., & Miyazawa, T. (2012). Significance of lutein in red blood cells of Alzheimer's disease patients. *JAD, 28*(3), 593–600.
79. Liu, C. B., Wang, R., Yi, Y. F., Gao, Z., & Chen, Y. Z. (2018). Lycopene Mitigates β-amyloid induced inflammatory response and inhibits NF-KB signaling at the choroid plexus in early stages of Alzheimer's disease rats. *Journal of Nutritional Biochemistry, 53*, 66–71.
80. Satou, T., Hanashima, Y., Mizutani, I., & Koike, K. (2018). The effect of inhalation of essential oil from Rosmarinus officinalis on Scopolamine-induced Alzheimer's type dementia model mice. *Flavour and Fragrance Journal, 33*(3), 230–234.
81. Postu, P. A., Sadiki, F. Z., El Idrissi, M., Cioanca, O., Trifan, A., Hancianu, M., & Hritcu, L. (2019). Pinus Halepensis essential oil attenuates the toxic Alzheimer's Amyloid Beta (1–42)-induced memory impairment and OS in the rat Hippocampus. *Biomedicine & Pharmacotherapy, 112*, 108673.
82. Dhillon, S. (2021). Aducanumab: First approval. *Drugs, 81*(12), 1437–1443.
83. Wagstaff, A. J., & McTavish, D. (1994). Tacrine: A review of its pharmacodynamic and pharmacokinetic properties, and therapeutic efficacy in Alzheimer's disease. *Drugs Aging, 4*, 510–540.
84. Almkvist, O., Jelic, V., Amberla, K., Hellström-Lindahl, E., Meurling, L., & Nordberg, A. (2010). Responder characteristics to a single oral dose of cholinesterase inhibitor: A double-blind placebo-controlled Study with Tacrine in Alzheimer Patients. *Dementia and Geriatric Cognitive Disorders, 12*, 22–32.
85. Gutzmann, H., Kühl, K.-P., Hadler, D., & Rapp, M. A. (2002). Safety and efficacy of Idebenone versus Tacrine in patients with Alzheimer's disease: Results of a randomized, double-blind, parallel-group multicenter study. *Pharmacopsychiatry, 35*, 12–18.
86. Raskind, M. A., Sadowsky, C. H., Sigmund, W. R., Beitler, P. J., & Auster, S. B. (1997). Effect of tacrine on language, praxis, and noncognitive behavioral problems in Alzheimer disease. *Archives of Neurology, 54*, 836–840.
87. Wong, W. J., Liu, H. C., Fuh, J. L., Wang, S. J., Hsu, L. C., Wang, P. N., & Sheng, W. Y. (1999). A double-blind, placebo-controlled study of tacrine in Chinese patients with Alzheimer's disease. *Dementia and Geriatric Cognitive Disorders, 10*, 289–294.
88. Jarrott, B. (2017). Tacrine: In vivo veritas. *Pharmacological Research, 116*, 29–31.
89. Lou, Y. H., Wang, J. S., Dong, G., Guo, P. P., Wei, D. D., Xie, S. S., Yang M. H., & Kong L. Y. (2015). The acute hepatotoxicity of tacrine explained by 1H NMR based metabolomic profiling. *Toxicology Research, 4*, 1465–1478.
90. Eagger, S. A., Levy, R., & Sahakian, B. J. (1991). Tacrine in Alzheimer's disease. *Lancet, 337*, 989–992.

91. Jacobson, S. A., & Sabbagh, M. N. (2008). Donepezil: Potential neuroprotective and disease-modifying effects. *Expert Opinion Drug Metabolism & Toxicology, 4*(10), 1363–1369.
92. Kumar, A., Gupta, V., & Sharma, S. (2022). *StatPearls.* StatPearls Publishing.
93. Homma, A., Imai, Y., Tago, H., Asada, T., Shigeta, M., Iwamoto, T., Takita, M., Arimoto, I., Koma, H., & Ohbayashi, T. (2008). Donepezil treatment of patients with severe Alzheimer's disease in a Japanese population: Results from a 24-week, double-blind, Placebo-Controlled, randomized trial. *Dementia and Geriatric Cognitive Disorders*, *25*, 399–407.
94. Howard, R. J., Juszczak, E., Ballard, C., Bentham, P., Brown, R., Bullock, R., Burns, A., Holmes, C., Jacoby, R., & Johnson, T. (2007). Donepezil for the treatment of agitation in Alzheimer's disease. *New England Journal of Medicine*, *357*, 1382–1392.
95. Jia, J., Wei, C., Jia, L., Tang, Y., Liang, J., Zhou, A., Li, F., Shi, L., & Doody, R. S. (2017). Efficacy and safety of Donepezil in Chinese patients with severe Alzheimer's disease: A randomized controlled trial. *Journal of Alzheimer's Disease*, *56*, 1495–1504.
96. Maher-Edwards, G., Dixon, R., Hunter, J., Gold, M., Hopton, G., Jacobs, G., Hunter, J., & Williams, P. (2011). SB-742457 and donepezil in Alzheimer disease: A randomized, placebo-controlled study. *International Journal of Geriatric Psychiatry*, *26*, 536–544.
97. Homma, A., Imai, Y., Tago, H., Asada, T., Shigeta, M,; Iwamoto, T., Takita, M., Arimoto, I., Koma, H., & Ohbayashi, T. (2008). Donepezil treatment of patients with severe Alzheimer's disease in a Japanese population: Results from a 24-week, double-blind, placebo-controlled, randomized trial. *Dementia and Geriatric Cognitive Disorders*, *25*, 399–407.
98. Hong, Y. J., Han, H. J., Youn, Y. C., Park, K. W., Yang, D. W., Kim, S., Kim, H. J., Kim, J. E., & Lee, J. H. (2019). Safety and tolerability of donepezil 23 mg with or without intermediate dose titration in patients with Alzheimer's disease taking donepezil 10 mg: A multicenter, randomized, open-label, parallel-design, three-arm, prospective trial. *Alzheimer's Research & Therapy*, *11*, 37.
99. Seltzer, B. (2007). Donepezil: An update. *Expert Opinion on Pharmacotherapy*, *8*, 1011–1023.
100. Farlow, M. R. (2003). Clinical pharmacokinetics of galantamine. *Clinical Pharmacokinetics*, *42*(15), 1383–1392.
101. Olin, J. T., & Schneider, L. S. (2002). Galantamine for dementia due to Alzheimer's disease. *Cochrane Database Systematic Reviews*, *3*, CD001747.
102. Tyagi, N., Goel, R., Mohd, J., Saini, R. K., & Singh, V. (2018). Effect of Galantamine with neuropharmacological benefits in Wistar Rats models of epilepsy and behaviour. *Journal of Pharmacology and Clinical Research*, *5*(3), 555664.
103. Hanafy, A. S., Farid, R. M., Helmy, M. W., & ElGamal, S. S. (2016). Pharmacological, toxicological and neuronal localization assessment of galantamine/chitosan complex nanoparticles in rats: Future potential contribution in Alzheimer's disease management. *Drug Delivery*, *23*, 3111–3122.
104. Misra, S., Chopra, K., Sinha, V. R., & Medhi, B. (2016). Galantamine-loaded solid–lipid nanoparticles for enhanced brain delivery: Preparation, characterization, in vitro and in vivo evaluations. *Drug Delivery*, *23*, 1434–1443.

105. Wahba, S. M., Darwish, A. S., & Kamal, S. M. (2016). Ceria-containing uncoated and coated hydroxyapatite-based galantamine nanocomposites for formidable treatment of Alzheimer's disease in ovariectomized albino-rat model. *Materials Science and Engineering C Materials for Biological Applications*, *65*, 151–163.
106. Kavanagh, S., Gaudig, M., Van Baelen, B., Adami, M., Delgado, A., Guzman, C., Jedenius, E., & Schäuble, B. (2011). Galantamine and behavior in Alzheimer disease: Analysis of four trials. *Acta Neurologica Scandinavica*, *124*, 302–308.
107. Tariot, P., Solomon, P., Morris, J., Kershaw, P., Lilienfeld, S., & Ding, C. (2000). The Galantamine USA-study group A 5-month, randomized, placebo-controlled trial of galantamine in AD. *Neurology*, *54*, 2269–2276.
108. Li, D, D., Zhang, Y. H., Zhang, W., & Zhao, P. (2019). Meta-analysis of randomized controlled trials on the efficacy and safety of Donepezil, Galantamine, Rivastigmine, and Memantine for the treatment of Alzheimer's disease. *Frontiers in Neuroscience*, *13*, 472.
109. Haake, A., Nguyen, K., Friedman, L., Chakkamparambil, B., & Grossberg, G. T. (2020). An update on the utility and safety of cholinesterase inhibitors for the treatment of Alzheimer's disease. *Expert Opinion Drug Safety*, *19*, 147–157.
110. Desai, P., Shete, H., Adnaik, R., Disouza, J., & Patravale, V. (2015). Therapeutic targets and delivery challenges for Alzheimer's disease. *World Journal of Pharmacology*, *4*(3), 236–264.
111 Birks, J. S., Chong, L. Y., & Evans, J. G. (2015). Rivastigmine for Alzheimer's disease. *Cochrane Database of Systematic Reviews*, *9*, CD001191.
112. Cummings, J., & Winblad, B. (2007). A rivastigmine patch for the treatment of Alzheimer's disease and Parkinson's disease dementia. *Expert Review of Neurotherapeutics*, *7*, 1457–1463.
113. Yunusa, I., Alsahali, S., Ranes, A., & Eguale, T. (2021). Comparative value of cholinesterase inhibitors and memantine in persons with moderate-to-severe Alzheimer's disease in the United States: A cost-effectiveness analysis. *Journal of Alzheimer's Disease Reports*, *5*, 705–713.
114. Feldman, H. H., & Lane, R. (2007). Rivastigmine: A placebo-controlled trial of twice daily and three times daily regimens in patients with Alzheimer's disease. *Journal of Neurology Neurosurgery and Psychiatry*, *78*, 1056–1063.
115. Mowla, A., Mosavinasab, M., Haghshenas, H., & Haghighi, A. B. (2007). Does Serotonin Augmentation have any effect on cognition and activities of daily living in Alzheimer's dementia? A double-blind, placebo-controlled clinical trial. *Journal of Clinical Psychopharmacology*, *27*, 484–487.
116. Winblad, B., Grossberg, G., Frölich, L., Farlow, M., Zechner, S., Nagel, J., & Lane, R. (2007). IDEAL: A 6-month, double-blind, placebo-controlled study of the first skin patch for Alzheimer disease. *Neurology*, *69*, S14–S22.
117. Mimica, N., & Presecki, P. (2009). Side effects of approved antidementives. *Psychiatria Danubina*, *21*, 108–113.
118. Lo, D., & Grossberg, G. T. (2011). Use of memantine for the treatment of dementia. *Expert Review of Neurotherapy*, *11*, 1359–1370.
119. Calhoun, A., King, C., Khoury, R., & Grossberg, G. T. (2018). An evaluation of memantine ER + donepezil for the treatment of Alzheimer's disease. *Expert Opinion Pharmacotherapy*, *19*, 1711–1717.

120. Guo, J., Wang, Z., Liu, R., Huang, Y., Zhang, N., & Zhang, R. (2020). Memantine, Donepezil, or combination therapy—What is the best therapy for Alzheimer's disease? A network meta-analysis. *Brain and Behavior, 10*, e01831.
121. Karthivashan, G., Ganesan, P., Park, S. Y., Kim, J. S., & Choi, D. K. (2018). Therapeutic strategies and nano-drug delivery applications in management of ageing Alzheimer's disease. *Drug Delivery, 25*(1), 307–320.
122. Bukhari, S. N. A. (2022). Nanotherapeutics for Alzheimer's disease with preclinical evaluation and clinical trials: Challenges, promises and limitations. *Current Drug Delivery, 19*(1), 17–31.
123. Khan, N. H., Mir, M., Ngowi, E. E., Zafar, U., Khakwani, M. M. A. K., Khattak, S., Zhai, Y. K., Jiang, E. S., Zheng, M., Duan, S. F., Wei, J. S., Wu, D. D., & Ji, X. Y. (2021). Nanomedicine: A promising way to manage Alzheimer's Disease. *Frontiers in Bioengeering and Biotechnology, 9*, 630055.
124. Puranik, N., Yadav, D., & Song, M. (2023). Advancements in the application of nanomedicine in Alzheimer's disease: A therapeutic perspective. *International Journal of Molecular Sciences, 24*(18), 14044.
125. Lohan, S., Raza, K., Mehta, S. K., Bhatti, G. K., Saini, S., & Singh, B. (2017). Anti-Alzheimer's potential of berberine using surface decorated multi-walled carbon nanotubes: A preclinical evidence. *International Journal of Pharmaceutics, 530*(1–2), 263–278.
126. Sorokina, S. A., Stroylova, Y. Y., Shifrina, Z. B., & Muronetz, V. I. (2016). Disruption of amyloid prion protein aggregates by Cationic Pyridylphenylene Dendrimers. *Macromolecular Bioscience, 16*(2), 266–275.
127. Ali, T., Kim, M. J., Rehman, S. U., Ahmad, A., & Kim, M. O. (2017). Anthocyanin-loaded PEG-Gold nanoparticles enhanced the neuroprotection of anthocyanins in an $A\beta_{1\text{-}42}$ Mouse model of Alzheimer's disease. *Molecular Neurobiology, 54*(8), 6490–6506.
128. Kuo, Y. C., Lin, C. Y., Li, J. S., & Lou, Y. I. (2017). Wheat germ agglutinin-conjugated liposomes incorporated with cardiolipin to improve neuronal survival in Alzheimer's disease treatment. *International Journal of Nanomedicine, 12*, 1757–1774.
129. Yang, L., Yin, T., Liu, Y., Sun, J., Zhou, Y., & Liu, J. (2016). Gold nanoparticle-capped mesoporous silica-based H_2O_2-responsive controlled release system for Alzheimer's disease treatment. *Acta Biomaterialia, 46*, 177–190.
130. Mirsadeghi, S., Shanehsazzadeh, S., Atyabi, F., & Dinarvand, R. (2016). Effect of PEGylated superparamagnetic iron oxide nanoparticles (SPIONs) under magnetic field on amyloid beta fibrillation process. *Materials Science and Engineering C Materials Biological Applications, 59*, 390–397.
131. Elnaggar, Y. S. R., Etman, S. M., Abdelmonsif, D. A., & Abdallah, O. Y. (2015). Intranasal Piperine-loaded Chitosan nanoparticles as brain-targeted therapy in Alzheimer's disease: Optimization, biological efficacy, and potential toxicity. *Journal of Pharmaceutical Sciences, 104*(10), 3544–3556.
132. Liu, H., Yu, L., Dong, X., & Sun, Y. (2017). Synergistic effects of negatively charged hydrophobic nanoparticles and (-)-epigallocatechin-3-gallate on inhibiting amyloid β-protein aggregation. *Journal of Colloid and Interface Science, 491*, 305–312.
133. Rassu, G., Soddu, E., Posadino, A. M., Pintus, G., Sarmento, B., Giunchedi, P., & Gavini, E. (2017). Nose-to-brain delivery of BACE1 siRNA loaded in solid lipid

nanoparticles for Alzheimer's therapy. *Colloids and Surfaces B Biointerfaces, 152,* 296–301.

134. Hernando, S., Gartziandia, O., Herran, E., Pedraz, J. L., Igartua, M., & Hernandez, R.M. (2016). Advances in nanomedicine for the treatment of Alzheimer's and Parkinson's diseases. *Nanomedicine, 11*(10), 1267–1285.
135. Ling, T. S., Chandrasegaran, S., & Xuan, L. Z. (2021). The potential benefits of nanotechnology in treating Alzheimer's disease. *Biomed Research International, 2021,* 1–9.
136. Spuch, C., Saida, O., & Navarro, C. (2012). Advances in the treatment of neurodegenerative disorders employing nanoparticles. *Recent Patents on Drug Delivery & Formulation, 6*(1), 2–18.
137. Fakhoury, M., Takechi, R., & Al-Salami, H. (2015). Drug permeation across the blood–brain barrier: Applications of nanotechnology. *British Journal of Medicine and Medical Research, i*(6), 547–556.
138. Leszek, J., Md, Ashraf, G., & Tse, W. H. (2017). Nanotechnology for Alzheimer disease. *Current Alzheimer Research, 14*(11), 1182–1189.
139. Kim, J. hoon., Hong, C. O., Koo, Y. chang., Choi, H. D., & Lee, K. W. (2012). Anti-glycation effect of gold nanoparticles on collagen. *Biological and Pharmaceutical Bulletin, 35*(2), 260–264.
140. Knop, K., Hoogenboom, R., Fischer, D., & Schubert, U.S. (2010). Poly(ethylene glycol) in drug delivery: Pros and cons as well as potential alternatives. *Angewandte Chemie International Edition, 49*(36), 6288–6308.
141. Altinoglu, G., & Adali, T., (2020). Alzheimer's disease targeted nano-based drug delivery systems. *Current Drug Targets, 21*(7), 628–646.
142. Carradori, D., Balducci, C., & Re, F. (2018). Antibody-functionalized polymer nanoparticle leading to memory recovery in Alzheimer's disease-like transgenic mouse model. *Nanomedicine, 14*(2), 609–618.
143. Sánchez-López, E., Ettcheto, M., & Egea, M. A. (2018). Memantine loaded PLGA PEGylated nanoparticles for Alzheimer's disease: In vitro and in vivo characterization. *Journal of Nanobiotechnology, 16*(1), 32.
144. Jeon, S. G., Cha, M. Y., & Kim, J. (2019). Vitamin D-binding protein-loaded PLGA nanoparticles suppress Alzheimer's disease-related pathology in 5XFAD mice. *Nanomedicine, 17,* 297–307.
145. Lauzon, M. A., Marcos, B., & Faucheux, N. (2018). Characterization of alginate/chitosan-based nanoparticles and mathematical modeling of their SpBMP-9 release inducing neuronal differentiation of human SH-SY5Y cells. *Carbohydrate Polymers, 181,* 801–811.
146. Hanafy, A. S., Farid, R. M., Helmy, M. W., & ElGamal, S. S. (2016). Pharmacological, toxicological and neuronal localization assessment of galantamine/chitosan complex nanoparticles in rats: Future potential contribution in Alzheimer's disease management. *Drug Delivery, 23,* 3111–3122.
147. Wang, Y., Yang, Y., & Hong, K. H. (2019). Design, synthesis and evaluation of a novel metal chelator as multifunctional agents for the treatment of Alzheimer's disease. *Bioorganic Chemistry, 87,* 720–727.

148. Kim, S. J., Son, T. G., & Park, H. R. (2008). Curcumin stimulates proliferation of embryonic neural progenitor cells and neurogenesis in the adult hippocampus. *Journal of Biological Chemistry*, *283*(21), 14497–14505.
149. Mishra, S., & Palanivelu, K. (2008). The effect of curcumin (turmeric) on Alzheimer's disease: An overview. *Annals of Indian Academy of Neurology*, *11*(1), 13.
150. Shen, Y., & Yu, L.C. (2008). Potential protection of curcumin against hypoxia-induced decreases in Beta-III Tubulin content in rat prefrontal cortical neurons. *Neurochemical Research*, *33*(10), 2112–2117.
151. Re, F., Airoldi, C., & Zona, C. (2010). Beta amyloid aggregation inhibitors: Small molecules as candidate drugs for therapy of Alzheimers disease. *Current Medicinal Chemistry*, *17*(27), 2990–3006.
152. Mohorko, N., Repovš, G., Popović, M., Kovacs, G. G., & Bresjanac, M. (2010). Curcumin labeling of neuronal fibrillar Tau inclusions in human brain samples. *Journal of Neuropathology & Experimental Neurology*, *69*(4), 405–414.
153. Yanagisawa, D., Shirai, N., & Amatsubo, T. (2010). Relationship between the tautomeric structures of curcumin derivatives and their Aβ-binding activities in the context of therapies for Alzheimer's disease. *Biomaterials*, *31*(14), 4179–4185.
154. Anand, P., Kunnumakkara, A. B., Newman, R. A., & Aggarwal, B. B. (2007). bioavailability of curcumin: Problems and promises. *Molecular Pharmaceutics*, *4*(6), 807–818.
155. Joseph, A., Wood, T., & Chen, C. C. (2018). Curcumin-loaded polymeric nanoparticles for neuroprotection in neonatal rats with hypoxic-ischemic encephalopathy. *Nano Research*, *11*(10), 5670–5688.
156. Muthukumaran, K., Kanwar, A., & Vegh, C. (2017). Ubisol-Q10 (a Nanomicellar water-soluble formulation of CoQ10) treatment inhibits Alzheimer-type behavioral and pathological symptoms in a double transgenic mouse (TgAPEswe, PSEN1dE9) model of Alzheimer's disease. *Journal of Alzheimer's Disease*, *61*(1), 221–236.
157. Gao, J., Chen, X., & Ma, T. (2020). <p>PEG-Ceramide nanomicelles induce autophagy and degrade Tau proteins in N2a cells</p>. *International Journal of Nanomedicine*, *15*, 6779–6789.
158. Mirzaie, Z., Ansari, M., Kordestani, S. S., Rezaei, M. H., & Mozafari, M. (2019). Preparation and characterization of curcumin-loaded polymeric nanomicelles to interference with amyloidogenesis through glycation method. *Biotechnology and Applied Biochemistry*, *66*(4), 537–544.
159. Aliev, G., Ashraf, G. M., & Tarasov, V. V. (2019). Alzheimer's disease – Future therapy based on dendrimers. *Current Neuropharmacology*, *17*(3), 288–294.
160. Gothwal, A., Kumar, H., & Nakhate, K. T. (2019). Lactoferrin coupled lower generation PAMAM dendrimers for brain targeted delivery of memantine in Aluminum-Chloride-induced Alzheimer's disease in mice. *Bioconjugate Chemistry*, *30*(10), 2573–2583.
161. Manning, F. C. (1994). Tacrine therapy for the dementia of Alzheimer's disease. *American Family Physician*, *50*(4), 819–826.
162. Serafín, V., Razzino, C. A., & Gamella, M. (2021). Disposable immunoplatforms for the simultaneous determination of biomarkers for neurodegenerative disorders using poly(amidoamine) dendrimer/gold nanoparticle nanocomposite. *Analytical and Bioanalytical Chemistry*, *413*(3), 799–811.

163. Aderibigbe, B., & Naki, T. (2018). Design and efficacy of nanogels formulations for intranasal administration. *Molecules*, *23*(6),1241.
164. Picone, P., Sabatino, M. A., & Ditta, L. A. (2018). Nose-to-brain delivery of insulin enhanced by a nanogel carrier. *Journal of Controlled Release*, *270*, 23–36.
165. Akel, H., Ismail, R., & Csóka, I. (2020). Progress and perspectives of brain-targeting lipid-based nanosystems via the nasal route in Alzheimer's disease. *European Journal Pharmaceutics and Biopharmacetics*, *148*, 38–53.
166. Sathya, S., Shanmuganathan, B., & Devi, K. P. (2020). Deciphering the anti-apoptotic potential of α-bisabolol loaded solid lipid nanoparticles against Aβ induced neurotoxicity in Neuro-2a cells. *Colloids and Surface B Biointerfaces*, *190*, 110948.
167. Arduino, I., Iacobazzi, R. M., & Riganti, C. (2020). Induced expression of P-gp and BCRP transporters on brain endothelial cells using transferrin functionalized nanostructured lipid carriers: A first step of a potential strategy for the treatment of Alzheimer's disease. *International Journal of Pharmacetics*, *591*, 120011.
168. Yasir, M., Chauhan, I., Haji, M. J., Tura, A. J., & Saxena, P. K. (2018). Formulation and evaluation of Glyceryl Behenate based solid lipid nanoparticles for the delivery of Donepezil to brain through nasal route. *Research Journal of Pharmacy and Technology*, *11*(7), 2836.
169. Sadegh Malvajerd, S., Izadi, Z., & Azadi, A. (2019). Neuroprotective potential of curcumin-loaded nanostructured lipid carrier in an animal model of Alzheimer's disease: Behavioral and biochemical evidence. *Journal of Alzheimer's Disease*, *69*(3), 671–686.
170. Micheli, M. R., Bova, R., Magini, A., Polidoro, M., & Emiliani, C. (2012). Lipid-based nanocarriers for CNS-targeted drug delivery. *Recent Patent CNS Drug Discovery*, *7*(1), 71–86.
171. Wong, H. L., Wu, X. Y., & Bendayan, R. (2012). Nanotechnological advances for the delivery of CNS therapeutics. *Advanced Drug Delivery Reviews*, *64*(7), 686–700.
172. Arora, S., Layek, B., & Singh, J. (2021). Design and validation of liposomal ApoE2 gene delivery system to evade blood–brain barrier for effective treatment of Alzheimer's disease. *Molecular Pharmaceutics*, *18*(2), 714–725.
173. Rinaldi, F., Hanieh, P., & Chan, L. (2018). Chitosan glutamate-coated niosomes: A Proposal for nose-to-brain delivery. *Pharmaceutics*,*10*(2), 38.
174. Kaur, A., Nigam, K., Srivastava, S., Tyagi, A., & Dang, S. (2020). Memantine nanoemulsion: A new approach to treat Alzheimer's disease. *Journal of Microencapsulation*, *37*(5), 355–365.
175. Md, S., Gan, S. Y., Haw, Y. H., Ho, C. L., Wong, S., & Choudhury, H. (2018). In vitro neuroprotective effects of naringenin nanoemulsion against β-amyloid toxicity through the regulation of amyloidogenesis and tau phosphorylation. *International Journal Biological Macromolecules*, *118*, 1211–1219.
176. Patil, R. P., Pawara, D. D., Gudewar, C. S., & Tekade, A. R. (2019). Nanostructured cubosomes in an *in situ* nasal gel system: an alternative approach for the controlled delivery of donepezil HCl to brain. *Journal of Liposome Research*, *29*(3), 264–273.
177. Kumar, S., Ganesh, L., & Santhoshkumar, J. (2019). Selenium nanoparticles as therapeutic agents in neurodegenerative diseases. In *Nanobiotechnology in neurodegenerative diseases* (pp. 209–224). Springer International Publishing.

178. Yin, T., Yang, L., Liu, Y., Zhou, X., Sun, J., & Liu, J. (2015). Sialic acid (SA)-modified selenium nanoparticles coated with a high blood–brain barrier permeability peptide-B6 peptide for potential use in Alzheimer's disease. *Acta Biomaterialia, 25*, 172–183.
179. Kwon, H. J., Cha, M. Y., & Kim, D. (2016). Mitochondria-targeting Ceria nanoparticles as antioxidants for Alzheimer's disease. *ACS Nano, 10*(2), 2860–2870.
180. Huo, X., Zhang, Y., Jin, X., Li, Y., & Zhang, L. (2019). A novel synthesis of selenium nanoparticles encapsulated PLGA nanospheres with curcumin molecules for the inhibition of amyloid β aggregation in Alzheimer's disease. *Journal of Photochemistry Photobiology B, 190*, 98–102.
181. Zhang, J., Liu, R., & Zhang, D. (2021). Neuroprotective effects of maize tetrapeptide-anchored gold nanoparticles in Alzheimer's disease. *Colloids Surfaces B Biointerfaces, 200*, 111584.
182. Cai, J., Dao, P., & Chen, H. (2020). Ultrasmall superparamagnetic iron oxide nanoparticles-bound NIR dyes: Novel theranostic agents for Alzheimer's disease. *Dyes and Pigments, 173*, 107968.

CHAPTER 9

MicroRNAs in AD Management: Significant Approaches as Diagnostic Biomarkers

AVIJIT BEJ[1], NANDITA MANDAL[2], ANWESHA DAS[3], and ARIJIT NANDI[3*]

[1]*Mahadeva Lal Schroff College of Pharmacy, Aurangabad, Bihar, India*

[2]*Department of Health and Family Welfare, Government of West Bengal, India*

[3]*Department of Pharmacy, Sanaka Educational Trust Group of Institutions (SETGOI), Malandighi, Durgapur, West Bengal, India*

*Corresponding author

ABSTRACT

Alzheimer's disease (AD) is a type of neurodegenerative condition and it is widely known as the most prevalent dementia. It is characterized by the existence of beta-amyloid (Aβ) plaques, neurofibrillary tangles, and alterations in the function of synaptic junction, all of which contribute to memory deterioration and behavioral alterations. The rise in the cases of AD places strain on the economy, families, and society. Despite the continuing endeavors to study the origins of AD, its reasons remain elusive, and there are currently no treatments with curative potential. Over the past few decades, notable advancements have been achieved in comprehending fundamental biology and molecular mechanisms, and creating biomarkers and therapeutic medications. Therefore, trustworthy biomarkers permitting advanced diagnosis and prognosis of AD are required to optimize the window of opportunity for treatment approaches. Recent research has

Cellular and Molecular Mechanisms in Alzheimer's Disease Therapeutics, Volume 1: Cellular Mechanisms. Prashant Tiwari, Sunil Kumar Kadiri, and Kunjbihari Sulakhiya (Eds.)

DOI: 10.1201/9781779641793-9

shown that microRNAs (miRNAs) are promising non-invasive, low-cost biomarkers for AD, easily found in blood and cerebrospinal fluid (CSF), among other biofluids. Furthermore, an increasing amount of research suggests that miRNAs control mechanisms related to synaptic plasticity and homeostasis, raising the possibility that they contribute to the initial dysfunction of synaptic junction in AD. This chapter discusses the miRNAs' diagnostic ability as AD biomarkers, reviews the latest studies on the roles of miRNAs in AD progression, and summarizes the methods currently employed to identify potential miRNA biomarkers in bodily fluids.

9.1 INTRODUCTION

9.1.1 ALZHEIMER'S DISEASE

Alzheimer's disease (AD) stands as a leading cause of dementia in the elderly population (60%) [1], with projections indicating that it will affect 80 million individuals by 2050 [2, 3]. Annually, approximately five million new AD cases are diagnosed, equating to one diagnosis in every 7 s [4]. Given its profound impact on patients and their families, a comprehensive understanding of AD's mechanisms and potential remedies holds significant importance [5]. Alzheimer's is a degenerative neurological disease that primarily affects the elderly, though it can afflict people of any age. The steady loss of cognitive faculties, including memory, logic, and language proficiency, sets it apart. AD is thought to arise from a combination of hereditary, environmental, and lifestyle factors, while the exact cause of the illness is yet unknown [6]. AD results from various factors, for example, alterations in proteins, oxidative stress, neuroinflammation, innate immunity-related dysfunctions, disrupted communication between neurons and glial cells, and exposure to neurotoxic substances. These elements lead to detrimental signaling via neuronal receptors, leading to the creation of amyloid-beta (Aβ) peptides and excessive tau protein phosphorylation [7]. The accumulation of abnormal protein aggregates in the brain, like tau tangles and Aβ plaques, which disrupt neuronal communication and result in cell death, is a hallmark of AD [8]. At present, while there is no known cure for AD, specific treatments can help manage symptoms and improve overall life quality. These therapies target behavioral and psychological issues, while medications are used to temporarily enhance cognitive function. To accomplish this objective, we need to improve our capability to identify individuals displaying moderate symptoms before the onset of dementia [9]. Various diagnostic criteria have been established, incorporating imaging techniques and biomarkers from cerebrospinal fluid (CSF), to establish a comprehensive classification system for AD [10]. Current research efforts in AD are focused on discovering a cure for this debilitating disorder and identifying biomarkers for early diagnosis and more effective therapies [11].

9.1.2 MICRORNAS

MicroRNAs (miRNA), also known as small RNA molecules, are a type of molecule that regulates the expression of mRNA after transcription through the pathway of RNA interference [12, 13]. They are generally 22 nucleotides long. RNA polymerase-II transcribes the genes that produce miRNA, producing a primary miRNA transcript (pri-miRNA) that typically has more than one kilobase pair (kbp) length and creates a stem-loop structure [14]. RNA polymerase III, however, can perform miRNA transcription in specific situations [15]. The microprocessor complex, which consists of RNase III endonucleases, Drosha and DCGR8, cuts the pri-miRNA within the nucleus to generate a pre-miRNA, which is a short RNA hairpin that is roughly 65 nucleotides long [16]. The pre-miRNA is moved to the cytoplasm, where it undergoes cleavage by an RNase-III endonuclease called Dicer near the terminal loop [17]. This process releases a duplex of small RNA that contains the sequences of mature miRNA, which are then incorporated into the RNA-induced silencing complex (RISC) [18]. The miRNA-induced silencing complex (miRISC) is created when a mature duplex's strands attach to a protein from the Argonaute (AGO) family, while the other is usually degraded. RISCs facilitate gene suppression by identifying the 3' untranslated region (3' UTR) of the target mRNA [11, 19]. Notably, the miRNA-AGO binding can influence gene expression. For instance, the engagement of miR-346 with *AGO2* targets the 5' untranslated region (5' UTR) of amyloid precursor protein (APP) mRNA, challenging the function of the iron response protein-1, which results in a higher rate of translation for APP mRNA [20]. Besides suppressing target gene expression through 3' UTR binding on mRNAs, there is growing witness indicating that miRNAs can exert non-canonical effects too by changing their interaction partners. For instance, MiR-181c is capable of targeting mitochondrial transcription [21]. MiRNAs can engage with proteins aside from Ago; research demonstrates that miR-let-7b functions as a signaling molecule, activating toll-like receptors (TLRs) [22]. In addition to binding with proteins, miRNAs can collaborate in interactions with non-coding RNAs like circular RNAs (circRNAs) and long non-coding RNAs (lncRNAs). Interestingly, miRNAs are found to perform the stimulation of gene expression, as observed with *miR-589* attaching to *cyclooxygenase-2 (COX-2)* transcript's promoter RNA, leading to an increase in transcription [11, 23].

9.1.3 MICRORNAS HELP WITH DIAGNOSTIC CHALLENGES IN AD

The main obstacle in AD research is identifying reliable predictive indicators for early diagnosis, well before dementia and loss of autonomy occur. This is crucial for enabling prompt intervention and care. The development of novel treatments to delay AD progression has underscored the importance of biomarker research. Discovering such a marker would be highly beneficial, facilitating quicker and more efficient medicine development [24]. It would support epidemiological and therapeutic studies, improve treatment monitoring, and be a critical indicator of disease progression. Much of this field's current work focuses on analyzing CSF and blood [25]. The biomarkers of blood provide a significant reservoir of proteomic species and genetic material for research, while CSF analysis has limitations in accessing DNA and RNA. At present, total tau (T-tau), Aβ 42, and hyperphosphorylated tau (tau-P) are the most researched biomarkers of CSF in specialized centers [26]. Researchers have extensively studied the biomarkers of CSF combined with imaging techniques for many years, leading to improved diagnostic accuracy. The diagnosis of probable AD currently depends on a combination of cognitive and clinical evaluations, structural and functional neuroimaging techniques (MRI and PET), and analysis of tau and Aβ levels in CSF. However, the sensitivity of this combined biomarker approach is approximately 93%, with a specificity of around 55% [27, 28]. Consequently, further research has focused on alternative biomarkers, with miRNAs being of particular interest for their potential use in screening, distinguishing between specific conditions, and tracking disease progression [29]. Several miRNAs, including miRNA-155, miRNA-125b, miRNA-9, and miRNA-146a have been seen to increase in both of CSF and extracellular fluid among the patients suffering from AD. These specific miRNAs arise from the pre-miRNA precursors' cleavage controlled by NF-κB and are involved in modulating inflammatory and innate immune reactions within the AD-affected brain [27]. The communication between cerebrospinal and extracellular fluid suggests a possible function for miRNAs in the central nervous system as paracrine messengers [30].

9.2 BIOGENESIS OF MIRNAS

Based on prior proteomic research, a single miRNA can affect numerous targets, playing a pivotal role in various animal developmental processes. In specific cases, these miRNAs are discharged into the extracellular fluid, showing potential as valuable biomarkers [31]. The developmental pathways of miRNA are associated with the beginning and advancement of various pathological conditions in both animals and humans. Initially, it was believed that the biogenesis pathway followed a uniform and linear process for all miRNAs. However, extensive research has revealed developmental and procedural variances that differentiate the biogenesis of various miRNAs [32]. In the majority of instances, miRNA transcription is facilitated by the intricate enzyme RNA polymerase II, leading to pri-miRNA creation. These structures typically feature single or multiple hairpin-like

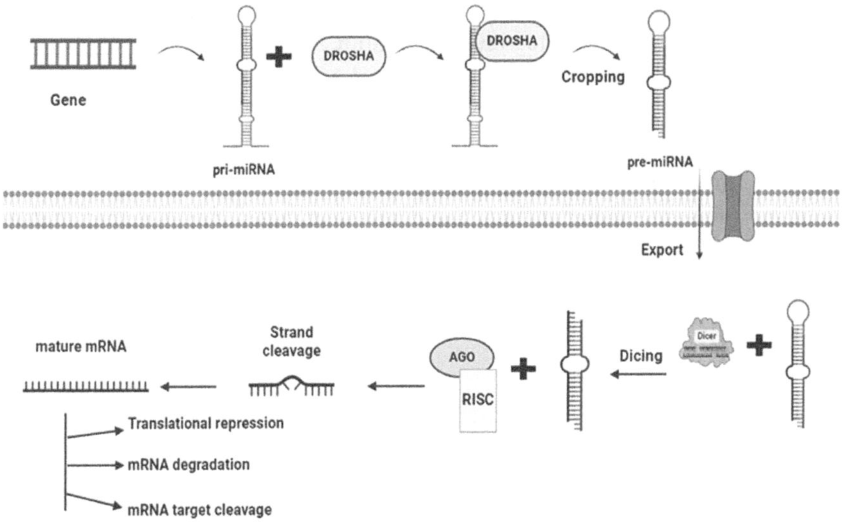

FIGURE 9.1 Biogenesis of microRNA.

formations with a terminal loop and stem. These pri-miRNAs possess polyadenylated characteristics, are 5' capped, and undergo splicing [33]. One primary miRNA can generate multiple functional miRNAs. Following this, two successive enzymatic processes occur in both the cytoplasm and nucleus, leading to the cleavage of pri-miRNA, and then, the subsequent generation of mature miRNA [34]. Frequently, miRNA transcription yields a single lengthy transcript known as a cluster, which shares the same seed region or family. There are two primary pathways for miRNA biogenesis: the canonical pathway and the non-canonical pathway [35].

9.2.1 CANONICAL PATHWAY

The canonical pathway is a linear process observed in the processing of nearly all miRNAs in mammals. In this pathway, primary miRNA (pri-miRNA) transcripts are processed with the assistance of RNA polymerase II and III (see Figure 9.1). Many pri-miRNAs undergo polyadenylation and capping during this process. Transcriptional regulation presents various regulatory possibilities, as RNA polymerase can recognize the specificity of different promoters and regulators in distinct ways. The "Drosha-DGCR8" microprocessor complex aids in cleaving the primary miRNA, leading to the formation of precursor hairpins. Subsequently, exportin-5-Ran-GTP assists in the transportation of pre-miRNA from the nucleus to the cytoplasm [36, 37]. In the cytoplasm, Dicer, an enzyme of the RNase III family, handles a double-stranded RNA (dsRNA), which is usually 20–25 nucleotides in length, originating from the pre-miRNA base [38]. Dicer cuts the pre-miRNA near the terminal loop in the stem, producing an intermediate miRNA duplex. The RISC loading process further refines the double-stranded miRNA into mature miRNA, facilitated by the AGO protein family. AGO proteins identify the guide strand for maturation as miRNA discards the passenger strand. The miRNA-loaded AGO proteins then dissociate from Dicer to form the RISC [39].

9.2.2 NON-CANONICAL PATHWAY

In addition to the canonical pathway, there exists another pathway independent of Dicer and Drosha for the biosynthesis of non-canonical miRNAs. These micro-processor-independent miRNAs are known as mirtrons and tailed mirtrons [40]. They are primarily formed through splicing followed by lariat debranching processes. Tailed mirtrons undergo further shortening by nuclear exosomes [41]. In this pathway, tRNAs (transfer RNA) mimic the structure of primary miRNA based on their sequence, forming distinct hairpin-like structures. Subsequently, nuclear exosomes cleave the tailed mirtrons. Occasionally, tRNAs can adopt hairpin-like structures based on their sequence, resembling pre-miRNAs. Typically, the formation of such structures is inhibited by the RNA chaperone La, which wraps around tRNA molecules. Certain small nucleolar RNAs (snoRNAs) can serve as substrates for Dicer. Both the C/D box and H/ACA snoRNA families possess properties that make them suitable substrates for Dicer. RNA polymerase II produces a hairpin-like structure at a specific transcription site, which then progresses into miRNA development. These resulting RNAs are referred to as pre-miRNA, undergo capping, and are subsequently exported through the exportin1 pathway. There exists a pathway independent of Dicer that revolves around the small hairpin-like structure of *mir-451*. Subsequently, *AGO-2* cleaves this structure, and poly A-specific ribonuclease (RNase) trims it [42, 43]. The 5p strand undergoes 3'–5' trimming, leading to full maturation [44]. Loss of *AGO-2* is embryonically lethal [45].

9.3 PATHOPHYSIOLOGY OF ALZHEIMER'S DISEASE

It is mainly a progressive irreversible brain disorder and the prime reason for dementia. AD is believed to arise from the abnormal accumulation of proteins in brain cells and their surrounding areas. Amyloid, one of those proteins, creates plaques surrounding the brain cells, on the other hand, another protein known as tau aggregates within the brain neurons. These proteins mainly target the prefrontal cortex and hippocampus, which are linked to emotional aspects, thinking, and memory [46]. As AD progresses throughout the brain, it impacts the additional areas of the brain and lobes. The cortex becomes thinner overall, leading to the loss of older memories, and the brain undergoes gradual shrinkage [47].

Certain hypotheses are correlated with this disease as discussed in the following sections.

9.3.1 METABOLISM OF AMYLOID PRECURSOR PROTEIN AND NEURITIC PLAQUES FORMATION

Senile plaque is one of the key pathological characteristics in AD caused by Aβ deposition. Aβ plays an essential role in the pathology of AD, as it accumulates in the brain, forming plaques that interfere with intercellular communication and ultimately cause cell death [48]. Originating from the APP, this process of amyloid pathogenesis initiates with the modified cleavage of APP, a crucial protein located on the cell membrane. This cleavage, mediated by γ-secretases and β-secretases (*BACE1*), leads to the creation of Aβ fibrils that are not soluble which then undergo oligomerization, migrate to synaptic clefts, and interfere with synaptic signaling. This protein fragment has adhesive properties, progressively aggregating into clusters, fibrils, and eventually plaques [49]. According to the amyloid hypothesis, this aggregation impairs cellular communication, induces inflammation, and culminates in cellular demise.

9.3.2 TAU PROTEIN EXCESSIVELY PHOSPHORYLATED LEADING TO THE FORMATION OF NEUROFIBRILLARY TANGLES

Tau protein is a highly soluble isoform predominantly present inside neurons within the central nervous system, notably in the cerebral cortex. Its main role revolves around maintaining the microtubule stability in axons, crucial for nutrient and substance transport within neurons. In humans, tau proteins are more prevalent in neurons than non-neuronal cells, which modulate axonal microtubule stability, primarily in axonal regions, offering stabilization and flexibility [50]. Apart from its microtubule-stabilizing role, tau also aids in recruiting signaling proteins and regulating microtubule-mediated axonal transport. When tau undergoes hyperphosphorylation, its capacity to bind to microtubules diminishes, resulting in their instability and eventual disintegration. Consequently, unbound tau aggregates into structures known as neurofibrillary tangles, disrupting various intracellular processes [48].

9.3.3 CHOLINERGIC HYPOTHESIS

The cholinergic system plays a vital role in the peripheral and central nervous systems. Acetylcholine serves as a primary brain neurotransmitter, operating across the basal ganglia, basal forebrain, and cortex. It is pivotal for several cognitive processes, such as memory, learning, and complex behaviors. Substantial deficiencies in the neocortex are observed concerning choline acetyltransferase (ChAT), the enzyme crucial for synthesizing Ach. Further research has revealed reduced choline absorption, diminished release of Ach, and the absence of cholinergic cell bodies in the basal frontal lobe, indicating a significant presynaptic cholinergic dysfunction [51]. Specifically, basal forebrain cholinergic neurons (BFCNs) significantly influence memory, learning, and cognitive abilities. In AD patients, the degeneration of BFCNs correlates with the severity of dementia. This degenerative process contributes to the loss of synaptic connections between the basal forebrain and associated brain regions, including the hippocampus and cortex, ultimately leading to memory impairment, a hallmark symptom of AD [52, 53]. Another crucial factor is the interaction between AChE and PS-1, an enzyme vital for Aβ production. AChE promotes the expression of PS-1 through direct binding, thereby elevating Aβ levels. It's widely recognized that the accumulation of Aβ accelerates cognitive deterioration in AD. Furthermore, unregulated alterations in the central cholinergic system could induce inflammation in neurons and phosphorylation of tau proteins, potentially resulting in neurotransmitter imbalance and neuronal demise [54].

9.3.4 OXIDATIVE STRESS HYPOTHESIS

According to AD's theory of oxidative stress, age-related cognitive decline and neurodegeneration are greatly accelerated by a rise in oxidative stress, which is characterized by the body's protective antioxidant systems and the formation of reactive oxygen species (ROS). The brain is particularly susceptible to oxidative stress due to its elevated oxygen consumption and abundant lipid content that is readily oxidized. Increased oxidative stress linked with AD is evidenced by heightened levels of end products of advanced glycation, as well as that for lipid peroxidation; oxidized proteins; and oxidative alterations in both nuclear and mitochondrial DNA [55]. This imbalance may cause tissue damage, which may lead to the production of harmful species and extremely reactive hydroxyl radicals. These substances may cause brain damage through a variety of pathways, including increased intracellular free calcium levels, the release of excitatory amino acids, and neurotoxicity. The reduction in antioxidants, such as glutathione, which is pivotal for scavenging ROS, further exacerbates the situation, creating a scenario where the rate of the production of ROS surpasses the antioxidant capacity, thereby promoting oxidative stress in the AD patients' brain tissue [56, 57].

9.4 ROLE OF MIRNAS IN ALZHEIMER'S DISEASE

Mature miRNAs are smaller RNA molecules, generally having a length of 20–25 nucleotides [58], which attach to the mRNA 3'UTR [14], thereby modulating gene expression after transcription. This binding can either initiate mRNA degradation or inhibit protein synthesis. While the primary action of miRNAs occurs intracellularly, they can also be detected outside of cells under conditions of cellular damage (due to factors like trauma, chronic inflammation, apoptosis, or necrosis), active release of protein-miRNA complexes (such as Ago2), or active release from microvesicles (MVs) (like microparticles, exosomes, or apoptotic bodies) [59, 60]. The stability of miRNAs is enhanced when they are encapsulated or bound to proteins, allowing for targeted analysis as potential biomarkers within exosomes or protein-bound fractions in bodily fluids [61, 62]. The levels of expression of many miRNAs and their corresponding target mRNAs undergo alterations during pathological conditions and the progression of AD, impacting critical processes essential for the development and advancement of the disease, such as the phosphorylation of tau, APP, and the production of Aβ (refer to Figure 9.2) [63, 64]. Given the dynamic behavior of miRNAs in AD, there have been attempts to identify and describe the varying levels of circulating miRNAs found in AD patients' body fluids, for example, serum, plasma, and CSF [65]. The miRNA participation in AD development, their presence in readily available bodily fluids, and their comparatively stable nature as compared to mRNA make them compelling for the target exploration for AD biomarkers [66]. Exosomes, considered as a potential origin of miRNAs, have attracted considerable attention because of their unique characteristics. They contain higher levels of miRNAs (3- to 4-fold greater than in serum) [67], and demonstrate non-random packaging mechanisms that reflect pathological conditions [68].

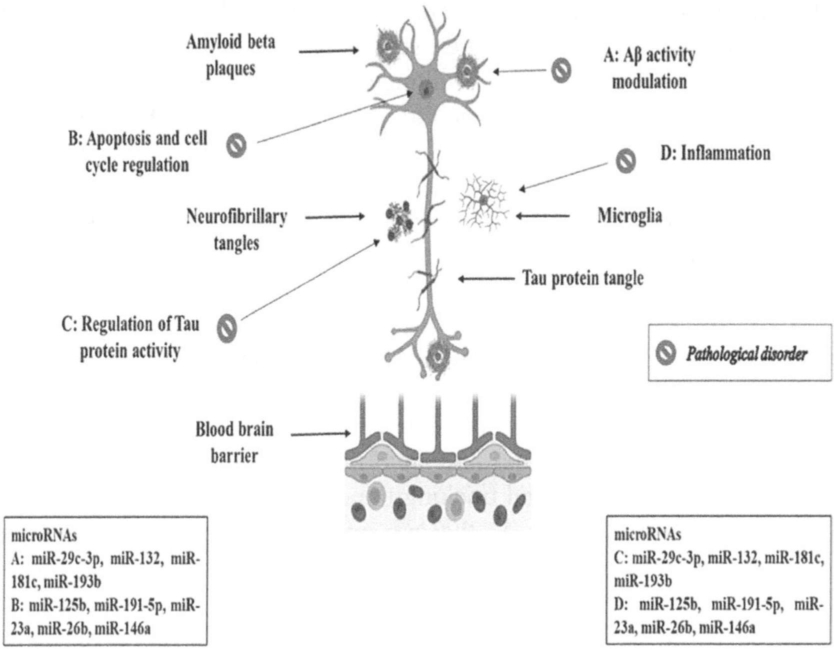

FIGURE 9.2 The impact of miRNAs on AD pathology is depicted in this illustration, showing how dysregulated miRNAs contribute to neuronal degeneration. These miRNAs influence cell cycle disruption, Aß metabolism regulation, tau-protein metabolism regulation, and neuroinflammation.

miRNA-based therapies that manipulate miRNA function by delivering artificial tiny RNA molecules have garnered significant interest and demonstrated potential in preliminary research, providing numerous benefits [69]. Firstly, miRNAs are natural components present in human cells, participating in biological processes *in vivo*. Secondly, unlike conventional treatments, miRNAs can also interact directly with downstream targets and regulate their expression. Additionally, the compact miRNA sizes allow for the creation of effective delivery methods [70]. Moreover, miRNAs can simultaneously regulate multiple genes in a particular pathway, resulting in more precise control. Artificial double-stranded oligonucleotides, functioning as miRNA mimetics, play a role in reinstating miRNA function in different diseases. Currently, clinical trials have investigated a couple of miRNA mimetics, namely MesomiR-1 (mimicking miR-16), and MRX34 (mimicking miR-34), for treating cancer [71, 72]. The significant therapeutic usages of miRNA-based therapies in the treatment of AD have not been fully investigated. While miRNAs offer promising therapeutic possibilities, several obstacles need attention, such as stability concerns, restricted cell membrane penetration, the challenge of precisely targeting tissues, and the presence of the blood–brain barrier (BBB) [73]. Unshielded RNAs can degrade easily in circulation and their size and negative charge pose challenges for crossing cell membranes [74]. Moreover, the BBB acts as an additional barrier that impedes the uptake and delivery of RNA into brain tissues [75].

9.5 MICRORNAS IN BIOFLUIDS AS DIAGNOSTIC TOOLS OF AD

miRNAs found in biofluids show potential as diagnostic tools for AD. These tiny RNA molecules are involved in AD's development and could help in detecting and tracking the disease early on. Research has pinpointed specific miRNAs that are either increased or decreased in mild cognitive impairment (MCI) and AD patients' blood serum, plasma, and CSF, indicating their usefulness as biomarkers (Figure 9.3) [76]. Although using miRNAs as diagnostic biomarkers for AD comes with challenges, they offer advantages over traditional biomarkers due to their stability and simplicity of detection through methods like real-time PCR and next-generation sequencing [11]. The combination of several miRNAs can accurately differentiate AD from control groups, suggesting potential improvements in AD's diagnosis, distinction, and prediction. In summary, miRNAs in biofluids show promise for improving AD diagnosis and management [11]. Advancements in AD diagnosis [11] have greatly improved thanks to the development of non-invasive neuroimaging techniques over the past few decades. These techniques allow for the real-time observation of brain structures. Examples include cutting-edge MRI, positron emission tomography (PET) for detecting metabolic changes, and amyloid imaging. These techniques not only assist in identifying brain diagnostic biomarkers of elderly individuals without cognitive impairments before symptoms appear, but also in tracking disease progression after symptoms develop [10]. Volumetric approaches of high-resolution subfields and diagnostic techniques for early changes in affected brain areas are particularly valuable because of their dependability and strong discriminating ability in the pre-dementia phase [77]. These tools make it easier to detect typical lesions associated with protein buildup and structural alterations in particular brain areas, forming the foundation of AD diagnosis [78, 79, 80]. In recent years, the progress in omics technologies, such as genomics, transcriptomics, proteomics, metabolomics, and secretomics, among others, has facilitated the thorough examination of different markers related to AD, covering both familial and sporadic instances. These methods analyze the fluid samples of humans, like blood, tears, urine, or saliva. Importantly, collecting these samples typically doesn't necessitate specialized training and offers one benefit, that is, their non-invasive nature, made possible by their simple accessibility. miRNAs offer distinct advantages compared to traditional biomarkers. Many research findings indicate that certain miRNAs found in AD patients' biofluids correlate with pathological alterations [81]. When attached to lipoproteins or enclosed in liposomes in bodily fluids, like plasma, serum, and CSF, these miRNAs display notable resilience, shielding them from deterioration and enabling them to withstand challenging environmental circumstances [82]. The significance of identifying and developing dependable and sensitive tools for preliminary AD detection lies in the potential advantages for patients. These advantages encompass prompt access to medical interventions aimed at slowing down disease advancement, thereby preserving cognitive function for longer periods and even enabling future planning [81]. Additionally, the types of biofluids and exosomes in which miRNA profiles are detected vary. This diversity is primarily explained by the selective transport mechanisms of miRNAs. Specific proteins, such as membrane-localized proteins and RNA-binding proteins, actively sort mature miRNAs into MVs, which are then released into the biofluid [83]. Exosomes originating from neurons can traverse the BBB and transport their contents into the CSF. Moreover, during coagulation, miRNAs may be passively released from platelets or apoptotic bodies. In particular, biomarkers found in exosomes may exhibit higher quality and greater diagnostic efficacy than those in biofluids [84, 85].

9.5.1 NON-INVASIVE BIOMARKERS

Non-invasive biomarkers used to detect AD early have become a focal point in research. These biomarkers, found in biofluids, like blood, saliva, breath, and urine, provide a less intrusive alternative to traditional methods, such as CSF sampling and imaging tests. Research indicates that biomarkers like Aβ, T-tau, phosphorylated tau (P-tau), and small metabolites can be identified in saliva, showing potential for future AD diagnostics [86]. Moreover, a wireless, handheld, non-invasive device has been created to detect biomarkers for both Parkinson's and ADs using electrical detection, presenting a more accurate and convenient approach for early detection [81]. These advancements in non-invasive biomarkers and technology hold promise for transforming the timely identification and management of AD. The composition of blood is more intricate than that of CSF, containing numerous cells and soluble substances that can create interferences. Additionally, the limited presence of brain-specific biomarkers in blood needs immensely sensitive detection methods. Variance in outcomes based on the approach used and conflicting results among studies are constraints to using blood biomarkers for AD diagnosis [87].

9.5.2 INVASIVE BIOMARKERS

In AD, invasive biomarkers often require more invasive procedures like lumbar puncture for collecting CSF or PET imaging. While these methods are presently considered the gold standard for the diagnosis of AD, they are limited by their invasiveness, high cost, and limited availability. The biomarkers of CSF, including P-tau, T-tau, and Aβ 42 have demonstrated excellent diagnostic accuracy for AD [86, 88]. However, the invasiveness of CSF collection via lumbar puncture restricts its common utilization in regular primary care settings. PET imaging can detect and identify the buildup of Aβ or tau in the brain, but it suffers from invasiveness, costliness, and limited accessibility. Non-invasive biomarkers, such as those based on blood, are being explored as alternatives to invasive techniques [86]. These biomarkers can exhibit AD-related alternation at preliminary or presymptomatic stages, allowing for earlier diagnosis and intervention. Recent years have witnessed increased interest in developing non-invasive AD biomarkers, including those based on blood, saliva, and the eyes (Figure 9.3) [89]. These non-invasive options offer a more accessible and cost-effective alternative to invasive methods, thereby enhancing early detection and AD management.

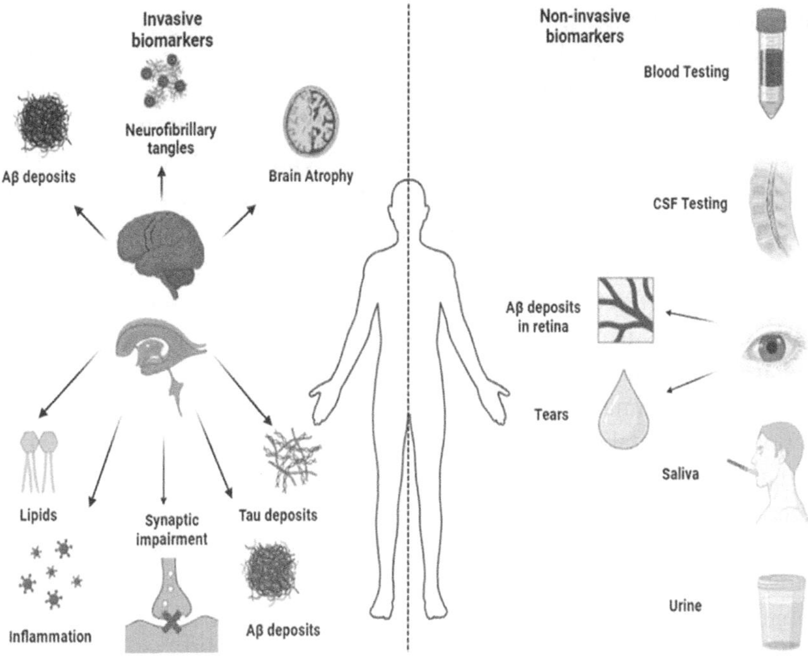

FIGURE 9.3 Schematic overview of non-invasive and invasive biomarkers.

9.6 SIGNIFICANT APPLICATION OF CLINICALLY IMPORTANT MIRNAS-BASED BIOMARKERS IN AD MANAGEMENT

The potential clinical applications of miRNA-based biomarkers in managing AD extend to their utility as diagnostic aids. Numerous miRNAs have been identified with notable efficacy and specificity, such as miR-19b-3p, miR-29c-3p, miR-29a, miR-455-3p, miR-324-3p, miR-106a-5p, and miR-107 [11]. When combined, these miRNAs can differentiate between control subjects and AD with 75–82% accuracy [11]. Although using miRNAs as AD biomarkers presents several challenges, their diagnostic potential is encouraging. Beyond diagnostics, miRNAs have demonstrated their ability to influence AD's onset and pathological progression [90]. They play crucial roles in five critical areas of AD: amyloid generation, tau pathology, neuroinflammation, synaptic malfunction, and neuronal demise. Monitoring changes in miRNA expression over time can also help guide or assess the impacts of therapeutic interventions [90]. Research into miRNA biomarkers for AD is a developing area. Levels of miRNAs in both the brain and circulation have been linked to AD (see Table 9.1). Importantly, miRNAs are directly associated with pathways believed to be crucial in AD's pathophysiology. Some specific uncommon types of AD that occur early in life result from genetic mutations accountable for processing the APP, which in turn affects brain Aβ 42 production [91]. While imaging techniques have demonstrated efficacy in early AD diagnosis, these methods often detect changes too late for implementing preventive measures [92]. Additionally, as AD progresses, significant morphological brain changes can occur, potentially reducing the effectiveness of pharmacological interventions. Therefore, early diagnosis using biomarkers is crucial for preventing or improving the treatment of the disease [93]. Recent advancements in biomarker technologies primarily target genomics, neuroimaging, epigenomics, transcriptomics, metabolomics, lipidomics, proteomics, and miRNAs [94, 95, 96]. Crucially, unlike numerous biomarkers found in CSF, miRNAs show abnormal expressions even before entering the CSF, making them highly effective diagnostic indicators [97]. Moreover, the methods for detecting miRNA biomarkers are more convenient and cost-effective for monitoring large populations. Given these advantages, miRNAs hold significant potential as biomarkers for diagnosing AD [5]. To authenticate miRNAs as indicators for the initiation of AD, examining longitudinal changes in biofluid miRNA levels among cohorts will provide valuable insights into whether alterations in miRNA levels could serve as predictors of AD onset.

Recently, there have been notable advancements in oligonucleotide therapy, opening new avenues for treating CNS diseases like AD. Encouraging preclinical research in animal models of AD sets the stage for forthcoming therapeutic strategies using oligonucleotides. Developing treatments with oligonucleotides for disorders, such as homozygous familial hypercholesterolemia [106, 107] and commencing Phase I clinical trials for oligonucleotide-based therapies for ALS suggest promising advances in therapeutic approaches. Additionally, ongoing research on circulating miRNAs as reliable biomarkers for AD progression could facilitate early AD diagnoses [108]. While advancements in oligonucleotide therapies hold promise for enhancing AD treatment strategies, it's important to note that miRNAs, despite their potential as therapeutic targets and biomarkers in AD, are not currently usable as diagnostic or therapeutic tools for neurodegenerative diseases [91]. To summarize, miRNA-based biomarkers hold promise for improving AD diagnosis and management by acting as diagnostic aids and regulators of disease advancement. Nonetheless, additional

TABLE 9.1 Chosen Studies Detailing microRNA Profiles in AD.

Tissue	Protocol Design	Results	References
Middle and superior temporal cortex	miRNA microarray analysis was confirmed through *in situ* hybridization and northern blot validation	miR-107 was lowered in case of AD when compared to the individuals that are nondemented without plaques.	[98]
CSF	The human miRNA microarray containing 728 miRNAs was validated using qRT-PCR	Decreases the generation of the levels of miR-27a-3p in the CSF when there is a comparison between AD patients and control subjects.	[99]
Serum	qRT-PCR conducts and validates the RNA sequencing	Only MiR-342-3p or 6 serum miRNAs were used to differentiate AD patients from healthy controls.	[100]
The mononuclear cells of blood	The validation of 462 miRNAs in humans was performed using qRT-PCR after microarray analysis	qRT-PCR confirmed that the levels of miR-34a and miR-181b were found to be higher in individuals with AD compared to the control group.	[101]
CA1 hippocampus	Northern blot validated the DNA array-based analysis	In AD, miR-125b, miR-128, and miR-9 were hiked when compared to the control group.	[102]
The white and gray matter sections of the middle and superior temporal cerebral cortex	Northern blot validated the human miRNA microarray-based analysis	Downregulation of white matter *miR-212* and gray matter miR-29 paralogs, and *miR-15/107* genes in AD and upregulation of *miR-424 gene* of the white matter in AD.	[103]
Frontal cortex	qRT-PCR validated the human miRNA microarray	Reduced levels of miR-29a in AD influence neurodegenerative mechanisms by increasing neuronal *NAV3* expression in the brains of affected individuals.	[104]
Plasma	qRT-PCR validated the human miRNA microarray-based analyses	Seven miRNA signatures (miR-7g, miR-7d, miR-142, miR-15b, miR-301a, miR-545, and miR-191) in plasma distinguishes between AD patients and control group.	[105]
Brain tissue and CSF derived from superior temporal lobe and extracellular fluid	Analysis using a LED northern dot-blot assay and fluorescent miRNA array	Increases in miR-9, miR-125b, miR-146, and miR-155 in AD, ECF, and CSF.	[30]

research is needed to tackle the constraints and discrepancies observed in the current research findings.

9.7 CONCLUSION

In summary, miRNAs show great potential as diagnostic biomarkers in the management of AD. Studies have pinpointed specific miRNAs that are dysregulated in AD, offering targets for early disease detection and monitoring. Employing miRNAs as biomarkers brings several advantages, including their stability in body fluids, like blood and CSF, their ability to mirror disease progression, and their suitability for non-invasive testing. The investigation of miRNA and circRNA holds significant promise for developing biomarkers that are non-invasive in AD. Continued investigation into these non-coding RNAs within the framework of AD offers the chance to revolutionize AD monitoring as well as diagnosing, along with the approaches of treatment, ultimately improving the management of disease and enhancing the well-being of those impacted. Continued research efforts and collaborative initiatives are vital to fully exploiting miRNAs' potential in enhancing timely AD detection and monitoring.

Conflict of interest statement: The authors declare no conflict of interest.

Consent for publication: Not applicable.

Funding: None.

Authors' contribution: Avijit Bej, Nandita Mandal, and Anwesha Das are responsible for data collection and manuscript preparation. Arijit Nandi is responsible for supervision and editing.

KEYWORDS

- **Alzheimer's disease**
- **biomarkers**
- **blood–brain barrier**
- **cerebrospinal fluid**
- **microRNAs**
- **neurodegenerative disorder**
- **synaptic junction**

REFERENCES

1. Hebert, L. E., Scherr, P. A., Bienias, J. L., Bennett, D. A., & Evans, D. A. (2003). Alzheimer disease in the US population: Prevalence estimates using the 2000 census. *Archives of Neurology*, *60*(8), 1119–1122. Doi: 10.1001/archneur.60.8.1119.
2. Plassman, B. L., et al. (2007). Prevalence of dementia in the United States: The aging, demographics, and memory study. *Neuroepidemiology*, *29*(1–2), 125–132. Doi: 10.1159/000109998.
3. Humpel, C. (2011). Identifying and validating biomarkers for Alzheimer's disease. *Trends in Biotechnology*, *29*(1), 26–32. Doi: 10.1016/j.tibtech.2010.09.007.
4. Lukiw, W. J., Andreeva, T. V., Grigorenko, A. P., & Rogaev, E. I. (2013). Studying micro RNA function and dysfunction in Alzheimer's disease. *Frontiers in Genetics*, *3*(1–13). Doi: 10.3389/fgene.2012.00327.
5. Pan, Y., et al. (2016). Dysregulation and diagnostic potential of microRNA in Alzheimer's disease. *Journal of Alzheimer's Disease*, *49*(1), 1–12. Doi: 10.3233/JAD-150451.
6. Farias, G., Cornejo, A., Jimenez, J., Guzman, L., & Maccioni, R. B. (2011). Mechanisms of Tau Self-aggregation and neurotoxicity. *Current Alzheimer Research*, *8*(6), 608–614. Doi: 10.2174/156720511796717258.
7. Farías, G., Ṕerez, P., & MacCioni, R. B. (2012). Platelet tau pattern correlates with cognitive status in alzheimer's disease. *Journal of Alzheimer's Disease*, *31*(1), 65–69. Doi: 10.3233/JAD-2012-120304.
8. Querfurth H. W., & LaFerla, F. M. (2010). Alzheimer's disease. *The New England Journal of Medicine*, *9*, 329–344.
9. Dickerson, B. C., *et al.*, (2005). Increased hippocampal activation in mild cognitive impairment compared to normal aging and AD. *Neurology*, *65*(3), 404–411. Doi: 10.1212/01.wnl.0000171450.97464.49.
10. Dubois, B., et al. (2007). Research criteria for the diagnosis of Alzheimer's disease: Revising the NINCDS-ADRDA criteria. *Lancet Neurology*, *6*(8), 734–746. Doi: 10.1016/S1474-4422(07)70178-3.
11. Wei, W., Wang, Z. Y., Ma, L. N., Zhang, T. T., Cao, Y., & Li, H. (2020). MicroRNAs in Alzheimer's disease: Function and potential applications as diagnostic biomarkers. *Frontiers in Molecular Neuroscience*, *13*, 1–16. Doi: 10.3389/fnmol.2020.00160.
12. Carthew, R. W., & Sontheimer, E. J. (2009). Origins and mechanisms of miRNAs and siRNAs. *Cell*, *136*(4), 642–655. Doi: 10.1016/j.cell.2009.01.035.
13. Patel, S., et al. (2021). Pyruvate kinase M2 in chronic inflammations: A potpourri of crucial protein–protein interactions. *Cell Biology and Toxicology*, *37*(5), 653–678. Doi: 10.1007/s10565-021-09605-0.
14. Ha, M., & Kim, V. N. (2014).Regulation of microRNA biogenesis. *Nature Molecular Cell Biology*, *15*(8), 509–524. Doi: 10.1038/nrm3838.
15. Borchert, G. M., Lanier, W., & Davidson, B. L. (2006). RNA polymerase III transcribes human microRNAs. *Nature Structural and Molecular Biology*, *13*(12), 1097–1101. Doi: 10.1038/nsmb1167.
16. Macias, S., Cordiner, R. A., & Cáceres, J. F. (2013). Cellular functions of the microprocessor. *Biochemical Society Transactions*, *41*(4), 838–843. Doi: 10.1042/BST20130011.

17. Song, M. S., & Rossi, J. J. (2017). Molecular mechanisms of Dicer: Endonuclease and enzymatic activity. *Biochemical Journal, 474*(10), 1603–1618. Doi: 10.1042/BCJ20160759.
18. Abe M., & Bonini, N. M. (2013). MicroRNAs and neurodegeneration: Role and impact. *Trends in Cell Biology, 23*(1), 30–36. Doi: 10.1016/j.tcb.2012.08.013.
19. Chauhan, H., Gupta, P., & Saxena, B. (2012). MicroRNAs as future treatment tools and diagnostic biomarkers in Alzheimer's disease. *IntechOpen*, 1–13. [Online]. Available: http://dx.doi.org/10.1039/C7RA00172J%0Ahttps://www.intechopen.com/books/advanced-biometric-technologies/liveness-detection-in-biometrics%0Ahttp://dx.doi.org/10.1016/j.colsurfa.2011.12.014
20. Long, J. M., Maloney, B., Rogers, J. T., & Lahiri, D. K. (2019). Novel upregulation of amyloid-β precursor protein (APP) by microRNA-346 via targeting of APP mRNA 5′-untranslated region: Implications in Alzheimer's disease. *Molecular Psychiatry, 24*(3) 345–363. Doi: 10.1038/s41380-018-0266-3.
21. Das, S., et al. (2012). Nuclear miRNA regulates the mitochondrial genome in the heart. *Circulation Research, 110*(12) 1596–1603. Doi: 10.1161/CIRCRESAHA.112.267732.
22. Lehmann, S. M., et al. (2012). An unconventional role for miRNA: Let-7 activates Toll-like receptor 7 and causes neurodegeneration. *Nature Neuroscience, 15*(6), 827–835. Doi: 10.1038/nn.3113.
23. Matsui, M., et al. (2013). Promoter RNA links transcriptional regulation of inflammatory pathway genes. *Nucleic Acids Research, 41*(22), 10086–10109. Doi: 10.1093/nar/gkt777.
24. El Kadmiri, N., Said, N., Slassi, I., El Moutawakil, B., & Nadifi, S. (2018). Biomarkers for Alzheimer disease: Classical and novel candidates' review. *Neuroscience, 370,* 181–190. Doi: 10.1016/j.neuroscience.2017.07.017.
25. Blennow, K., & Zetterberg, H. (2015). The past and the future of Alzheimer's disease CSF biomarkers-a journey toward validated biochemical tests covering the whole spectrum of molecular events. *Frontiers in Neuroscience, 9,* 1–8. Doi: 10.3389/fnins.2015.00345.
26. Vemuri, P., et al. (2010). Effect of apolipoprotein E on biomarkers of amyloid load and neuronal pathology in Alzheimer disease. *Annals of Neurology, 67*(3), 308–316. Doi: 10.1002/ana.21953.
27. Femminella, G. D., Ferrara, N., & Rengo, G. (2015). The emerging role of microRNAs in Alzheimer's disease. *Frontiers in Physiology, 6,* 1–5. doi: 10.3389/fphys.2015.00040.
28. Singh, D., et al. (2023). Design, synthesis, anticancer and antimicrobial studies of 2-Phenylthiazolidin-4-one glycinamide conjugates. *Chemistry Europe, 8*(42), 1–13. Doi: 10.1002/slct.202301806.
29. Jack C. R., & Holtzman, D. M. (2013). Biomarker modeling of alzheimer's disease. *Neuron, 80*(6), 1347–1358. Doi: 10.1016/j.neuron.2013.12.003.
30. Alexandrov, P. N., Dua, P., Hill, J. M., Bhattacharjee, S., Zhao, Y., & Lukiw, W. J. (2012). MicroRNA (miRNA) speciation in Alzheimer's disease (AD) cerebrospinal fluid (CSF) and extracellular fluid (ECF). *International Journal of Biochemistry and Molecular Biology, 3*(4), 365–373.
31. O'Brien, J., Hayder, H., Zayed, Y., & Peng, C. (2018). Overview of microRNA biogenesis, mechanisms of actions, and circulation. *Frontiers in Endocrinology, 9,* 1–12. Doi: 10.3389/fendo.2018.00402.

32. Obernosterer, G., Martinez, J., & Alenius, M. (2007). Locked nucleic acid-based in situ detection of microRNAs in mouse tissue sections. *Nature Protocols*, *2*(6), 1508–1514. Doi: 10.1038/nprot.2007.153.
33. Kim, V. N., Han, J., & Siomi, M. C. (2009). Biogenesis of small RNAs in animals. *Nature Reviews Molecular Cell Biology*, *10*(2), 126–139. Doi: 10.1038/nrm2632.
34. Siomi, H., & Siomi, M. C. (2010). Posttranscriptional regulation of MicroRNA biogenesis in animals. *Molecular Cell*, *38*(3), 323–332. Doi: 10.1016/j.molcel.2010.03.013.
35. Tanzer, A., & Stadler, P. F. (2004). Molecular evolution of a microRNA cluster. *Journal of Molecular Biology*, *339*(2), 327–335. Doi: 10.1016/j.jmb.2004.03.065.
36. Chong, M. M. W., Zhang, G., Cheloufi, S., Neubert, T. A., Hannon, G. J., & Littman, D. R. (2010). Canonical and alternate functions of the microRNA biogenesis machinery. *Genes and Development*, *24*(19), 1951–1960. Doi: 10.1101/gad.1953310. and.
37. Rathod, B., et al. (2024). Design and synthesis of novel quinazolinone-based pyruvate kinase M2 activators as selective inhibitors of oral cancer cells. *Journal of Molecular Structure*, *1304*, 137595–137607. Doi: 10.1016/j.molstruc.2024.137595.
38. Grishok, A., et al. (2001). Genes and mechanisms related to RNA interference regulate expression of the small temporal RNAs that control C. elegans developmental timing. *Cell*, *106*(1), 23–34. Doi: 10.1016/S0092-8674(01)00431-7.
39. Winter, J., Jung, S., Keller, S., Gregory, R. I., & Diederichs, S. (2009). Many roads to maturity : microRNA biogenesis pathways and their regulation. *Nature Cell Biology*, *11*(3), 228–234.
40. Ruby, J. G., Jan, C. H., & Bartel, D. P. (2007). Intronic microRNA precursors that bypass Drosha processing. *Nature*, *448*, 83–87. Doi: 10.1038/nature05983.
41. Flynt, A. S., Greimann, J. C., Chung, W., Lima, C. D., & Lai, E. C. (2010). MicroRNA biogenesis via Splicing and exosome-mediated trimming in Drosophila. *Molecular Cell*, *38*(6), 900–907. Doi: 10.1016/j.molcel.2010.06.014.
42. Singh, D., et al. (2024). Phenylthiazolidin-4-one piperazine conjugates: Design, synthesis, anticancer and antimicrobial studies. *Results in Chemistry*, *7*, 2023, 101237–101248. Doi: 10.1016/j.rechem.2023.101237.
43. Treiber, T., Treiber, N., & Meister, G. (2019). Regulation of microRNA biogenesis and its crosstalk with other cellular pathways. *Nature Reviews Molecular Cell Biology*, *20*, 5–20. doi: 10.1038/s41580-018-0059-1.
44. Cheloufi, S., Dos Santos, C. O., Chong, M. M. W., & Hannon, G. J. (2010). A Dicer-independent miRNA biogenesis pathway that requires Ago catalysis. *Nature*, *465*, 584–590. Doi: 10.1038/nature09092.
45. Yoda, M., Cifuentes, D., Izumi, N., Sakaguchi, Y., Suzuki, T., & Giraldez, A. J. (2013). Poly(A)-Specific Ribonuclease mediates 30-end trimming of Argonaute2-cleaved precursor MicroRNAs. *Cell Reports*, *5*(3), 715–726. Doi: 10.1016/j.celrep.2013.09.029.
46. Scheltens, P., et al. (2021). Alzheimer 's disease. *The Lancet*, *397*, 1577–1590. Doi: 10.1016/S0140-6736(20)32205-4.
47. Blinkouskaya, Y., & Weickenmeier, J. (2021). Brain shape changes associated with cerebral atrophy in healthy aging and Alzheimer's disease. *Frontiers in Mechanical Engineering*, *7*, 1–17. Doi: 10.3389/fmech.2021.705653.

48. Thakur, A. K., Kamboj, P., Goswami, K., & Ahuja, K. (2018). Pathophysiology and management of alzheimer's disease : An overview. *Journal of Analytical & Pharmaceutical Research, 7*(2), 226–235. Doi: 10.15406/japlr.2018.07.00230.
49. Tiwari, S., Atluri, V., Kaushik, A., Yndart, A., & Nair, M. (2019). Alzheimer's disease : Pathogenesis, diagnostics, and therapeutics. *International Journal of Nanomedicine, 14*, 5541–5554.
50. Michalicova, A., Majerova, P., & Kovac, A. (2020). Tau Protein and Its role in blood – brain barrier dysfunction. *Frontiers in Molecular Neuroscience, 13*, 1–13. Doi: 10.3389/fnmol.2020.570045.
51. Stanciu, G. D., et al. (2020). Alzheimer's disease pharmacotherapy in relation to cholinergic system involvement. *Biomolecules, 10*(40), 1–21.
52. Latina, V., et al. (2018). NGF-dependent changes in Ubiquitin Homeostasis Trigger early cholinergic degeneration in cellular and animal AD-model. *Frontiers in Cellular Neuroscience, 12*, 1–30. Doi: 10.3389/fncel.2018.00487.
53. Chowdhury, A., Patel, S., Sharma, A., Das, A., Meshram, P., & Shard, A. (2020). A perspective on environmentally benign protocols of thiazole synthesis. *Chemistry of Heterocyclic Compounds, 56*(4), 455–463. Doi: 10.1007/s10593-020-02680-x.
54. Campanari, M.-L., Garcia-Ayllona, M.-S., Belbin, O., Galceran, J., Lleo, A., & Saez-Valero, J. (2014). Acetylcholinesterase Modulates Presenilin-1 Levels and γ-Secretase activity. *Journal of Alzheimer's Disease*, 1–14. Doi: 10.3233/JAD-140426.
55. Gella, A., & Durany, N. (2009). Oxidative stress in Alzheimer disease. *Cell Adhesion & Migration, 3*(1), 88–93. Doi: 10.4161/cam.3.1.7402.
56. Chattaraj, B. et al. (2022). In vitro antilithiatic activity of Enhydra fluctuans Lour. on calcium oxalate crystallisation and the elucidation of the possible mechanism involved therein through in silico studies. *Frontiers in Pharmacology, 13*, 1–29 .
57. Huang, W. E. N. J., Zhang, X. I. A., & Chen, W. E. I. W. E. I. (2016).Role of oxidative stress in Alzheimer ' s disease. *Biomedical Reports, 4*, 519–522. Doi: 10.3892/br.2016.630.
58. Wahid, F., Shehzad, A., Khan, T., & Young, Y. (2010). MicroRNAs : Synthesis, mechanism, function, and recent clinical trials. *BBA - Molecular Cell Research, 1803*(11), 1231–1243. Doi: 10.1016/j.bbamcr.2010.06.013.
59. Creemers, E. E., Tijsen, A. J., & Pinto, Y. M. (2012). Circulating MicroRNAs: Novel biomarkers and extracellular communicators in cardiovascular disease? *Journl of American Heart Association, 110*, 483–495. Doi: 10.1161/CIRCRESAHA.111.247452.
60. Redis, R. S., Calin, S., Yang, Y., You, M. J., & Calin, G. A. (2012). Cell-to-cell miRNA transfer : From body homeostasis to therapy. *Pharmacology and Therapeutics, 136*(2), 169–174. Doi: 10.1016/j.pharmthera.2012.08.003.
61. Geekiyanage, H., Rayatpisheh, S., Wohlschlegel, J. A., Brown, R., & Ambros, V. (2020). Extracellular microRNAs in human circulation are associated with miRISC complexes that are accessible to anti-AGO2 antibody and can bind target mimic oligonucleotides oligonucleotides. *Applied Biological Sciences*, 1–11. Doi: 10.1073/pnas.2008323117.
62. Nur, N., Binti, S., Mohamed, N., Nazatul, W., & Shahidan, S. (2020). Non-exosomal and exosomal circulatory MicroRNAs : Which are more valid as biomarkers ? *Frontiers in Pharmacology, 10*, 1–22. Doi: 10.3389/fphar.2019.01500.

63. Dehghani, R., Rahmani, F., & Rezaei, N. (2017). MicroRNA in Alzheimer's disease revisited : implications for major neuropathological mechanisms. *Reviews in the Neurosciences*, *29*(2), 161–182.
64. Nandi, A., Das, A., Dey, Y. N., & Roy, K. K. (2023). The Abundant Phytocannabinoids in rheumatoid arthritis: Therapeutic targets and molecular processes identified using integrated bioinformatics and network pharmacology. *Life*, *13*(3), 1–19. Doi: 10.3390/life13030700.
65. Kumar S., & Reddy, P. H. (2016). Are circulating microRNAs peripheral biomarkers for Alzheimer ' s disease ? *Biochimica et Biophysica Acta - Molecular Basis of Disease*, *1862*(9), 1617–1627. Doi: 10.1016/j.bbadis.2016.06.001.
66. Nagaraj, S., Zoltowska, K. M., Laskowska-kaszub, K., & Wojda, U. (2019). microRNA diagnostic panel for Alzheimer's disease and epigenetic trade-off between neurodegeneration and cancer. *Ageing Research Reviews*, *49*, 125–143. Doi: 10.1016/j.arr.2018.10.008.
67. Xia, X., Wang, Y., Huang, Y., Zhang, H., Lu, H., & Zheng, J. C. (2019). Exosomal miRNAs in central nervous system diseases: biomarkers, pathological mediators, protective factors and therapeutic agents. *Progress in Neurobiology*, *183*, 101694–101715. Doi: 10.1016/j.pneurobio.2019.101694.
68. Saeedi, S., Israel, S., Nagy, C., & Turecki, G. (2019). The emerging role of exosomes in mental disorders," *Translational Psychiatry*, *9*(1), 122–132. Doi: 10.1038/s41398-019-0459-9.
69. Rupaimoole R., & Slack, F. J. (2017). MicroRNA therapeutics: Towards a new era for the management of cancer and other diseases. *Nature Reviews Drug Discovery*, *16*(3), 203–221. Doi: 10.1038/nrd.2016.246.
70. Wang, L., Shui, X., Diao, Y., Chen, D., Zhou, Y., & Lee, T. H. (2023). Potential implications of miRNAs in the pathogenesis, diagnosis, and therapeutics of Alzheimer's disease. *International Journal of Molecular Sciences*, *24*, 16259–16282.
71. Winkle, M., El-Daly, S. M., Fabbri, M., & Calin, G. A. (2021). Noncoding RNA therapeutics — Challenges and potential solutions. *Nature Reviews Drug Discovery*, *20*(8), 629–651. Doi: 10.1038/s41573-021-00219-z.
72. Nandi, A., Auti, P. S., Jagtap, U. A., & Paul, A. T. (2024). Investigating the role of indole and quinazolinone-based hybrid analogues with ketoamide fragment and alkyl extension for potential PL inhibition. *Journal of Molecular Structure*, *1301*, 137337. Doi: 10.1016/j.molstruc.2023.137337.
73. Diener, C., Keller, A., & Meese, E. (2022). Emerging concepts of miRNA therapeutics : From cells to clinic. *Trends in Genetics*, *38*(6), 613–626. Doi: 10.1016/j.tig.2022.02.006.
74. Yu, A. M., Choi, Y. H., & Tu, M. J. (2020). RNA drugs and RNA targets for small molecules: Principles, progress, and challenges. *Pharmacological Reviews*, *72*(4), 862–898. Doi: 10.1124/pr.120.019554.
75. Dammes, N., & Peer, D. (2020). Paving the road for RNA therapeutics. *Trends in Pharmacological Sciences*, *41*(10), 755–775. Doi: 10.1016/j.tips.2020.08.004.
76. Martinez, B., & Peplow, P. V. (2017). MicroRNAs as diagnostic and therapeutic tools for Alzheimer's disease : Advances and limitations. *Neural Regeneration Research*, *14*(2), 242–255. Doi: 10.4103/1673-5374.244784.

77. Mueller, S. G., Schuff, N., Yaffe, K., Madison, C., Miller, B., & Weiner, M. W. (2010). Hippocampal atrophy patterns in mild cognitive impairment and Alzheimer's disease. *Human Brain Mapping, 31*, 1339–1347. Doi: 10.1002/hbm.20934.
78. La, R., et al. (2013). NeuroImage : Clinical Hippocampal sub field volumetry in mild cognitive impairment , Alzheimer's disease and semantic dementia. *NeuroImage: Clinical Journal, 3*, 155–162. Doi: 10.1016/j.nicl.2013.08.007.
79. Pluta, J., Yushkevich, P., Das, S., & Wolk, D. (2012). In vivo analysis of Hippocampal subfield atrophy in mild cognitive impairment via semi-automatic segmentation of T2-weighted MRI. *Journal of Alzheimer's Disease, 31*, 85–99. Doi: 10.3233/JAD-2012-111931.
80. Auti, P. S., Nandi, A., Kumari, V., & Paul, A. T. (2022). Design, synthesis, biological evaluation and molecular modelling studies of oxoacetamide warhead containing indole-quinazolinone based novel hybrid analogues as potential pancreatic lipase inhibitors. *Royal Society of Chemistry, 46*(24), 11661. Doi: 10.1039/d2nj01210c.
81. Auso, E., Violeta, G.-V., & Esquiva, G. (2020). Biomarkers for Alzheimer's disease early diagnosis. *Journal of Personalized Medicine, 10*, 114–139.
82. Keller, A., et al., (2016). Validating Alzheimer's disease micro RNAs using next-generation sequencing. *Alzheimer's & Dementia, 1*, 1–12. Doi: 10.1016/j.jalz.2015.12.012.
83. Li, D., et al. (2018). Effect of regulatory network of exosomes and microRNAs on neurodegenerative diseases. *Chinese Medical Journal, 131*(18), 2216–2225. Doi: 10.4103/0366-6999.240817.
84. Groot, M., & Lee, H. (2020). Sorting mechanisms for MicroRNAs into extracellular vesicles and their associated diseases. *Cells*, 9(Dm), 1044–1059.
85. Upadhyay, D. B., et al. (2024). Indole clubbed 2,4-thiazolidinedione linked 1,2,3-triazole as a potent antimalarial and antibacterial agent against drug-resistant strain and molecular modeling studies. *Archiv der Pharmazie* (December 2023), 1–20. Doi: 10.1002/ardp.202300673.
86. Paraskevaidi, M., Allsop, D., Karim, S., Martin, F. L., & Crean, S. (2020). Diagnostic biomarkers for Alzheimer's disease using non - invasive specimens. *Journal of Clinical Medicine, 9*, 1–19. Doi: 10.3390/jcm9061673.
87. Lewczuk, P., *et al.*, (2018). Plasma neurofilament light as a potential biomarker of neurodegeneration in Alzheimer's disease. *Alzheimer's Research & Therapy, 10*, 1–10.
88. Chattaraj, B., et al. (2023). Enhydra fluctuans Lour. aqueous extract inhibited the growth of calcium phosphate crystals: An in vitro study. *Food Chemistry Advances, 2*, (August 2022), 100287. Doi: 10.1016/j.focha.2023.100287.
89. Klyucherev, T. O., et al. (2022). Advances in the development of new biomarkers for Alzheimer ' s disease. *Translational Neurodegeneration, 11*, 1–24. Doi: 10.1186/s40035-022-00296-z.
90. Eddin, A. N., et al. (2023). Cerebrospinal fluid microRNAs as potential biomarkers in Alzheimer ' s disease. *Frontiers in Aging Neuroscience*, 1–15. Doi: 10.3389/fnagi.2023.1210191.
91. Bekris, L. M., & Leverenz, J. B. (2015). The biomarker and therapeutic potential of miRNA in Alzheimer ' s disease. *Neurodegenerative Disease Management, 5*(1), 61–74. Doi: 10.2217/nmt.14.52.

92. Marx, J., (2007). Alzheimer's disease: A new take on tau. *Science, 316*(5830), 1416–1417. Doi: 10.1126/science.316.5830.1416.
93. Craft, S. (2006). Insulin resistance syndrome and Alzheimer disease: Pathophysiologic mechanisms and therapeutic implications. *Alzheimer Disease and Associated Disorders, 20*(4), 298–301. Doi: 10.1097/01.wad.0000213866.86934.7e.
94. Fehlbaum-Beurdeley, P., et al. (2012). Validation of AclarusDx™, a blood-based transcriptomic signature for the diagnosis of Alzheimer's disease. *Journal of Alzheimer's Disease, 32*(1), 169–181. Doi: 10.3233/JAD-2012-120637.
95. Xu, X. H., Huang, Y., Wang, G., & Di Chen, S. (2012). Metabolomics: A novel approach to identify potential diagnostic biomarkers and pathogenesis in Alzheimer's disease," *Neuroscience Bulletin, 28*(5), 641–648. Doi: 10.1007/s12264-012-1272-0.
96. Nandi, A., Nigar, T., Das, A., & Dey, Y. N. (2023). Network pharmacology analysis of Plumbago zeylanica to identify the therapeutic targets and molecular mechanisms involved in ameliorating hemorrhoids. *Journal of Biomolecular Structure and Dynamics, 0*(0), 1–15. Doi: 10.1080/07391102.2023.2280681.
97. Ishikawa, M., Tajima, Y., Murayama, M., Senoo, Y., Maekawa, K., & Saito, Y. (2013). Plasma and serum from nonfasting men and women differ in their lipidomic profiles. *Biological and Pharmaceutical Bulletin, 36*(4), 682–685. Doi: 10.1248/bpb.b12-00799.
98. Wang, W., et al. (2008). The expression of MicroRNA miR-107 decreases early in Alzheimer's disease and may accelerate disease progression through regulation of$^{N}_{L}$-site amyloid precursor protein- cleaving Enzyme 1. *The Journal of Neuroscience, 28*(5), 1213–1223. Doi: 10.1523/JNEUROSCI.5065-07.2008.
99. Frigerio, C. S., et al. (2013). Reduced expression of hsa-miR-27a-3p in CSF of patients with Alzheimer disease. *Neurology, 81*(24), 2103–2106. Doi: 10.1212/01.wnl.0000437306.37850.22.
100. Tan, L. et al. (2014). Circulating miR-125b as a biomarker of Alzheimer's disease. *Journal of the Neurological Sciences, 336*(1–2), 52–56. Doi: 10.1016/j.jns.2013.10.002.
101. Schipper, H. M., Maes, O. C., Chertkow, H. M., & Wang, E. (2007). MicroRNA expression in Alzheimer blood mononuclear cells. *Gene Regulation and Systems Biology, 1*, 263–274. Doi: 10.4137/grsb.s361.
102. Lukiw, W. J. (2007). Micro-RNA speciation in fetal, adult and Alzheimer's disease hippocampus. *Molecular Neuroscience, 18*(3), 297–300. Doi: 10.1097/WNR.0b013e3280148e8b.
103. Wang, W. X., Huang, Q., Hu, Y., Stromberg, A. J., & Nelson, P. T. (2011). Patterns of microRNA expression in normal and early Alzheimer's disease human temporal cortex: White matter versus gray matter. *Acta Neuropathologica, 121*(2), 193–205. Doi: 10.1007/s00401-010-0756-0.
104. Shioya, M., et al. (2010). Aberrant microRNA expression in the brains of neurodegenerative diseases: MiR-29a decreased in Alzheimer disease brains targets neurone navigator 3," *Neuropathology and Applied Neurobiology, 36*(4), 320–330. Doi: 10.1111/j.1365-2990.2010.01076.x.
105. Kumar, P., et al. (2013). Circulating miRNA biomarkers for Alzheimer's disease. *PLoS ONE, 8*(7), 1–10. Doi: 10.1371/journal.pone.0069807.

106. Li, N., Li, Q., Tian, X. Q., Qian, H. Y., & Yang, Y. J. (2014). Mipomersen is a promising therapy in the management of hypercholesterolemia: A meta-analysis of randomized controlled trials. *American Journal of Cardiovascular Drugs, 14*(5), 367–376. Doi: 10.1007/s40256-014-0077-0.
107. Cuchel, M., et al. (2014). Homozygous familial hypercholesterolaemia: New insights and guidance for clinicians to improve detection and clinical management. A position paper fromthe Consensus Panel on Familial Hypercholesterolaemia of the European Atherosclerosis Society. *European Heart Journal, 35*(32), 2146–2157. Doi: 10.1093/eurheartj/ehu274.
108. Miller, T. M., et al. (2013). An antisense oligonucleotide against SOD1 delivered intrathecally for patients with SOD1 familial amyotrophic lateral sclerosis : A phase 1 , randomised , fi rst-in-man study. *Lancet Neurology, 12*, 435–442. Doi: 10.1016/S1474-4422(13)70061-9.

CHAPTER 10

Cellular Symphony: Harmonizing Therapeutic Approaches for Alzheimer's Disease Through Targeted Intervention Strategies

SHUBHRAT MAHESHWARI[1], ANAS ISLAM[2], PURBA MANDAL[2], ADITYA SINGH[2*], ASAD AHMAD[2], MOHAMMAD IRFAN[2], BADRUDDEEN[2], MOHAMMAD AHMAD[2], and JUBER AKHTAR[2*]

[1]*Bioorganic and Medicinal Chemistry Research Laboratory, Department of Pharmaceutical Sciences, Sam Higginbottom University of Agriculture, Technology and Sciences, Prayagraj, Uttar Pradesh, India*

[2]*Department of Pharmaceutics, Parul Institute of Pharmacy and Research, Parul University, Waghodia, Vadodara, Gujarat, India*

*Corresponding author

ABSTRACT

Alzheimer's disease (AD) presents a formidable challenge in the landscape of neurodegenerative disorders, characterized by progressive cognitive decline and functional impairment. Despite extensive research efforts, therapeutic options for AD remain limited, with available treatments offering only symptomatic relief. The shifting focus toward early disease detection reflects a paradigmatic change toward preventive strategies, emphasizing the importance of personalized medicine in AD management. Recent advancements in AD genetics and molecular characteristics have profound implications for both clinical research and personalized

Cellular and Molecular Mechanisms in Alzheimer's Disease Therapeutics, Volume 1: Cellular Mechanisms. Prashant Tiwari, Sunil Kumar Kadiri, and Kunjbihari Sulakhiya (Eds.)

DOI: 10.1201/9781779641793-10

medicine, enabling risk assessment and individualized preventive strategies. However, the ethical implications of disclosing genetic risk information remain subjects of ongoing debate. The integration of genetic and molecular insights into personalized therapies heralds a new frontier in AD management, with the potential to revolutionize treatment and prevention strategies. Yet, this transformative potential is accompanied by a myriad of ethical dilemmas and practical challenges. Against the backdrop of burgeoning scientific knowledge and technological advancements, navigating the ethical landscape in AD care and research requires robust ethical frameworks to guide decision-making practices. Moreover, as personalized medicine in AD becomes increasingly feasible, it is imperative to examine the societal impact and distributional justice implications of such approaches. This comprehensive exploration of the ethical considerations surrounding personalized medicine in AD aims to inform ethical decision-making practices that prioritize the well-being and autonomy of individuals affected by this devastating disease. By synthesizing current research, ethical frameworks, and practical insights, this study contributes to the ongoing dialogue surrounding the ethical dimensions of personalized medicine in AD.

10.1 INTRODUCTION

Alzheimer's disease (AD) is a significant neurodegenerative disorder characterized by the gradual decline of cognitive abilities and functional abilities [1]. As the predominant form of dementia globally, its impact on individuals, families, and societies is profound, manifesting as disability and dependence among older populations worldwide [2]. Despite extensive research efforts, therapeutic options for AD remain limited, with available treatments offering only modest relief from symptoms. Current AD treatment approaches often concentrate on cholinergic neurotransmission modulation to alleviate late-stage symptoms. However, just one drug received conditional approval to slow the course of the illness in its early stages, and its therapeutic benefits are still debatable [3]. However, the shifting focus of research toward early disease detection reflects a paradigmatic change from curative to preventive strategies [4]. This transition underscores the growing recognition of the importance of personalized medicine in AD management, tailoring interventions to individuals' genetic and molecular profiles [5]. AD has a complex genetic profile that includes a wide range of genes that are risk factors for late-onset AD and uncommon detrimental variants in early-onset forms. Recent advancements in AD genetics hold significant importance for clinical research and personalized treatment. It allows for the evaluation of disease risk and the creation of tailored preventative interventions. However, the moral consequences of revealing genetic risk information to mentally capable persons either in research trials or requesting direct-to-consumer genetic testing continue to be a topic of continuous discussion [6]. As we delve into the intricate ethical considerations surrounding AD, it becomes evident that the integration of genetic and molecular insights into personalized therapies heralds a new frontier in AD management [7]. The advent of precision medicine holds promise for tailoring interventions to individuals' unique genetic makeup and molecular characteristics, potentially revolutionizing the landscape of AD treatment and prevention. However, this transformative potential is accompanied by a myriad of ethical dilemmas and practical challenges that warrant careful consideration [8]. Against the backdrop of burgeoning scientific knowledge and technological advancements, the ethical discourse surrounding AD encompasses multifaceted issues, ranging from the complexities of genetic risk assessment to the implications of disclosing predictive genetic information. The intersection of genetic determinism, autonomy, privacy, and equity underscores the need for robust ethical frameworks to guide clinical practice, research endeavors, and policy formulation in the field of AD. Moreover, as we navigate the ethical terrain of customizing therapies based on individual genetics and molecular profiles, it is imperative to critically examine the societal impact and distributional justice implications of personalized medicine in AD [9]. In this chapter, we embark on a comprehensive exploration of the ethical considerations underpinning the customization of therapies according to individual genetics and molecular characteristics of AD [10]. By synthesizing current research, ethical frameworks, and practical insights, we aim to elucidate the complexities of navigating the ethical landscape in AD care and research [11]. We aim to contribute to the ongoing discussion on the ethical aspects of individualized treatment for Alzheimer's illness through careful analysis and deep reflection. Our goal is to provide insights that can guide ethical decision-making, with a focus on prioritizing the well-being and autonomy of individuals affected by this debilitating condition [12].

10.2 ALZHEIMER'S DISEASE: GENETICS AND MOLECULAR CHARACTERISTICS

AD involves a complex brain disorder causing long-term memory loss, cognitive issues, and behavioral changes. These symptoms gradually lead to a loss of independence [13]. AD is characterized histopathologically by two main features: the presence of improperly phosphorylated Tau protein within cells, which results in the creation of neurofibrillary tangles (NFTs) (Figure 10.1). Familial AD, which represents a minority of cases, is associated with mutations in genes encoding amyloid precursor protein (APP), Presenilin 1 (PSEN1), and Presenilin 2 (PSEN2), disrupting the physiological processing of the Aβ peptide. Mutations in PSEN1 and PSEN2 result in severe forms of AD with an early start, and changes in Aβ production caused by APP mutations. The apolipoprotein E (APOE) gene's ε4 allele is consistently associated in sporadic AD, with a three-to-eight-fold increased risk [14]. APOE impacts Aβ removal, tau-induced neuron degeneration, and microglial cell responses, all contributing to AD development [15]. The identification of a wide array of genes associated with AD has significantly advanced our understanding of its pathogenesis, providing insights into potential early biomarkers and therapeutic targets. These genes, collectively, have the potential to explain a substantial portion of late-onset Alzheimer's disease (LOAD) cases, shedding light on novel disease-related mechanisms [16]. The amyloid hypothesis is a key concept in understanding AD pathophysiology, suggesting that it implicates genes involved in pathological Aβ processing [17]. The most important component in this cascade is the disintegrin and ADAM10 (metalloproteinase domain-containing protein, which controls α-secretase activity and modifies the Aβ42/Aβ40 ratio [18]). Phosphatidylinositol-binding clathrin assembly protein (PICALM) and clustered mitochondria protein homolog (CLU) contribute to neuritic plaque formation and amyloid fibril inhibition, respectively. However, while crucial, the amyloid hypothesis alone fails to fully elucidate the complexity of AD [19].

Beyond Aβ pathology, the neuroimmunomodulation hypothesis has emerged, highlighting the role of immune-related genes and glial cell responses in AD pathogenesis. Genes such as transcription factor PU.1 (SPI1) and myeloid cell surface antigen CD33 modulate microglial activation and cytokine production, contributing to neuroinflammation and Tau protein dysregulation [20]. Similarly, genes involved in Tau metabolism (shown in Table 10.1), including triggering receptor expressed

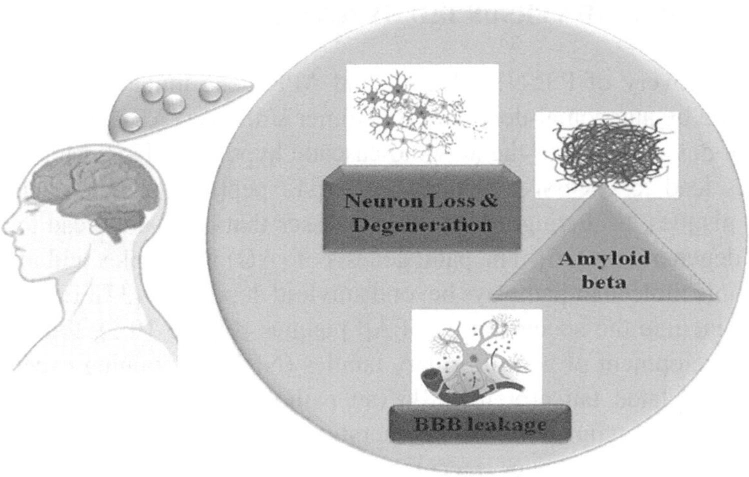

FIGURE 10.1 Process of Alzheimer's disease.

on myeloid cells 2 (TREM2) and bridging integrator 1 (BIN1), further exacerbate neurodegeneration through aberrant signaling pathways [21]. The infectious hypothesis suggests a connection between pathogens and AD pathology. Genes like paired immunoglobulin-like type 2 receptor alpha (PILRA) and apolipoprotein E (APOE) ε4 allele may mediate viral interactions and microglial dysfunction, exacerbating neuroinflammation and amyloid aggregation, moreover, metabolic pathways [22].

TABLE 10.1 Genes Involved in Molecular Mechanisms Implicated in AD.

Gene/Location	References
ABCA7 (19p13.3)	[23]
CD2A (6p12.3)	[24]
ADAM10 (15q21.3)	[25]
SORL1 (11q24.1)	[26]
CR1 (1q32.2)	[27]
CD33 (19q13.3)	[28]
MAPT (17q21.21)	[29]
CLU (8p21.1)	[30]
TOMM40 (19q13.32)	[31]
INSR (19p13.2)	[32]
PILRA (7q22.1)	[33]
APH1B (15q22.2)	[34]

10.3 AMYLOID VERSUS TAU HYPOTHESES

The discovery of PSEN1, PSEN2, and APP, gene mutations in families affected by autosomal-dominant Alzheimer's disease (ADAD) was crucial in the development of the amyloid cascade hypothesis [35]. Mutations in genes lead to excessive synthesis of Aβ42 peptides, which accumulate into plaques, triggering a series of processes that ultimately lead to nerve cell degeneration [36]. The pathogenesis of (AD) is complex and involves multiple molecular pathways beyond amyloid deposition [37]. Early symptoms include the accumulation of Aβ plaques (Figure 10.2), followed by the development of neurofibrillary tangles (NFTs) containing excessively phosphorylated tau protein [38]. Tau pathology is more closely linked to neuronal dysfunction and degeneration than synaptic dysfunction and inflammation. Moreover, recent studies suggest that Aβ-induced capillary constriction may contribute to decreased cerebral blood flow, further exacerbating neuronal dysfunction [39]. Neuronal hyperactivity is another early feature of AD, driven by the accumulation of soluble Aβ [40]. This hyperactivity may be targeted therapeutically by modulating excitatory amino acid transporters, offering a potential strategy to mitigate early disease progression. Additionally, immune response and inflammation play critical roles in AD pathology, with microglial activation by Aβ deposits initiating an inflammatory cascade that promotes further Aβ accumulation and tau pathology [41]. Detractors of the amyloid cascade theory contend that Aβ can be an indirect consequence rather than the fundamental catalyst of AD [42]. The study suggests that AD is primarily caused by dysfunction in the metabolism of amyloid precursor protein (APP), with tau-related abnormalities playing a crucial role [43]. Injection of tau from AD-affected brains into transgenic mice with high Aβ levels leads to tau seed accumulation and tau pathology, indicating a crucial role for tau in the disease progression [44]. Epigenetic modifications, like histone acetylation, may contribute to the development of AD [45]. Histone 3 lysine 9 acetylation (H3K9ac) marks have been associated with tau pathology, indicating early chromatin reorganization induced by tau protein, independent of Aβ deposition [46]. APOE isoforms influence Aβ deposition, tau pathology, microglial activation, and lipid metabolism, highlighting its polyvalent role in AD pathogenesis [47]. Research on AD has advanced mostly because of our growing understanding of how hereditary factors influence the disease. Early genome-wide association studies (GWAS) highlighted the importance of the immunological pathway in AD pathogenesis by identifying several genetic risk factors (including BIN1, CD33, CLU, CR1, and PICALM) for the disease [48]. Treating AD via targeting APOE-related pathways may be a promising therapeutic approach [49]. Further investigations using bigger datasets identified additional genes that are significantly associated with immune response on a genome-wide scale, including ABCA7, CD2AP, EPHA1, and MS4A. The discovery of loss-of-function mutations in TREM2 is particularly intriguing since it offers researchers precise targets to explore the involvement of the immune response in AD pathogenesis [50]. TREM2, along with SPI1 (encoding PU.1), ADAM10, and ACE, has shown genome-wide significant associations in large GWAS, underscoring their importance in AD pathogenesis. While ADAM10 is involved in α-secretase activity for APP processing and ACE is implicated in Aβ degradation, the precise mechanisms linking these genes to microglial dysfunction remain unclear. Excessive ACE production in microglia and macrophages in AD [51] mice models significantly reduced soluble Aβ42 and amyloid plaque levels, suggesting potential therapeutic intervention. The activation of p38MAPK signaling in microglia, which is initiated by Aβ plaques, stimulates the secretion of results in proinflammatory cytokine production, inflammation, and tau phosphorylation [52]. Changes in PPAR/p38MAPK signaling have also been associated with the dysregulation of TREM2, which influences the reprogramming of microglial metabolism from oxidative phosphorylation to glycolysis in response to Aβ plaques. Conversely, metabolic reprogramming is impaired in iPSC-derived microglia derived from patients with pathogenic TREM2 mutations; this suggests a possible association between dysfunctional TREM2 proteins and aberrant microglial metabolism in AD [53]. Microglial transmembrane proteins CD33 and TREM2 play contrasting roles in modulating microglial phagocytosis in response to Aβ plaques. While CD33 activation inhibits phosphatidylinositol-3 kinase (PI3K), dampening microglial phagocytosis, TREM2 activation promotes phagocytosis by activating PI3K [54]. Research has indicated that inhibiting the expression of TREM2 may have adverse effects on amyloid levels, whereas downregulation of CD33 may have positive implications in this regard [55]. TREM2 ablation in AD mouse models intensifies neurodegeneration and inflammation, contrary to the correlation between elevated TREM2 levels and reduced cognitive decline in AD and MCI. Research indicates that MS4A4A may indirectly regulate TREM2 expression, with a prevalent variant linked to elevated sTREM2 levels in cerebrospinal fluid [56]. Elevated levels of sTREM2 may confer protection against AD, according to Mendelian randomization analyses; this underscores the potential therapeutic importance of targeting MS4A4A [57]. The interaction between microglia and the autophagy pathway is crucial in AD pathology, with deficiencies in the lysosome and autophagosome pathways identified in iPSC-derived neurons from patients with pathogenic mutations in PSEN1 and APP [58]. Dysfunctional autophagy pathways in microglia may influence their response to Aβ plaques, as demonstrated by the importance of autophagy regulators like ATG5 in Aβ clearance [59]. The lysosome and autophagosome dysfunction-related TMEM106B has been linked to a higher risk of AD, especially when combined with APOE. Recent findings suggest that TMEM106B may regulate neuronal proportion in adults, with impaired lysosomal function contributing to abnormal accumulation of toxic components in neurons [60].

The landscape of therapeutic options for neurodegenerative diseases, including AD, remains limited, with currently available treatments offering only symptomatic relief rather than disease modification [61]. The two main pharmacological therapies for AD are memantine, an NMDA receptor modulator, and acetylcholinesterase inhibitors (AChE-I), which include galantamine, donepezil, and rivastigmine. Although these drugs may provide modest improvements in cognitive and functional symptoms, they do not significantly alter the overall course of the disease or its outcome [62]. Similarly, amyotrophic lateral sclerosis (ALS) has only one approved drug therapy, riluzole, which extends lifespan by a few months compared with placebo [63]. Currently, there are no authorized therapy options available for additional neurodegenerative illnesses such as frontotemporal dementia (FTD). Although epidemiological studies have linked many preventative strategies, such as aerobic physical exercise, to a decreased incidence of AD, there is still a paucity of trustworthy data from prospective controlled trials. However, using aerobic exercise as a kind of intervention has been able to change cognitive deficiencies in individuals suffering from moderate cognitive impairment (MCI) to some extent, suggesting that this might be a way to stop the deterioration from continuing [64]. Despite improvements in the early identification of the characteristic pathophysiological indications of AD, such as those achieved by biomarker-based diagnostic techniques, there are still very few treatment and prevention options available [65]. The "therapeutic and preventive vacuum" raises questions about the benefits and potential drawbacks of preclinical and early predementia diagnosis, as well as the appropriate conditions for conducting biomarker-based early diagnostic procedures [66]. An argument supporting the use of early diagnostic biomarkers is that it allows patients to contemplate the consequences of the diagnosis and how it will affect their lives [67]. This enables people to possibly modify their life goals while their cognitive and functional abilities still allow for independent decision-making [68]. Nevertheless, there are possible adverse medical and psychological consequences linked to the detection of biomarkers in asymptomatic persons. Predictive accuracy of biomarker-supported diagnostic criteria is crucial, as false-positive diagnoses may emotionally affect individuals with a poor prognosis, leading to increased stress and depressive reactions. Individuals diagnosed with predementia AD should receive psychological support to manage their emotional burden [69]. There are apprehensions regarding the possible exploitation of predictive biomarker data by entities like insurance companies and employers, which may be interested in precise future mental health and cognitive performance prognoses. Privacy concerns regarding the use of biomarker and genetic information in medical records also need to be addressed [70].

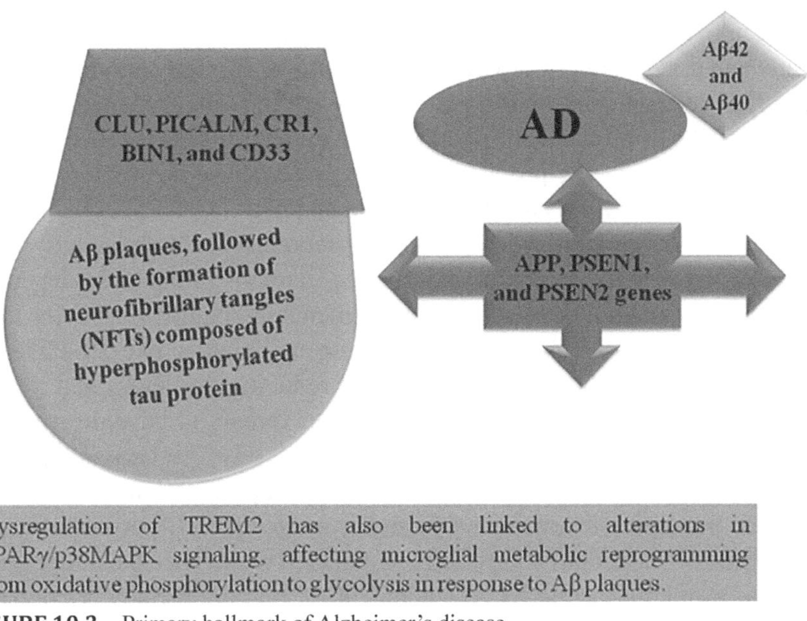

FIGURE 10.2 Primary hallmark of Alzheimer's disease.

10.4 THERAPEUTIC APPROACHES FOR ALZHEIMER'S DISEASE THROUGH TARGETED INTERVENTION STRATEGIES

Aβ peptides, Aβ42 and Aβ40, which are the main constituents of amyloid plaques in (AD) patients, are produced by the breakdown of amyloid precursor protein (APP) by β-Secretase (BACE), sometimes referred to as beta-site amyloid precursor protein-cleaving enzyme (BACE1) [71]. One intriguing treatment approach to lower Aβ generation and lessen AD pathogenesis is to inhibit BACE1 activity. Various classes of BACE1 inhibitors, including non-peptidic and peptidomimetic agents, have been developed and evaluated for their efficacy and safety profiles. One notable BACE1 inhibitor is KMI-429, which incorporates a hydroxymethylcarbonyl (HMC) isostere in its structure. KMI-429 has demonstrated significant BACE1-blocking activity *in vivo* and *in vitro* models, showing dose-dependent reductions in soluble APP and Aβ levels. Moreover, its structural properties, including favorable blood–brain barrier permeation and low molecular mass, hold promise for improved CNS drug delivery [72]. However, challenges such as selectivity toward BACE1 and potential off-target effects remain significant hurdles in the development of BACE1 inhibitors. BACE2, an isoform of BACE1, shares structural similarities but exhibits differential tissue expression patterns [73]. Strategies to enhance selectivity toward BACE1 over BACE2 are actively pursued to minimize unwanted side effects. Apart from preventing BACE1, another possible therapy for AD is to focus on γ-secretase, an intricate enzyme that breaks down APP to produce Aβ peptides. To regulate the APP processing and precisely modify the production of Aβ, γ-Secretase modulators (GSMs) and inhibitors (GSIs) have been developed [74].

BIIB042, an orally administered γ-secretase modulator, has shown promising results in preclinical studies, demonstrating dose-dependent reductions in Aβ42 levels and improvements in cognitive behavior in the animal model with AD [75]. γ-secretase modulators like NGP555 and SGSM-36 have shown effectiveness in reducing Aβ levels and altering Aβ aggregate conformation, but challenges remain to be addressed [76]. Off-target effects, limited selectivity, and adverse reactions observed in clinical trials underscore the need for further optimization of drug candidates. Strategies to enhance selectivity, improve BBB permeability, and mitigate off-target effects are actively pursued to advance the development of safer and more effective AD therapeutics. Immunotherapy is a potential technique for preventing and treating (AD) [77]. It uses the body's immune response to target amyloid-beta (Aβ) peptides, which are key to AD pathogenesis. This method includes two forms of immunotherapy: active and passive [78]. Active immunotherapy stimulates the immune system to respond to Aβ peptides, whereas passive immunotherapy uses externally generated antibodies to target Aβ [79].

Synthetic Aβ-42 fragments are utilized in active immunotherapy to stimulate T and B cells, thereby activating microglia and generating a cellular and humoral response, potentially enhancing the elimination of Aβ plaques and reducing AD damage [80]. However, challenges such as the risk of generating an immune response and adverse effects, such as meningoencephalitis, have been associated with this approach. Alternative strategies, including the incorporation of Aβ fragments with carrier proteins, have shown promise in reducing AD pathology without triggering an immune response [81]. Passive immunotherapy involves the administration of externally produced antibodies that target Aβ peptides. These antibodies can effectively bind to Aβ plaques, preventing their accumulation and promoting their clearance [82]. Passive immunotherapy has shown significant improvements in cognitive function and reductions in Aβ peptide levels in animal models of AD. Additionally, genetically engineered antibodies, such as BAM-10, have been developed to enhance their therapeutic potential. Targeting Aβ plaques more effectively has been made possible by the use of techniques like magnetic resonance imaging-guided transcranial focused ultrasound (MRIgFUS), which improves the delivery of antibodies across the blood–brain barrier (BBB) [83].

Furthermore, the complement system has been implicated in AD pathology, with complement proteins found to be associated with amyloid deposits in the brain. Strategies aimed at modulating complement activation may offer therapeutic potential in AD [84]. Furthermore, the physiological mechanisms responsible for removing Aβ peptides, including the glymphatic system, lysosome-induced degradation, and ubiquitin-proteasome pathway, are essential factors in the development of AD [85]. Facilitating these clearance routes may aid in preventing the buildup of harmful Aβ peptides and relieving the pathology associated with AD [86]. The discovery of the amyloid peptide fragment KLVFF as an aggregation inhibitor in 1996 marked a significant milestone in the development of peptide-based ligands targeting Aβ (amyloid-beta) peptides, which are central to AD pathology [87]. Subsequent research efforts have focused on identifying peptides derived from the amyloid sequence itself that exhibit high affinity for Aβ peptides and have the potential to influence fibril formation dynamics. Through systematic testing and optimization, the pentapeptide KLVFF was identified as a potent inhibitor of Aβ peptide aggregation. This sequence, along with variations such as KLXXF, has been shown to reduce the toxicity of Aβ peptides and modulate the kinetics of fibril aggregation [88].

Further modifications to the KLVFF sequence, such as the addition of repeating oligoproline units, have been explored to enhance the inhibitory effects on Aβ aggregation. These modifications disrupt the formation of β-sheets and impair the stacking capability of amyloid peptides, thereby preventing the growth of larger fibrils [89]. Additionally, peptide sequences with similar binding characteristics to KLVFF have been developed through amino acid shuffling and structural modifications, highlighting the importance of hydrophobic interactions with amyloid binding. Design criteria for amyloid aggregation inhibitors include factors such as binding affinity, stability, and permeability across biological barriers like the blood–brain barrier (BBB) [90]. Peptides derived from the KLVFF sequence, such as OR2, have been engineered with terminal charged residues and glycine spacers to enhance solubility and impair fibril formation. Additionally, modifications such as replacing amino acids with their D-enantiomers have been employed to improve proteolytic stability and reduce immunogenicity. The "retro inverso" form of peptide inhibitors, like RI-OR2, has shown promise in blocking Aβ oligomerization and protecting neuronal cells from Aβ-induced toxicity [91].

Computer-aided drug design (CADD) techniques have been utilized to design and screen peptide inhibitors with enhanced antiaggregatory properties [92]. By analyzing molecular dynamics simulations and employing fluorescence assays, researchers have identified inhibitors capable of disrupting interactions between Aβ monomers and inhibiting fibril formation. These inhibitors, along with variations such as D-amino acid substitutions, are being further tested *in vitro* and *in vivo* to evaluate their therapeutic potential for AD. Peptide inhibitors offer advantages over monoclonal antibodies (MAbs) in terms of production cost, versatility, and safety [93]. Strategies to enhance the BBB permeability of peptide inhibitors, such as incorporating cell-penetrating peptides (CPPs), have been explored to improve their delivery to the central nervous system. Additionally, bi-specific MAbs targeting both amyloid peptides and BBB characteristics have been investigated to enhance drug delivery for AD treatment. Peptide inhibitors known as autacoid local injury antagonist amides (ALIAmides) have recently become a new and innovative treatment approach for (AD). Compounds like palmitoyl ethanol amide (PEA) have anti-inflammatory properties and have shown efficacy in mitigating neuroinflammation associated with AD. PEA and its derivatives have demonstrated neuroprotective effects in animal models of spinal cord injury and traumatic brain injury by modulating inflammatory pathways and promoting the release of neurotrophic factors. Furthermore, PEA has been shown to alleviate amyloid peptide-induced astrogliosis and promote angiogenesis, offering potential therapeutic benefits in AD [94]. ALIAmides, a treatment for AD, is a promising approach that targets neuroinflammatory pathways and promotes neuroprotection.

Active immunotherapy with AN1792 was a pioneering approach in AD treatment, but it faced significant setbacks. A clinical investigation in 2003 was halted when 6% of patients developed meningoencephalitis, leading to neurological and cognitive impairment in some cases [95]. This adverse reaction highlighted the challenges associated with immune-based therapies for AD. Further analysis of the epitopes showed that the immune response to AN1792 was specifically targeted at the N-terminal residues of the Aβ peptide. Studies conducted on mice demonstrated that antibodies directed against the Aβ N-terminal region effectively suppressed neuronal toxicity and the production of fibrils, indicating a promising therapeutic approach [96]. A monoclonal antibody called bepineuzumab was created with the express purpose of targeting the Aβ N-terminal region in AD treatment [97]. However, phase 1 clinical trials reported MRI-related abnormalities, including vasogenic edema, which resolved over time. Vasogenic edema is caused by changes in vascular permeability due to the binding of bapineuzumab to Aβ in blood vessel walls. Despite lower doses of bapineuzumab showing improved cognitive scores, higher doses were associated with MRI abnormalities. Beyond immunotherapy, the relationship between diabetes and AD has garnered attention. Disrupted insulin signaling, a common feature of both conditions, has led to the concept of type 3 diabetes, where AD is considered a form of diabetes affecting the brain [98]. Activation of glucagon-like peptide (GLP-1) has shown potential in decreasing the buildup of Aβ and the harmful effects on the nervous system in animal experiments, indicating a possible treatment approach. Statins, commonly used for cholesterol management, have been shown to protect against AD by activating microglia to break down Aβ peptides. The complex nature of statins' role in treating AD is highlighted by contradictory findings regarding their cognitive impacts [99].

Ibuprofen, NSAIDs, or nonsteroidal anti-inflammatory drugs have shown long-term benefits in (AD) patients by reducing Aβ levels in neurons [100] and improving cognitive function in animal experiments. Its mechanism involves inhibiting nicotinamide adenine dinucleotide phosphate oxidase activation and reducing oxidative stress. However, other NSAIDs like celecoxib [101], and naproxen did not show significant cognitive improvement in clinical trials. Innovative approaches use piezoelectric materials. These nanosheets exhibit piezoelectric properties and induce oxidative stress, leading to the destabilization of Aβ aggregates and minimizing their neurotoxic potential. This novel approach represents a potential avenue for future AD therapies [102]. The herbal and phytoconstituent industries have shown promise in providing potential benefits for various disorders, including neurological conditions like AD [103]. Proline, a β-sheet breaker, has been studied for its ability to break amyloid protofibrils and disrupt β-sheet structures [104, 105]. It promotes the production of 310 helices, loosens interatomic arrangements, impairs hydrogen bonds and salt bridges, and interacts with charged residues [106].

10.5 CONCLUSION

Advancements in understanding the genetic and molecular basis of AD are driving rapid progress in research and therapy development. Although there has been notable progress in understanding the intricate pathophysiological processes of AD, the task of converting this knowledge into successful therapies still poses a difficult obstacle. Current therapeutic options offer only symptomatic relief, highlighting the urgent need for disease-modifying interventions. The advent of precision medicine holds promise for tailoring AD therapies to individuals' unique genetic and molecular profiles, ushering in a new era of personalized medicine. By leveraging genetic and biomarker data, clinicians may identify individuals at heightened risk for AD and implement targeted preventive strategies to delay or mitigate disease progression. Moreover, personalized therapies may optimize treatment efficacy while minimizing adverse effects, enhancing overall patient outcomes. However, the pursuit of personalized medicine in AD is accompanied by ethical, social, and practical considerations that must be carefully navigated. Ethical frameworks must address issues surrounding genetic testing, informed consent, privacy, equity, and the responsible use of predictive information. The equitable distribution of personalized therapies and access to genetic testing must be ensured to mitigate disparities in healthcare delivery.

KEYWORDS

- **Alzheimer's disease**
- **dementia**
- **DTC genetic**
- **molecular characteristics**
- **neurodegenerative disorder**
- **psychological**

REFERENCES

1. Alberti, S., Dormann, D., & et al. (2019). Liquid–liquid phase separation in disease. *Annual Review of Genetics*, *53*, 171–94.
2. Ambadipudi, S., Biernat, J., Riedel, D., & et al. (2017). Liquid–liquid phase separation of the microtubule-binding repeats of the Alzheimer-related protein Tau. *Nature Communications*, *8*(1), 275.
3. Andersen, K., Launer, L. J., Dewey, M. E., & et al. (1999). Gender differences in the incidence of AD and vascular dementia: The EURODEM studies. *Neurology*, *53*(9), 1992–1997.
4. Auluck, P. K., Chan, H. E., Trojanowski, J. Q., & et al. (2002). Chaperone suppression of α-synuclein toxicity in a Drosophila model for Parkinson's disease. *Science*, *295*(5556), 865–868.
5. Babinchak, W. M., Haider, R., Dumm, B. K., & et al. (2019). The role of liquid–liquid phase separation in aggregation of the TDP-43 low-complexity domain. *Journal of Biological Chemistry*, *294*(16), 6306–6317.
6. Bateman, R. J., Xiong, C., Benzinger, T. L., & et al. (2012). Clinical and biomarker changes in dominantly inherited Alzheimer's disease. *The New England Journal of Medicine*, *367*, 795–804.
7. Beckmann, N. D., & et al. (2018). *Multiscale approaches to complex human diseases* (Doctoral dissertation, Icahn School of Medicine at Mount Sinai).

8. Bo, Y., Zhang, X., Wang, Y., & et al. (2017). The n-3 polyunsaturated fatty acids supplementation improved the cognitive function in the Chinese elderly with mild cognitive impairment: A double-blind randomized controlled trial. *Nutrients*, *9*(1), 54.
9. Brody, K. M., Taylor, J. M., Wilson, G. R., & et al. (2008). Regional and cellular localisation of Parkin co-regulated gene in developing and adult mouse brain. *Brain Research*, *1201*, 177–186.
10. Buell, A. K., Dhulesia, A., White, D. A., & et al. (2012). Detailed analysis of the energy barriers for amyloid fibril growth. *Angewandte Chemie International Edition*, *51*(21), 5247–5251.
11. Butterworth, R. F., Héroux, M., & et al. (1989). Effect of pyrithiamine treatment and subsequent thiamine rehabilitation on regional cerebral amino acids and thiamine-dependent enzymes. *Journal of Neurochemistry*, *52*(4), 1079–1084.
12. Camino, J. D., Gracia, P., Chen, S. W., & et al. (2020). The extent of protein hydration dictates the preference for heterogeneous or homogeneous nucleation generating either parallel or antiparallel β-sheet α-synuclein aggregates. *Chemical Science*, *11*(43), 11902–10914.
13. Chami, L., Checler, F., & et al. (2012). BACE1 is at the crossroad of a toxic vicious cycle involving cellular stress and β-amyloid production in Alzheimer's disease. *Molecular Neurodegeneration*, *7*(1), 1–5.
14. Davis, A. A., Leyns, C. E., Holtzman, D. M., & et al. (2018). Intercellular spread of protein aggregates in neurodegenerative disease. *Annual Review of Cell and Developmental Biology*, *34*, 545–568.
15. de Araújo, F. M., Cuenca-Bermejo, L., Fernandez-Villalba, E., & et al. (2022). Role of microgliosis and NLRP3 inflammasome in Parkinson's disease pathogenesis and therapy. *Cellular and Molecular Neurobiology*, *42*(5), 1283–1300.
16. Dorsey, M. P., Nguelifack, B. M., Yates, E. A., & et al. (2019). Colorimetric detection of mutant β-amyloid (1–40) membrane-active aggregation with biosensing vesicles. *ACS Applied Bio Materials*, *2*(11), 4966–4977.
17. Erkkinen, M. G., Kim, M. O., Geschwind, M. D., & et al. (2018). Clinical neurology and epidemiology of the major neurodegenerative diseases. *Cold Spring Harbor Perspectives in Biology*, *10*(4), a033118.
18. Fenimore, P. W., Frauenfelder, H., McMahon, B. H., & et al. (2002). Slaving: Solvent fluctuations dominate protein dynamics and functions. *Proceedings of the National Academy of Sciences*, *99*(25), 16047–16051.
19. Fitzpatrick, A. W., Debelouchina, G. T., Bayro, M. J., & et al. (2013). Atomic structure and hierarchical assembly of a cross-β amyloid fibril. *Proceedings of the National Academy of Sciences*, *110*(14), 5468–5473.
20. Flach, T. L., Ng, G., Hari, A., & et al. (2011). Alum interaction with dendritic cell membrane lipids is essential for its adjuvanticity. *Nature Medicine*, *17*(4), 479–487.
21. Fu, R. H., Liu, S. P., Huang, S. J., & et al. (2013). Aberrant alternative splicing events in Parkinson's disease. *Cell Transplantation*, *22*(4), 653–661.
22. Galvagnion, C., Brown, J. W., Ouberai, M. M., & et al. (2016). Chemical properties of lipids strongly affect the kinetics of the membrane-induced aggregation of α-synuclein. *Proceedings of the National Academy of Sciences*, *113*(26), 7065–7070.

23. Gazit, E., & et al. (2002). The "correctly folded" state of proteins: Is it a metastable state? *Angewandte Chemie International Edition*, *41*(2), 257–259.
24. Glenner, G. G., Wong, C. W., & et al. (1984). Alzheimer's disease: Initial report of the purification and characterization of a novel cerebrovascular amyloid protein. *Biochemical and Biophysical Research Communications*, *120*(3), 885–890.
25. Goh, K. I., Cusick, M. E., Valle, D., & et al. (2007). The human disease network. *Proceedings of the National Academy of Sciences*, *104*(21), 8685–8690.
26. Goodman, R. A., Lochner, K. A., Thambisetty, M., & et al. (2017). Prevalence of dementia subtypes in United States Medicare fee-for-service beneficiaries, 2011–2013. *Alzheimer's & Dementia*, *13*(1), 28–37.
27. Gorelick, P. B., Furie, K. L., Iadecola, C., & et al. (2017). Defining optimal brain health in adults: A presidential advisory from the American Heart Association/American Stroke Association. *Stroke*, *48*(10), e284–e303.
28. Greenamyre, J. T., & et al. (1986). The role of glutamate in neurotransmission and in neurologic disease. *Archives of Neurology*, *43*(10), 1058–1063.
29. Grossberg, G. T., Manes, F., Allegri, R. F., & et al. (2013). The safety, tolerability, and efficacy of once-daily memantine (28 mg): A multinational, randomized, double-blind, placebo-controlled trial in patients with moderate-to-severe Alzheimer's disease taking cholinesterase inhibitors. *CNS Drugs*, *27*, 469–478.
30. Guo, J. L., Lee, V. M., & et al. (2014). Cell-to-cell transmission of pathogenic proteins in neurodegenerative diseases. *Nature Medicine*, *20*(2), 130–138.
31. Gupta, P. P., Pandey, R. D., Jha, D., & et al. (2015). Role of traditional nonsteroidal anti-inflammatory drugs in Alzheimer's disease: A meta-analysis of randomized clinical trials. *American Journal of Alzheimer's Disease & Other Dementias®*, *30*(2), 178–182.
32. Gwozdzinska, P., & et al. (2018). Hypercapnia impairs ENaC cell surface expression and function by promoting phosphorylation and polyubiquitination of ENaC beta-subunit in alveolar epithelial cells. *Frontiers in Immunology*, *8*, 591.
33. Hanna, K., Giebel, C., Tetlow, H., & et al. (2022). Emotional and mental wellbeing following COVID-19 public health measures on people living with dementia and carers. *Journal of Geriatric Psychiatry and Neurology*, *35*(3), 344–352.
34. Hidalgo, C. A., Blumm, N., Barabási, A. L., & et al. (2009). A dynamic network approach for the study of human phenotypes. *PLoS Computational Biology*, *5*(4), e1000353.
35. Hol, E. M., Roelofs, R. F., Moraal, E., & et al. (2003). Neuronal expression of GFAP in patients with Alzheimer pathology and identification of novel GFAP splice forms. *Molecular Psychiatry*, *8*(9), 786–796.
36. Honig, L. S., Vellas, B., Woodward, M., Boada, M., Bullock, R., Borrie, M., Hager, K., Andreasen, N., Scarpini, E., Liu-Seifert, H., & Case, M. (2018). Trial of solanezumab for mild dementia due to Alzheimer's disease. *New England Journal of Medicine*, *378*(4), 321–330.
37. Hornberger, M., Piguet, O., & et al. (2012). Episodic memory in frontotemporal dementia: A critical review. *Brain*, *135*(3), 678–692.
38. Hoyer, W., Cherny, D., Subramaniam, V., & et al. (2004). Impact of the acidic C-terminal region comprising amino acids 109–140 on α-synuclein aggregation *in vitro*. *Biochemistry*, *43*(51), 16233–16242.

39. Hu, G., Agarwal, P., & et al. (2009). Human disease-drug network based on genomic expression profiles. *PLoS One*, *4*(8), e6536.
40. Huang, L. K., Chao, S. P., & Hu, C. J. (2020). Clinical trials of new drugs for Alzheimer disease. *Journal of Biomedical Science*, *27*(1), 1–3.
41. Hussain, R., Zubair, H., Pursell, S., & et al. (2018). Neurodegenerative diseases: Regenerative mechanisms and novel therapeutic approaches. *Brain Sciences*, *8*(9), 177.
42. Jarrett, J. T., Lansbury, P. T., & et al. (1992). Amyloid fibril formation requires a chemically discriminating nucleation event: Studies of an amyloidogenic sequence from the bacterial protein OsmB. *Biochemistry*, *31*(49), 12345–12352.
43. Jimenez, J. L., Guijarro, J. I., Orlova, E., & et al. (1999). Cryo-electron microscopy structure of an SH3 amyloid fibril and model of the molecular packing. *The EMBO Journal*, *18*(4), 815–821.
44. Kakio, A., Nishimoto, S. I., Yanagisawa, K., & et al. (2002). Interactions of amyloid β-protein with various gangliosides in raft-like membranes: Importance of GM1 ganglioside-bound form as an endogenous seed for Alzheimer amyloid. *Biochemistry*, *41*(23), 7385–7390.
45. Kennedy, M. E., Stamford, A. W., Chen, X., Cox, K., Cumming, J. N., Dockendorf, M. F., Egan, M., Ereshefsky, L., Hodgson, R. A., Hyde, L. A., & Jhee, S. (2016). The BACE1 inhibitor verubecestat (MK-8931) reduces CNS β-amyloid in animal models and in Alzheimer's disease patients. *Science Translational Medicine*, *8*(363), 363ra150.
46. Kitwood, T., & et al. (1997). The experience of dementia. *Aging and Mental Health*, *1*(1), 13–22.
47. Knowles, T. P., Waudby, C. A., Devlin, G. L., & et al. (2009). An analytical solution to the kinetics of breakable filament assembly. *Science*, *326*(5959), 1533–1537.
48. Kosik, K. S., Joachim, C. L., Selkoe, D. J., & et al. (1987). Microtubule-associated protein tau (?) is a major antigenic component of paired helical filaments in Alzheimer disease. *Alzheimer Disease & Associated Disorders*, *1*(3), 203.
49. Lee, L. K., Shahar, S., Chin, A. V., & et al. (2013). Docosahexaenoic acid-concentrated fish oil supplementation in subjects with mild cognitive impairment (MCI): A 12-month randomised, double-blind, placebo-controlled trial. *Psychopharmacology*, *225*, 605–612.
50. Li, D., Kaner, R. B., & et al. (2007). How nucleation affects the aggregation of nanoparticles. *Journal of Materials Chemistry*, *17*(22), 2279–2282.
51. Li, G. W., Xie, X. S., & et al. (2011). Central dogma at the single-molecule level in living cells. *Nature*, *475*(7356), 308–315.
52. Librizzi, F., Rischel, C., & et al. (2005). The kinetic behavior of insulin fibrillation is determined by heterogeneous nucleation pathways. *Protein Science*, *14*(12), 3129–3134.
53. Linse, S., & et al. (2017). Monomer-dependent secondary nucleation in amyloid formation. *Biophysical Reviews*, *9*, 329–338.
54. Littlejohns, T. J., Henley, W. E., Lang, I. A., & et al. (2014). Vitamin D and the risk of dementia and Alzheimer disease. *Neurology*, *83*(10), 920–928.

55. Loscalzo, J., Kohane, I., Barabasi, A. L., & et al. (2007). Human disease classification in the postgenomic era: A complex systems approach to human pathobiology. *Molecular Systems Biology*, *3*(1), 124.
56. Marques, O., Outeiro, T. F., & et al. (2012). Alpha-synuclein: From secretion to dysfunction and death. *Cell Death and Disease*, *3*(7), e350.
57. Marsh, S. E., Blurton-Jones, M., & et al. (2017). Neural stem cell therapy for neurodegenerative disorders: The role of neurotrophic support. *Neurochemistry International*, *106*, 94–100.
58. Montero-Odasso, M., Pieruccini-Faria, F., Bartha, R., Black, S. E., & et al. (2017). Motor phenotype in neurodegenerative disorders: Gait and balance platform study design protocol for the Ontario neurodegenerative research initiative (ONDRI). *Journal of Alzheimer's Disease*, *59*(2), 707–721.
59. Necula, M., Chirita, C. N., Kuret, J., & et al. (2003). Rapid anionic micelle-mediated α-synuclein fibrillization *in vitro*. *Journal of Biological Chemistry*, *278*(47), 46674–46680.
60. Niedzielska, E., Smaga, I., Gawlik, M., & et al. (2016). Oxidative stress in neurodegenerative diseases. *Molecular Neurobiology*, *53*, 4094–4125.
61. Northrop, N. A., Yamamoto, B. K., & et al. (2015). Methamphetamine effects on blood–brain barrier structure and function. *Frontiers in Neuroscience*, *9*, 69.
62. Novak, P., Schmidt, R., Kontsekova, E., & et al. (2017). Safety and immunogenicity of the tau vaccine AADvac1 in patients with Alzheimer's disease: A randomised, double-blind, placebo-controlled, phase 1 trial. *The Lancet Neurology*, *16*(2), 123–134.
63. Orgeta, V., Leung, P., del-Pino-Casado, R., Qazi, A., Orrell, M., Spector, A. E., & Methley, A. M. (2022). Psychological treatments for depression and anxiety in dementia and mild cognitive impairment. *Cochrane Database of Systematic Reviews*, *4*(4), CD009125.
64. Panza, F., Solfrizzi, V., Seripa, D., Imbimbo, B. P., Lozupone, M., Santamato, A., Tortelli, R., Galizia, I., Prete, C., Daniele, A., & Pilotto, A. (2016). Tau-based therapeutics for Alzheimer's disease: Active and passive immunotherapy. *Immunotherapy*, *8*(9), 1119–1134.
65. Pedersen, J. T., & Sigurdsson, E. M. (2015). Tau immunotherapy for Alzheimer's disease. *Trends in Molecular Medicine*, *21*(6), 394–402.
66. Perczel, A., Hudáky, P., Pálfi, V. K., & et al. (2007). Dead-end street of protein folding: Thermodynamic rationale of amyloid fibril formation. *Journal of the American Chemical Society*, *129*(48), 14959–14965.
67. Perry, D., Sperling, R., Katz, R., Berry, D., Dilts, D., Hanna, D., Salloway, S., Trojanowski, J. Q., Bountra, C., Krams, M., & Luthman, J. (2015). Building a roadmap for developing combination therapies for Alzheimer's disease. *Expert Review of Neurotherapeutics*, *15*(3), 327–333.
68. Pfrieger, F. W., & et al. (2021). Neurodegenerative diseases and cholesterol: Seeing the field through the players. *Frontiers in Aging Neuroscience*, *13*, 766587.
69. Pohjasvaara, T., Erkinjuntti, T., Ylikoski, R., & et al. (1998). Clinical determinants of poststroke dementia. *Stroke*, *29*(1), 75–81.

70. Politi, C., Ciccacci, C., Novelli, G., & et al. (2018). Genetics and treatment response in Parkinson's disease: An update on pharmacogenetic studies. *Neuromolecular Medicine, 20,* 1–7.
71. Pooler, A. M., Polydoro, M., Wegmann, S., & et al. (2013). Propagation of tau pathology in Alzheimer's disease: Identification of novel therapeutic targets. *Alzheimer's Research & Therapy, 5*(5), 1–8.
72. Pronchik, J., He, X., Giurleo, J. T., & et al. (2010). *In vitro* formation of amyloid from α-synuclein is dominated by reactions at hydrophobic interfaces. *Journal of the American Chemical Society, 132*(28), 9797–9803.
73. Przedborski, S., Vila, M., Jackson-Lewis, V., & et al. (2003). Series Introduction: Neurodegeneration: What is it and where are we? *The Journal of Clinical Investigation, 111*(1), 3–10.
74. Qiu, C., Kivipelto, M., Von Strauss, E., & et al. (2022). Epidemiology of Alzheimer's disease: Occurrence, determinants, and strategies toward intervention. *Dialogues in Clinical Neuroscience, 11*(2), 111–128.
75. Rai, S. N., Singh, P., & et al. (2020). Advancement in the modelling and therapeutics of Parkinson's disease. *Journal of Chemical Neuroanatomy, 104,* 101752.
76. Rezazadeh, M., Khorrami, A., Yeghaneh, T., & et al. (2016). Genetic factors affecting late-onset Alzheimer's disease susceptibility. *Neuromolecular Medicine, 18,* 37–49.
77. Rizek, P., Kumar, N., Jog, M. S., & et al. (2016). An update on the diagnosis and treatment of Parkinson disease. CMAJ, *188*(16), 1157–1165.
78. Salloway, S., Sperling, R., Fox, N. C., & et al. (2014). Two phase 3 trials of bapineuzumab in mild-to-moderate Alzheimer's disease. *New England Journal of Medicine, 370*(4), 322–333.
79. Schapira, A. H., & et al. (2015). Glucocerebrosidase and Parkinson disease: Recent advances. *Molecular and Cellular Neuroscience, 66,* 37–42.
80. Singh, A., Ansari, V. A., Mahmood, T., & et al. (2022). Neurodegeneration: Microglia: Nf-Kappab signaling pathways. *Drug Research, 72*(9), 496–499.
81. Singh, A., Maheshwari, S., & et al. (2023). Dendrimers for neuro targeting. *International Journal of Pharma Professional's Research (IJPPR), 14*(1), 124–130.
82. Smith, C. J., Anderton, B. H., Davis, D. R., & et al. (1995). Tau isoform expression and phosphorylation state during differentiation of cultured neuronal cells. *FEBS Letters, 375*(3), 243–248.
83. Stegmaier, P., Krull, M., Voss, N., & et al. (2010). Molecular mechanistic associations of human diseases. *BMC Systems Biology, 4,* 1–9.
84. Stephenson, J., Nutma, E., van der Valk, P., & et al. (2018). Inflammation in CNS neurodegenerative diseases. *Immunology, 154*(2), 204–219.
85. Tarpey, P., Parnau, J., Blow, M., & et al. (2004). Mutations in the DLG3 gene cause nonsyndromic X-linked mental retardation. *The American Journal of Human Genetics, 75*(2), 318–324.
86. Thrash-Williams, B., Karuppagounder, S. S., Bhattacharya, D., & et al. (2016). Methamphetamine-induced dopaminergic toxicity prevented owing to the neuroprotective effects of salicylic acid. *Life Sciences, 154,* 24–29.
87. Törnquist, M., Michaels, T. C., Sanagavarapu, K., & et al. (2018). Secondary nucleation in amyloid formation. *Chemical Communications, 54*(63), 8667–8684.

88. Varma, H., Lo, D. C., Stockwell, B. R., & et al. (2008). High throughput screening for neurodegeneration and complex disease phenotypes. *Combinatorial Chemistry and High Throughput Screening, 11*(3), 238–248.
89. Vassar, R. (2014). BACE1 inhibitor drugs in clinical trials for Alzheimer's disease. *Alzheimer's Research & Therapy, 6*(9), 1–4.
90. Vidal, M., Cusick, M. E., Barabási, A. L., & et al. (2011). Interactome networks and human disease. *Cell, 144*(6), 986–998.
91. Wang, C., Shah, N., Thakur, G., & et al. (2010). α-Synuclein in α-helical conformation at air–water interface: Implication of conformation and orientation changes during its accumulation/aggregation. *Chemical Communications, 46*(36), 6702–6704.
92. Wilson, G. R., Wang, H. X., Egan, G. F., & et al. (2010). Deletion of the Parkin co-regulated gene causes defects in ependymal ciliary motility and hydrocephalus in the quaking viable mutant mouse. *Human Molecular Genetics, 19*(8), 1593–1602.
93. Wimo, A., Jönsson, L., Bond, J., Prince, M., Winblad, B., & Alzheimer Disease International. (2013). The worldwide economic impact of dementia 2010. *Alzheimer's & Dementia, 9*(1), 1–11.e3.
94. Woolley, J. D., Khan, B. K., Murthy, N. K., & et al. (2011). The diagnostic challenge of psychiatric symptoms in neurodegenerative disease: Rates of and risk factors for prior psychiatric diagnosis in patients with early neurodegenerative disease. *The Journal of Clinical Psychiatry, 72*(2), 4437.
95. Xie, L., Li, J., Xie, L., Bourne, P. E., & et al. (2009). Drug discovery using chemical systems biology: Identification of the protein-ligand binding network to explain the side effects of CETP inhibitors. *PLoS Computational Biology, 5*(5), e1000387.
96. Xue, W. F., Homans, S. W., Radford, S. E., & et al. (2008). Systematic analysis of nucleation-dependent polymerization reveals new insights into the mechanism of amyloid self-assembly. *Proceedings of the National Academy of Sciences, 105*(26), 8926–8931.
97. Yang, H., Chen, J., Tang, S., & et al. (2009). New drug R&D of traditional Chinese medicine: Role of data mining approaches. *Journal of Biological Systems, 17*(03), 329–347.
98. Zhu, H., Ding, Q., & et al. (2015). Lower expression level of two RAGE alternative splicing isoforms in Alzheimer's disease. *Neuroscience Letters, 597*, 66–70.
99. Rachmian, N., Medina, S., Cherqui, U., Akiva, H., Deitch, D., Edilbi, D., Croese, T., Salame, T. M., Ramos, J. M., Cahalon, L., & Krizhanovsky, V. (2024). Identification of senescent, TREM2-expressing microglia in aging and Alzheimer's disease model mouse brain. *Nature Neuroscience, 27*(6), 1116–1124.
100. Hou, K., Pan, H., Shahpasand-Kroner, H., Hu, C., Abskharon, R., Seidler, P., Mekkittikul, M., Balbirnie, M., Lantz, C., Sawaya, M. R., & Dolinsky, J. L. (2024). D-peptide-magnetic nanoparticles fragment tau fibrils and rescue behavioral deficits in a mouse model of Alzheimer's disease. *Science Advances, 10*(18), eadl2991.
101. Seo, D. O., & Holtzman, D. M. (2024). Current understanding of the Alzheimer's disease-associated microbiome and therapeutic strategies. *Experimental & Molecular Medicine, 56*(1), 86–94.
102. Yang, P., Huang, Q., Zhang, J., Li, Y., Gao, H., & Gu, Z. (2024). Natural polyphenolic nanodots for Alzheimer's disease treatment. *Advanced Materials, 36*(3), 2308393.

103. Sachdeva, P., Narayanan, K. B., Sinha, J. K., Gupta, S., Ghosh, S., Singh, K. K., Bhaskar, R., Almutary, A. G., Zothantluanga, J. H., Kotta, K. K., & Nelson, V. K. (2024). Recent advances in drug delivery systems targeting insulin signalling for the treatment of Alzheimer's disease. *Journal of Alzheimer's Disease, 98*(4), 1169–1179.
104. Alhujaily, M. (2024). Molecular assessment of methylglyoxal-induced toxicity and therapeutic approaches in various diseases: Exploring the interplay with the glyoxalase system. *Life*, *14*(2), 263.
105. JunYi, L., Wang, Y., Liu, B., Dong, X., Cai, W., Zhang, N., & Zhang, H. (2023). Deciphering the significance of metal ions in the pathogenesis of Alzheimer's disease and forging pathways towards prospective therapeutic strategies. *Authorea*. https://www.doi.org/10.22541/au.169753470.03178440/v1.
106. Liang, J., Liu, B., Dong, X., Wang, Y., Cai, W., Zhang, N., & Zhang, H. (2023). Decoding the role of gut microbiota in Alzheimer's pathogenesis and envisioning future therapeutic avenues. *Frontiers in Neuroscience*, *17*, 1242254.

CHAPTER 11

Exploring the Role of the Gene Mutation in the Management of Alzheimer's Disease

VARSHA YADAV, SINCHAN DAS, and ARJUN PATRA*

Department of Pharmacy, Guru Ghasidas University, Bilaspur, Chhattisgarh, India

*Corresponding author

ABSTRACT

Alzheimer's disease (AD) is a progressive neurodegenerative condition marked by cognitive decline, memory loss, and behavioral issues. While most cases are sporadic, a portion of familial AD is closely associated with specific genetic mutations, particularly in the *APP*, *PSEN1*, and *PSEN2* genes. These mutations affect the processing of amyloid precursor protein, resulting in the abnormal buildup of amyloid-beta plaques and tau-related pathology. This chapter delves into the mechanistic roles of these genetic mutations in the development of AD and their implications for treatment strategies. We discuss how genetic profiling can improve early diagnosis, assess risk, and facilitate personalized treatment plans. Recent advancements in genomic technologies have led to the discovery of new mutations and polymorphisms, such as *APOE ε4*, which influence the susceptibility and progression of the disease. Additionally, gene-editing technologies like CRISPR/Cas9 present promising opportunities for correcting harmful genetic variants, while approaches like antisense oligonucleotides and RNA interference aim to address abnormal gene expression. The chapter also highlights the importance of integrating genetic mutation information

Cellular and Molecular Mechanisms in Alzheimer's Disease Therapeutics, Volume 1: Cellular Mechanisms.
Prashant Tiwari, Sunil Kumar Kadiri, and Kunjbihari Sulakhiya (Eds.)

DOI: 10.1201/9781779641793-11

into the design of clinical trials, the development of biomarkers, and pharmacogenomic strategies. By understanding the relationships between genotype and phenotype, healthcare providers can customize interventions to enhance treatment effectiveness and reduce side effects. Ethical issues related to genetic testing and counseling are also discussed, underscoring the necessity for informed consent and psychosocial support. In summary, genetic mutations are crucial not only to the cause of Alzheimer's disease but also in influencing its clinical management. Leveraging genetic insights has the potential to revolutionize precision medicine, paving the way for more effective and personalized treatments for this challenging disorder.

11.1 INTRODUCTION

Nearly 44% of people worldwide are experiencing psychological instability of several kinds. Psychological disease is believed to severely compromise everyone's health. The number of dementia patients is expected to be 139 million in 2050 due to population growth. As per the International Classification of Diseases, one way to describe dementia is a combination of brain diseases, usually of a chronic or of progressive nature, that disrupts various enhanced cortical abilities, like memory, reasoning, direction, understanding, computation, ability to learn, and language and comprehension. There is no clouding of consciousness that exists. Motivational decline, social behavior, and emotional regulation often occur before or along with impairments in cognitive function. This syndrome can be brought on by various illnesses affecting the brain. Alzheimer's disease (AD) is the most common cause of dementia, accounting for 50–70% of cases of the condition. The pathophysiology includes amyloid-beta 42 (Aβ 42) and phosphorylated tau formation and accumulation, and there is also excessive glial cell activation. Moreover, two crucial aspects of AD are reduced neurotrophin signaling and compromised synapse function [1–3].

There are basically two recognized types of AD: "sporadic" and "familial." Among all the cases of AD, less than 1% are familial. Their early onset—occurring before the age of 60, substantial mendelian transmission—predominantly autosomal dominant, and family aggregation are their defining characteristics. The most prevalent type of AD is sporadic. The condition often manifests after the age of 60, has a minimal familial aggregation, and lacks a clear method of transmission [4]. According to estimates, a 65-year-old with a first-degree relative with AD has a 5-fold elevated risk of developing AD by the time they are 87 years old (49 vs. 10%). The risk varies according on ethnicity; some research indicates that communities of African, Americans, and Caribbean Hispanics may be more at risk [5–8].

The pathophysiology, etiology, and causation of each synaptic dysfunction and cognitive impairment remain poorly understood despite much research. Furthermore, the current treatment choices have negative consequences, like disorientation, feeling dizzy, constipation, depression, and diarrhea, and are only symptomatic and supportive. Despite the abundance of information regarding the molecular causes of AD, it has been challenging to create effective disease-modifying medications. The required level of efficacy has not been explored in preventing the creation, accumulation, and toxicity of AD. This challenged the amyloid beta theory and supported the need for different approaches for treatment. Clustered regularly interspaced short palindromic repeats/CRISPR associated proteins 9 system (*CRISPR/Cas9*) gene editing is a relatively new and highly significant strategy that has attracted attention because of its possible advantages for managing and treating AD. The comparatively simple, affordable, and accurate nature of this new technology has raised attention in its application to neurodegenerative disorders. This may be applied directly to treatment or could help create more accurate animal models that accurately represent non-verbal developmental disorders in humans [9, 10].

DNA methylation is another relatively new and highly significant approach for the supervision and treatment of AD epigenetic alterations. The chromatin shape and dynamics are caused by DNA methylation, primarily at the cytosine and adenine sites. The bases in rRNAs and tRNAs that are methylated, pseudo-uridylated, and thiouridylated are the most crucial for survival among all epigenetic changes. These alterations are crucial for preserving DNA's double-helical pairing and guarding against base modification's mutagenic consequences. By preserving genetic material in its unaltered form, they also enable RNA to remember the biochemical variety that is necessary for their function. A limited range of DNA modifications has merely emerged during development and has been employed to ascertain biological activities. Neuronal degeneration can advance as a result of changes in the pattern of methylation of specific genes, which affects neuronal survival and regeneration [11–13].

11.2 THE GENETICS OF ALZHEIMER'S DISEASE

Families with AD persons may have an autosomal dominant disorder, which facilitates the identification of disease genes. Three genes are the source of fully penetrant mutations that cause early-onset AD and are referred to as causative genes. The genes that encode presenilin 1 *(PSEN1)*, presenilin 2 *(PSEN2)*, and amyloid precursor protein *(APP)* on chromosome 21q21 and chromosome 14q24 are the ones that cause the disease. Approximately 5% of AD cases are resulting from changes in these three genes. Research is currently being conducted to identify the other three probable AD-susceptibility genes. Number of genes, each accounting for a relatively tiny percentage of instances, will be found to impart risk. About half of the entire genetic risk is accounted for by the four genes (*APP, PSEN1, PSEN2,* and *APOE*) associated with AD [14, 15]. The genes that are responsible for autosomal dominant disorder have been summarized in Figure 11.1.

Protein	Gene symbol	Locus	Age at onset (Yr)
Presenilin 1	PSEN1	14q24.3	30-58
Presenilin 2	PSEN2	1q31-q42	45-88
Amloid Precursor Protein	APP	21q21.2	40-60
Apolipoprotein E	APOE	19q13.2	40-90

FIGURE 11.1 Genes confirmed in Alzheimer's disease.

11.2.1 AMYLOID PRECURSOR PROTEIN (APP)

A 4.2 kDa polypeptide was identified in 1984 by Glenner and Wong from fibrils found in amyloid plaques linked to AD and cerebrovascular amyloidoses. The partly beta-pleated sheet shape of this protein led to its naming as the beta-amyloid protein (Aβ). The 28-amino acid sequence of the proteins from Down syndrome and AD is the same [16].

Numerous cells produce *APP*, a transmembrane glycoprotein which is subsequently metabolized by the endosomal-lysosomal or secretory pathways. Ten isoforms, containing varying number of amino acids are produced via alternative splicing of *APP*. The *APP* isoforms with 695, 751, and 770 amino acids are the predominant types. Neuronal tissue expresses the 695 amino acid variant more frequently than other tissues. While the exact physiological roles of *APP* are yet unknown, its prevalent expression throughout development in several adult tissues is responsible for basic functions in cell physiology. *APP* is important in neurite outgrowth, static cell-substrate adhesion, synaptic plasticity synaptogenesis, and maintenance of neuronal cell survival. Additionally, *APP* also controls cell motility. Similar to growth hormones, the soluble secreted form *APP*, or sAPPa, supports development of neural stem cells in the embryo [17–20].

Currently, there are >30 distinct *APP* missense mutations known to exist, 25 of which are harmful and cause most occurrences of AD. AD Mutation Database and the NCBI database both provide information about *APP* mutations [21].

Two recessive mutations in *APP* have been found to be the reason to cause disease solely in homozygous state: A673V in one family and a trinucleotide deletion called E693D that segregates in the family of Japanese origin with diminished Aβ 40 and Aβ 42 but no alterations in their ratio. Since three copies of *APP* are provided by an additional 21 chromosomes in cases of Down syndrome, AD in these circumstances results from an overdose effect. Moreover, *APP* and AD have copy number variant (CNV) alterations. However, locus duplications are extremely uncommon and might not be fully penetrated, which could suggest the presence of previously unidentified protective genetic factors. In an Icelandic investigation, *A673T* (an APP variant) protected against AD and cognitive impairment brought on by aging. Compared to the AD group, mutation of this *APP* variant, *A673T* was substantially more prevalent in the elderly. This alteration reduces the production of amyloidogenic peptides *in vitro* by 40%. This finding offers preliminary support for the theory that inhibiting *APP* β-cleavage could offer protection from the illness. According to reports, this *A673T* substitution is recessive for AD and presents itself relatively early in life in one Italian family. However, in examining the mutation's protective effect, the researchers discovered three homozygous *A673T* carriers in Icelandic samples; none had a history of dementia, and one had passed away at the age of 88. The other two were presently 67 and 83 years old, respectively [22–26].

11.2.2 PSEN1 GENE

One year of following the initial *APP* mutation's discovery, a second area for AD linkage which is located at 14q24 was identified. An extremely instructive Mexican household was included in this investigation.

Since the 14q24.3 locus was found in two different study teams' series, the 14q24.3 gene is a key locus for familial AD. The chromosome 14 gene was cloned in 1995 and given the name presenilin1 (*PSEN1*) [27–29].

PSEN1 mutations are the main cause for early onset of AD. There are various *PSEN1* mutations known to exist. Severe AD is brought on by *PSEN1* mutations, which have 100% penetration and beginning at the age of 30. However, there is a large range in onset age; in other *PSEN1*-AD patients having a mean onset at the age of 58 years. Individuals that carry mutations in *PSEN1* gene inherit by autosomal dominant pattern. Progressive dementia and parkinsonism characterize the clinical picture, however patients with *PSEN1*-associated language dysfunction, gait abnormalities, and reflex alterations also exhibit diversity in their phenotype. Spastic paraparesis and other symptoms of atypical AD, such as ataxia and mental loss T2 signal alterations, frontotemporal atrophy are developed by some patients. It was demonstrated that a distinct and variant neuropathology, known as cotton wool plaques, was present in a few *PSEN1* AD pedigrees with spastic paraparesia. These plaques are big and noncored and lack substantial neuritic dystrophy. *PSEN1* mutations may have full penetrance or incomplete penetrance [28, 30–34].

There has been reported to be over 180 mutations in *PSEN1*. The bulk of the mutations are missense mutations which alter the amino acid composition of the PSEN1 protein and seem to raise the proportion of Aβ 42 peptides [21].

Certain types of mutations in *PSEN1*, like the *A431E* variant discovered by Yescas et al. among families who are prone to early onset of AD in Jalisco, Mexico, and the one reported in unrelated Caribbean Hispanic families, which was subsequently validated by other authors, have a founder impact. This discovery facilitates the timely diagnosis and genetic counseling [35–37]. The *PSEN1* L166P mutation manifests in adolescence, which is an unusual late age for it to manifest. Studies conducted *in vitro* reveal that this mutation results in abnormally high amounts of Aβ 42 synthesis along with defective notch signaling. In contrast to *PSEN2* mutations, which are uncommon, *PSEN1* mutations are frequent [38, 39].

The pre-dementia clinical signs and symptoms of the E280A mutation in 449 carriers of familial early-onset AD were investigated in a retrospective cohort study considering pre-mild cognitive impairment (Pre-MCI) and MCI, which have been suggested to represent phases prior to AD. They found that *PSEN1* E200A carriers had detectable cognitive impairment almost 20 years before dementia manifested itself. They arrived at the conclusion that the onset and development of pre-dementia phases should be taken into account while researching familial AD and when implementing therapeutic measures [40].

The stability of γ-secretase complex depends on *PSEN1*, nicastrin, anterior pharynx deficient 1 (*Aph-1*), and presenilin enhancer 2 (*PEN-2*) (Figure 11.2). One can find *PSEN1* on chromosome 14q24.2. The 467 amino acid protein that is encoded by this gene's 12 exons is expected to cross the membrane six to ten times. An essential membrane protein, γ-secretase is normally located at the cell surface but can additionally exist in the golgi apparatus, endoplasmic reticulum, and mitochondria [41–43].

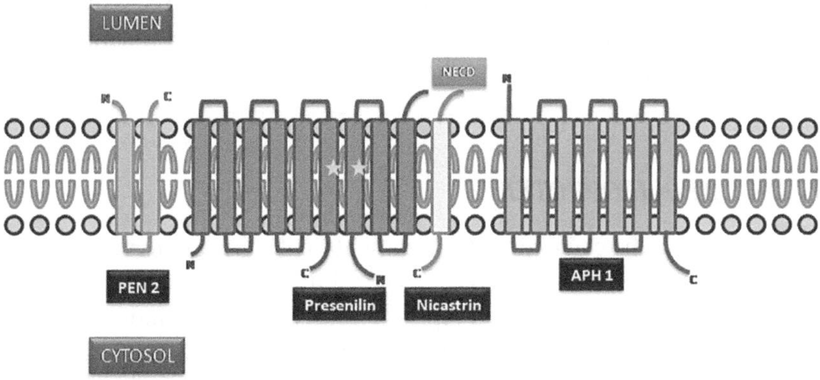

FIGURE 11.2 Schematic representation of subunits of the γ-secretase complex.

11.2.3 PSEN2 GENE

The year 1995 saw the discovery of a putative gene on chromosome 1 in a Volga German AD clan. Later on, presenilin 2 (*PSEN2*) was created by cloning this gene. Both at the sequence and protein levels, *PSEN1* and *PSEN2* exhibit significant sequence homology [44, 45]. *PSEN2* encodes a 448-amino acid protein and has 12 exons arranged into 10 translated exons. According to predictions, the PSEN2 protein has nine trans-membrane domains [43]. The γ-secretase, an unusual aspartyl protease, contains *PSEN2*, which is involved in the splitting of Aβ. In both humans and animals, mutations in *PSEN2* increase Aβ 42 to Aβ 40 ratio. It has been shown that distinct presenilin mutations have varying effects on the processing of *APP* at the γ-secretase site. It's unclear exactly how

PSEN2 convinces production of higher Aβ in AD patients. According to a recent study, extracellular signal-regulated kinase is activated by mutant presenilin 2 in a species-dependent manner, hence increasing β-secretase activity [41, 42, 46].

Rare missense mutations in the *PSEN2* are mostly responsible for the early development of AD, at least in populations of Caucasians. Compared to certain family members who have PSEN1 mutations, these family members often had an age of onset between 45 and 88 years. With *PSEN2*, patients within the same family have quite different onset ages, but with *PSEN1*, family members who are affected by the condition have very similar onset ages. The *PSEN2* gene may be less susceptible to missense mutations than the *PSEN1* gene [47].

A genome-wide linkage analysis in 2012 identified a recessive inheritance *LOAD* gene that maps to locus 8p22-p21.2 in an inbreeding group. These findings lend credence to the idea that recessive genes play a part in late onset of AD [48].

11.2.4 APOLIPOPROTEIN E

In 1993, research revealed that the patients with the late development of AD had an amplified occurrence of the ε4 allele and that the *APOE* had a high propensity to bind with Aβ amyloid. The 299-amino-acid protein known as *APOE* is concerned in the transport and biotransformation of lipids. Ninety percent of the *APOE* in the blood is produced by the liver. In addition, *APOE* is produced by glia, macrophages, and neurons in CNS, as well as in other organs [49–51].

In the brain, astrocytes create the majority of *APOE*, with microglia and in certain situations, neurons producing it to a lower degree. Among the essential lipoproteins in lipoprotein complexes, *APOE* controls lipid biotransformation by facilitating the transfer, distribution, and transfer of lipids via APOE receptors and proteins between different types of tissues and cells linked to lipolysis and lipid transfer [52, 53].

Four exons of the *APOE* gene on chromosome 19q13.32 encode a protein of 299 amino acids. The two single nucleotide polymorphisms (SNPs) in exon four, which encodes three different protein isoforms (E2, E3, and E4), describe the three APOE alleles (ε2, ε3, and ε4). APOE*E3 is the most prevalent APOE isoform. In all populations, the APOE ε3 allele is found in 50–90% of individuals, while 3.35% of people have APOE ε3/ε4 and 1–5% of people have APOE ε2. It has been established that APOE ε4 enlarges the incidence of sporadic late development of AD and in certain populations has also been connected to autosomal dominant family illness [54–58].

Compared to 20–25% of controls, 50% of AD patients take the APOE ε4 allele. The threat of late development of AD increases around three times for one copy of APOEε4 and twelve times for two copies. Moreover, APOE ε4 raises the chances of development of late onset of AD patients by roughly 10–20 years when compared to noncarriers. Conversely, persons with PSEN1 mutation in the large Colombian lineage and families with *APP* gene mutations experienced an earlier onset of dementia as a result of APOE ε4. In contrast to those with APOE ε4, those with APOE ε2 are less likely to have late onset of AD [59, 60].

Numerous genetic epidemiology studies that include individuals in their 60s or 70s who have one or two APOE ε4s substantially confirm an enhanced risk of developing AD. However, the effect varies with gender and is age-specific, peaking at about 70 years of age. Some aged people with the two APOE ε4s emerge to have avoided its impact. Reportedly, compared to APOE ε3, the isolated APOE ε4 isoform exhibits a higher affinity for Aβ which suggests that it may function as a pathologic chaperone to stabilize the Aβ fibril's β-sheet structure and hinder its clearance. Exquisite studies using transgenic and *APOE* deletion animal models demonstrate that APOE ε4 notably increases the magnitude of fibrillar Aβ deposits in the encephalon in contrast to the actions of APOE ε3. However, the genotype APOE ε3 is neither required nor adequate for AD to arise. Therefore, the risk for AD can be altered by additional genetic or environmental variables, either independently or in combination with APOE ε4 [61–67].

11.2.5 MICROTUBULE-ASSOCIATED PROTEIN T

This protein participates in the process of development and maintenance of neuronal polarity and facilitates microtubule assembly and stability. PNS-τ, fetal-τ, τ-A, τ-B, τ-C, τ-D (382 amino acids τ-E, and τ-F are the eight isoforms of protein τ. Histologically defining filamentous deposits (neurofibrillar tangles) occur in AD, Pick's disease, and progressive supranuclear palsy. The filaments of neurofibrillar tangles are made of protein τ in its insoluble, hyperphosphorylated form. It was not evident for a long time if illness comes first and then protein τ malfunction.

This was settled when it was discovered that inherited FTD and parkinsonism (FTDP-17), are caused due to the mutations in the τ gene on chromosome 17q21 (MAPT). More than 100 families currently have 32 distinct variants discovered. Roughly 50% of the identified mutations cause the protein to become less capable of interacting with microtubules and more likely to form aberrant filaments. The usual ratio of protein τ isoforms is disrupted by the other alterations. The amount of neurofibrillar tangles is correlated with the severity of neurodegeneration, despite the fact that protein τ mutations are not yet linked to AD. In AD, neurofibrillar tangles might represent a "downstream" response to the pathologic processes that the disease initiates. In a triple transgenic mouse model of AD carrying mutant genes for protein τ (*P301L*), PS-1 (*M146V*), and *APP* (K670N/M671L), Aβ deposition takes place before the tangle pathology. Likewise, research conducted on twin transgenic mice that produce the APP mutation found in Sweden and the τ protein that carries three of the mutations found in FTDP-17 demonstrates that Aβ promotes τ's phosphorylation and abnormal aggregation [68–74].

11.2.6 OTHER SUSCEPTIBILITY GENES

The *α-T catenin* gene, which is encoded by the *VR22* gene, is a recognized PS-1 interactor and β-catenin's binding partner. Variants of *VR22* affect elevated plasma Aβ 42 levels. The insulin-degrading enzyme, which is encoded by the *IDE* gene, is essential for microglial cells and neurons to break down and eliminate Aβ. Recently, there has been a suggestion that pathologic polymorphisms with elevated plasma Aβ 42 levels that are present in the area harboring *IDE* may influence late-onset AD. The urokinase-type plasminogen activator, which transforms plasminogen into plasmin, is encoded by the *PLAU* gene. Plasmin is upregulated in response to Aβ aggregates, leading to an increase in plasmin, which breaks down Aβ. It has recently been proposed that a single missense mutation (P141L) in the urokinase-type plasminogen activator increases Aβ 42 in AD with a late onset [75–78].

At first, the genes on chromosome 12 that encode for LDL receptor-related protein (LRP) and α2-macroglobulin (α2M) were thought to be suspects. Later, it was suggested that the susceptibility factor for late development of AD is a genetic difference in the transcriptional factor LBP-1c/CP2/LSF. Serum protease inhibitor α2M is also articulated in the brain and has been associated to AD due to its capacity to mediate Aβ breakdown and clearance. It is thought that α2M polymorphisms caused by an amino acid substitution or deletion allele (V1000I) account for its correlation with AD. New population-based studies have been published more recently, and they show that α2M deletion/insertion polymorphisms do not increase the chance of developing AD. Therefore, it's probable that the chromosomal 12 locus linked to AD does not match to α2M, but rather to a nearby gene [79–82].

There are many glyceraldehyde-3-phosphate dehydrogenase (*GAPD*) gene members identified, one of which is located in the 12p candidate area linked to late-onset AD. In addition to being essential for the synthesis of cellular energy, *GAPD* is involved in neuronal death and neurodegenerative illnesses like AD. It is recognized to bind Aβ and *APP* [83–85].

In the brain, the low-density LRP receptor is highly expressed and has multiple functions. Dystrophic neuritis, reactive astrocytes in AD brain, senile plaques, and endocytosis of released APP are all associated with LRP. It was recently demonstrated that LRP functions as a substrate for β-secretase and interacts with it on the cell surface. There is evidence of a genetic relationship between AD and two distinct LRP polymorphisms [86–88].

A highly expressed protein called ubiquilin 1 (*UBQLN1*) stimulates the manufacture of uncleaved PS-1 and PS-2, causing them to accumulate. On chromosome 9q22, the *UBQLN1* gene is one of the several potential genes for AD that are situated close to a known linkage peak. *UBQLN1* variants considerably raise the risk of having AD, potentially through affecting the brain's alternative splicing of this gene [89].

A zinc metalloproteinase known as nerprilysin (*NEP*) is an Aβ-degrading enzyme expressed in the encephalon. Decreased expression of *NEP* causes building up of Aβ, that aid in the prognosis of AD. *NEP* is encoded by a gene found on chromosome 3q25. There are at least 10 SNPs in the *NEP* that are linked to sporadic AD, indicating that genetic differences inside or very near the *NEP* may affect an individual's susceptibility to AD [90–92].

The enzyme that initiates the initial cleavage of *APP* to produce Aβ is called β-secretase (*BACE1*). On chromosome 11q23, the gene encoding *BACE1* is situated close to an area with elevated lod scores for AD. This gene is a risk factor of AD. In Chinese populations, the variant C786G of *BACE1* gene is a risk factor for AD [93].

Tumour necrosis factor-a (*TNF-a*) is a suitable gene for AD since it can mediate neurotoxicity. The chromosomal region 6p21, which exhibits genetic linkage relationship with AD, is close to the gene producing *TNF-a*. Recent research, however, indicates that 308 adenine/guanine (A/G) polymorphism is not linked to late-onset AD and *TNF-a* may solely function as the cause for vascular dementia [94, 95].

Acute phase protein α1-antichymotrypsin (*ACT*) inhibits serine proteases. *ACT* is mostly synthesized in liver and brain. The liver and other tissues, including the brain, are where *ACT* is mostly synthesized. A significant part of senile plaques, *ACT* might function as a pathologic chaperone, encouraging the assembly of Aβ into neurotoxic fibrils. Higher concentrations of *ACT* are observed in peripheral blood and brain of AD patients. One can find the *ACT* gene on chromosome 14q32. It has adenine/thymine (A/T) transversion. This causes the signal peptide sequence's codon 15 to have either a threonine or an alanine. It also has typical bi-allelic polymorphisms. Only *APOE ε4* carriers are linked to an elevated risk for AD due to the alanine allele. However, these results are not supported by other research. According to more recent research, regardless of the occurrence of *APOE ε4*, the T allele of the *ACT* location 51 guanine/thymine (G/T) polymorphism is highly linked to early-onset AD. In patients with the *APOE ε4* gene, the *ACT* TT genotype is also linked to a quicker rate of cognitive loss when the disease develops [96–99].

11.3 GENETIC POLYMORPHISMS

Within the field of genomics, polymorphism (variation in certain gene sequence) can be found in some individuals or groups. Patients diagnosed with AD have been found to have several genetic abnormalities or disruptions, such as mitochondrial mutations, epigenetic modifications, malfunctioning mutations, and SNPs. Jiang et al.'s recent work demonstrated the likelihood that genetic variation participated in AD development. Additionally, the same study revealed a high enrichment of AD genetic variants in immune system pathways [100–102].

Effective pharmaceutical therapies of AD are currently deficient. The initial course of therapy for mild-to-moderate AD comprises both acetylcholinesterase inhibitors (AChEI) and non-pharmacological treatments, such as MST (multidimensional stimulation therapy), which aims to delay and prevent cognitive impairment that is frequently seen in AD patients. Nevertheless, the end result of certain medicines may be altered by variances in individual genetic polymorphisms. For instance, variations in *CYPD26* and *ABCA1* can have an impact on AChEI therapy; however, ApoE-4 and SNAP-25 therapies can impact MST [103, 104]. Table 11.1 discusses the relationship between medication response and gene polymorphism.

TABLE 11.1 Association of Gene Polymorphism with Therapy Response.

Gene	Polymorphism	Therapy Type	Patient Population	Response Rate	Significance (P Value)	Comments
Gene A	SNP rs123456	Chemotherapy	Breast cancer patients	75%	0.01	Improved response with polymorphism [104–111]
Gene B	Insertion/deletion	Immunotherapy	Lung cancer patients	60%	0.05	Polymorphism linked to adverse effects [112–117]
Gene C	SNP rs789101	Targeted therapy	Colon cancer patients	80%	<0.001	Strong association with positive response [118–125]

11.4 DNA METHYLATION/HYDROXYMETHYLATION IN ALZHEIMER'S DISEASE

In DNA methylation, a group of proteins known as DNA methyltransferases (DNMTs) methylate the nucleobases in DNA. In CpGs and CpHs (H = A, T, and C), the majority of the methylated nucleobases are cytosines, which combine to generate 5-methylcytosine (5mC). The majority of CpG dinucleotides are enriched at five mC. While CpGs receive most of the emphasis when discussing DNA methylation, CpHs methylation has also been linked to neural gene expression. It is determined that another nucleobase is methylated in adding up to 5 mC. Deoxyadenosine methylation on N6 (m6dA) in mouse prefrontal cortex neurons is associated with memory formation and gene expression. DNMTs consist of three different potential enzymes: DNA methyltransferase 1, 3A, and 3B. DNA hydroxymethylation is the process of hydroxymethylating cytosine to produce 5-hydroxymethylcytosine (5 hmC). Ten-eleven translocase (TET) isoforms convert 5 mC to 5 hmC. The acronyms CpG and CpH stand for cytosine and other nucleobases that are phosphate-divided. The main groove of the DNA helix, where numerous DNA-binding proteins come into touch, is where the methyl/hydroxymethyl groups are found. Consequently, genomic imprinting, X-chromosome inactivation, and gene silence can all result from DNA methylation. DNA hydroxymethylation may either increase or suppress gene expression by attaching to a gene's regulatory regions (Table 11.2), however its precise roles are still unknown [126–133].

TABLE 11.2 Alzheimer's Disease-related Unambiguous Gene Methylation and Hydroxymethylation: The Potential Gene Strategy.

Gene	Methylation Status	Hydroxymethylation Status	Role in Alzheimer's Disease	Potential Strategy
APP	Hypermethylated	Low hydroxymethylation	Amyloid precursor protein, implicated in plaque formation	Target methylation to reduce APP expression [134–140]
PSEN1	Hypomethylated	High hydroxymethylation	Part of the gamma-secretase complex; involved in amyloid production	Modulate hydroxymethylation to alter expression [141–154]
BACE1	Variable methylation	Moderate hydroxymethylation	Beta-secretase, key in amyloid beta formation	Control methylation for gene regulation [155–158]
MAPT (Tau)	Hypermethylated	Low hydroxymethylation	Encodes tau protein; involved in neurofibrillary tangles	Reduce methylation to control tau aggregation [159]
APOE	Low methylation	Variable hydroxymethylation	Associated with increased AD risk, especially APOE ε4 variant	Precision methylation targeting by allele [160]
TREM2	Hypermethylated	Low hydroxymethylation	Immune receptor affecting microglial function	Target methylation for immune modulation [161–164]

11.5 CONCLUSION

AD is the main cause of dementia in elderly, which poses a significant and growing clinical difficulty in regard to diagnosis and management. This chapter highlights the genes recognized to be implicated in AD and its role in genetics in piecing together the complex AD puzzle. According to estimates, 70% of the possibility of having AD is genetic. Early-onset autosomal dominant AD occurs by mutations in the genes encoding APP, presenilin 1, and presenilin 2, which account for only 1% of AD cases but whose recognition has been essential to comprehending the molecular mechanisms of AD. Apolipoprotein E (*APOE)* gene ε-4 allele has been identified as a major genetic susceptibility factor for the more common complex late-onset AD. Genetic polymorphisms should be taken into account to attain cost- and prognosis-effective and efficient treatment outcomes. As a result, for successful treatment of AD, the genetic polymorphisms of patient should be considered. Polymorphisms in AD patients, either directly altering any of the amino acids in a protein or manifesting as SNPs, can affect how differently people respond to the same medication during treatment. Apart from genetic polymorphisms, DNA methylation is a new approach of treatment. In the genome of mammal, DNA methylation is an epigenetic alteration that results in 5-methylcytosine by transferring a methyl group to the C5 site of cytosine. An intriguing field of study has been the relationship between DNA methylation and the maturation of the nervous system, as well as the development of neurodegenerative illnesses like AD. Moreover, mutations affecting DNA methylation have an impact on neurodevelopmental processes and may accelerate the onset of a number of neurodegenerative illnesses. In order to identify new indicators for early diagnosis and potential pharmacological targets, as well as to unearth likely mechanisms influencing the development and the disease's progression, epigenetic alterations in neurodegenerative disorders are extensively researched across a diversity of populations. Although the field of epigenetics in AD has yielded many exciting results, the overall picture remains unsatisfactory; larger longitudinal cohorts are essential in order to fully comprehend the epigenetic modifications of AD and potentially identify new treatment approaches for the disease in the future.

KEYWORDS

- **amyloid precursor protein**
- ***APOE* gene**
- **DNA methylation**
- **genetic polymorphisms**
- ***PSEN1* gene**
- ***PSEN2* gene**

60. Corder, E. H., Saunders, A. M., Pericak-Vance, M. A., & Roses, A. D. (1995). There is a pathologic relationship between ApoE-ε4 and Alzheimer's disease. *Archives of Neurology, 52*(7), 650–650.
61. Selkoe, D. J. (2005). Defining molecular targets to prevent Alzheimer disease. *Archives of Neurology, 62*(2), 192–195.
62. Farrer, L. A., Cupples, L. A., Haines, J. L., Hyman, B., Kukull, W. A., Mayeux, R., & Van Duijn, C. M. (1997). APOE and Alzheimer Disease Meta Analysis Consortium Effects of age, sex, and ethnicity on the association between apolipoprotein E genotype and Alzheimer disease. A meta-analysis. *JAMA, 278*(16), 1349–1356.
63. Tokuda, T., Calero, M., Matsubara, E., Vidal, R., Kumar, A., Permanne, B., & Ghiso, J. (2000). Lipidation of apolipoprotein E influences its isoform-specific interaction with Alzheimer's amyloid β peptides. *Biochemical Journal, 348*(2), 359–365.
64. Wisniewski, T., & Frangione, B. (1992). Apolipoprotein E: A pathological chaperone protein in patients with cerebral and systemic amyloid. *Neuroscience Letters, 135*(2), 235–238.
65. Cedazo-Mínguez, A., Wiehager, B., Winblad, B., Hüttinger, M., & Cowburn, R. F. (2001). Effects of apolipoprotein E (apoE) isoforms, β-amyloid (Aβ) and apoE/Aβ complexes on Protein Kinase C-α (PKC-α) translocation and amyloid precursor protein (APP) processing in human SH-SY5Y neuroblastoma cells and fibroblasts. *Neurochemistry International, 38*(7), 615–625.
66. Holtzman, D. M., Bales, K. R., Tenkova, T., Fagan, A. M., Parsadanian, M., Sartorius, L. J., & Paul, S. M. (2000). Apolipoprotein E isoform-dependent amyloid deposition and neuritic degeneration in a mouse model of Alzheimer's disease. *Proceedings of the National Academy of Sciences, 97*(6), 2892–2897.
67. Kamboh, M. I. (2004). Molecular genetics of late-onset Alzheimer's disease. *Annals of Human Genetics, 68*(4), 381–404.
68. St George-Hyslop, P. H., & Petit, A. (2005). Molecular biology and genetics of Alzheimer's disease. *Comptes Rendus Biologies, 328*(2), 119–130.
69. Heutink, P. (2000). Untangling tau-related dementia. *Human Molecular Genetics, 9*(6), 979–986.
70. Goedert, M., & Spillantini, M. G. (2000). Tau mutations in frontotemporal dementia FTDP-17 and their relevance for Alzheimer's disease. *Biochimica et Biophysica Acta (BBA)-Molecular Basis of Disease, 1502*(1), 110–121.
71. Goedert, M., & Jakes, R. (2005). Mutations causing neurodegenerative tauopathies. *Biochimica et Biophysica Acta (BBA)-Molecular Basis of Disease, 1739*(2-3), 240–250.
72. D'Souza, I., & Schellenberg, G. D. (2005). Regulation of tau isoform expression and dementia. *Biochimica et Biophysica Acta (BBA)-Molecular Basis of Disease, 1739*(2-3), 104–115.
73. Oddo, S., Caccamo, A., Kitazawa, M., Tseng, B. P., & LaFerla, F. M. (2003). Amyloid deposition precedes tangle formation in a triple transgenic model of Alzheimer's disease. *Neurobiology of Aging, 24*(8), 1063–1070.
74. Perez, M., Ribe, E., Rubio, A., Lim, F., Moran, M. A., Ramos, P. G., & Avila, J. (2005). Characterization of a double (amyloid precursor protein-tau) transgenic: Tau phosphorylation and aggregation. *Neuroscience, 130*(2), 339–347.

75. Ertekin-Taner, N., Ronald, J., Asahara, H., Younkin, L., Hella, M., Jain, S., & Younkin, S. (2003). Fine mapping of the α-T catenin gene to a quantitative trait locus on chromosome 10 in late-onset Alzheimer's disease pedigrees. *Human Molecular Genetics, 12*(23), 3133–3143.
76. Vekrellis, K., Ye, Z., Qiu, W. Q., Walsh, D., Hartley, D., Chesneau, V., & Selkoe, D. J. (2000). Neurons regulate extracellular levels of amyloid β-protein via proteolysis by insulin-degrading enzyme. *Journal of Neuroscience, 20*(5), 1657–1665.
77. Ertekin-Taner, N., Allen, M., Fadale, D., Scanlin, L., Younkin, L., Petersen, R. C., & Younkin, S. G. (2004). Genetic variants in a haplotype block spanning IDE are significantly associated with plasma Aβ42 levels and risk for Alzheimer disease. *Human Mutation, 23*(4), 334–342.
78. Ertekin-Taner, N., Ronald, J., Feuk, L., Prince, J., Tucker, M., Younkin, L., & Younkin, S. G. (2005). Elevated amyloid β protein (Aβ42) and late onset Alzheimer's disease are associated with single nucleotide polymorphisms in the urokinase-type plasminogen activator gene. *Human Molecular Genetics, 14*(3), 447–460.
79. Lambert, J. C., Goumidi, L., Vrièze, F. W. D., Frigard, B., Harris, J. M., Cummings, A., & Chartier-Harlin, M. C. (2000). The transcriptional factor LBP-1c/CP2/LSF gene on chromosome 12 is a genetic determinant of Alzheimer's disease. *Human Molecular Genetics, 9*(15), 2275–2280.
80. Qiu, Z., Strickland, D. K., Hyman, B. T., & Rebeck, G. W. (1992). α2-Macroglobulin enhances the clearance of endogenous soluble β-amyloid peptide via low-density lipoprotein receptor-related protein in cortical neuron. *Journal of Neurochemistry, 73*(4), 1393–1398.
81. Gibson, A. M., Singleton, A. B., Smith, G., Woodward, R., McKeith, I. G., Perry, R. H., & Morris, C. M. (2000). Lack of association of the α2-macroglobulin locus on chromosome 12 in AD. *Neurology, 54*(2), 433–433.
82. Zhang, P., Yang, Z., Wan, C. L., Zheng, W. D., Zhang, C. F., Li, S., & Wang, L. (2004). Neither the tumor necrosis factor alpha-308 A/G polymorphism nor the alpha2-macroglobulin polymorphism was associated with late-onset Alzheimer's disease in the Chinese population. *Yi Chuan xue bao= Acta Genetica Sinica, 31*(1), 1–6.
83. Li, Y., Nowotny, P., Holmans, P., Smemo, S., Kauwe, J. S., Hinrichs, A. L., & Grupe, A. (2004). Association of late-onset Alzheimer's disease with genetic variation in multiple members of the GAPD gene family. *Proceedings of the National Academy of Sciences, 101*(44), 15688–15693.
84. Schulze, H., Schuler, A., Stüber, D., Döbeli, H., Langen, H., & Huber, G. (1993). Rat brain glyceraldehyde-3-phosphate dehydrogenase interacts with the recombinant cytoplasmic domain of Alzheimer's β-amyloid precursor protein. *Journal of Neurochemistry, 60*(5), 1915–1922.
85. Oyama, R., Yamamoto, H., & Titani, K. (2000). Glutamine synthetase, hemoglobin α-chain, and macrophage migration inhibitory factor binding to amyloid β-protein: Their identification in rat brain by a novel affinity chromatography and in Alzheimer's disease brain by immunoprecipitation. *Biochimica et Biophysica Acta (BBA)-Protein Structure and Molecular Enzymology, 1479*(1-2), 91–102.

86. Deane, R., Wu, Z., Sagare, A., Davis, J., Du Yan, S., Hamm, K., & Zlokovic, B. V. (2004). LRP/amyloid β-peptide interaction mediates differential brain efflux of Aβ isoforms. *Neuron, 43*(3), 333–344.
87. Von Arnim, C. A., Kinoshita, A., Peltan, I. D., Tangredi, M. M., Herl, L., Lee, B. M., & Hyman, B. T. (2005). The low density lipoprotein receptor-related protein (LRP) is a novel β-secretase (BACE1) substrate. *Journal of Biological Chemistry, 280*(18), 17777–17785.
88. Gläser, C., Schulz, S., Handschug, K., Huse, K., & Birkenmeier, G. (2004). Genetic and functional characteristics of the human in vivo LRP1/A2MR receptor suggested as a risk marker for Alzheimer's disease and other complex (degenerative) diseases. *Neuroscience Research, 50*(1), 85–101.
89. Bertram, L., Hiltunen, M., Parkinson, M., Ingelsson, M., Lange, C., Ramasamy, K., & Tanzi, R. E. (2005). Family-based association between Alzheimer's disease and variants in UBQLN1. *New England Journal of Medicine, 352*(9), 884–894.
90. Carson, J. A., & Turner, A. J. (2002). β-Amyloid catabolism: roles for neprilysin (NEP) and other metallopeptidases? *Journal of Neurochemistry, 81*(1), 1–8.
91. Helisalmi, S., Hiltunen, M., Vepsäläinen, S., Iivonen, S., Mannermaa, A., Lehtovirta, M., & Soininen, H. (2004). Polymorphisms in neprilysin gene affect the risk of Alzheimer's disease in Finnish patients. *Journal of Neurology, Neurosurgery & Psychiatry, 75*(12), 1746–1748.
92. Shi, J., Zhang, S., Tang, M., Ma, C., Zhao, J., Li, T., & Zhao, Z. (2005). Mutation screening and association study of the neprilysin gene in sporadic Alzheimer's disease in Chinese persons. *The Journals of Gerontology Series A: Biological Sciences and Medical Sciences, 60*(3), 301–306.
93. Kan, R., Wang, B., Zhang, C., Jin, F., Yang, Z., Ji, S., & Wang, L. (2005). Genetic association of BACE1 gene polymorphism C786G with late-onset Alzheimer's disease in Chinese. *Journal of Molecular Neuroscience, 25*, 127–131.
94. Floden, A. M., Li, S., & Combs, C. K. (2005). β-Amyloid-stimulated microglia induce neuron death via synergistic stimulation of tumor necrosis factor α and NMDA receptors. *Journal of Neuroscience, 25*(10), 2566–2575.
95. Fung, H. C., Chen, C. M., Wu, Y. R., Hsu, W. C., Ro, L. S., Lin, J. C., & Lee-Chen, G. J. (2005). Heat shock protein 70 and tumor necrosis factor alpha in Taiwanese patients with dementia. *Dementia and Geriatric Cognitive Disorders, 20*(1), 1–7.
96. DeKosky, S. T., Ikonomovic, M. D., Wang, X., Farlow, M., Wisniewski, S., Lopez, O. L., & Kamboh, M. I. (2003). Plasma and cerebrospinal fluid α1-antichymotrypsin levels in Alzheimer's disease: Correlation with cognitive impairment. *Annals of Neurology: Official Journal of the American Neurological Association and the Child Neurology Society, 53*(1), 81–90.
97. Kamboh, M. I., Sanghera, D. K., Ferrell, R. E., & DeKosky, S. T. (1995). A4POE* 4-associated Alzheimer's disease risk is modified by α1–antichymotrypsin polymorphism. *Nature Genetics, 10*(4), 486–488.
98. Muller, U., Bodeker, R. H., Gerundt, I., & Kurz, A. (1996). Lack of association between alpha 1-antichymotrypsin polymorphism, Alzheimer's disease, and allele epsilon 4 of apolipoprotein E. *Neurology, 47*(6), 1575–1577.
99. Licastro, F., Chiappelli, M., Grimaldi, L. M., Morgan, K., Kalsheker, N., Calabrese, E., & Canal, N. (2005). A new promoter polymorphism in the alpha-1-antichymotrypsin

gene is a disease modifier of Alzheimer's disease. *Neurobiology of Aging, 26*(4), 449–453.

100. Harold, D., MacGregor, S., Patterson, C. E., Hollingworth, P., Moore, P., Owen, M. J., & Jones, L. (2006). A single nucleotide polymorphism in CHAT influences response to acetylcholinesterase inhibitors in Alzheimer's disease. *Pharmacogenetics and Genomics, 16*(2), 75–77.
101. Minocherhomji, S., Tollefsbol, T. O., & Singh, K. K. (2012). Mitochondrial regulation of epigenetics and its role in human diseases. *Epigenetics, 7*(4), 326–334.
102. Jiang, Q., Jin, S., Jiang, Y., Liao, M., Feng, R., Zhang, L., & Hao, J. (2017). Alzheimer's disease variants with the genome-wide significance are significantly enriched in immune pathways and active in immune cells. *Molecular Neurobiology, 54*, 594–600.
103. Guerini, F. R., Farina, E., Costa, A. S., Baglio, F., Saibene, F. L., Margaritella, N., & Clerici, M. (2016). ApoE and SNAP-25 polymorphisms predict the outcome of multidimensional stimulation therapy rehabilitation in Alzheimer's disease. *Neurorehabilitation and Neural Repair, 30*(9), 883–893.
104. Sumirtanurdin, R., Thalib, A. Y., Cantona, K., & Abdulah, R. (2019). Effect of genetic polymorphisms on Alzheimer's disease treatment outcomes: an update. *Clinical Interventions in Aging,* 631–642.
105. Varsaldi, F., Miglio, G., Scordo, M. G., Dahl, M. L., Villa, L. M., Biolcati, A., & Lombardi, G. (2006). Impact of the CYP2D6 polymorphism on steady-state plasma concentrations and clinical outcome of donepezil in Alzheimer's disease patients. *European Journal of Clinical Pharmacology, 62*, 721–726.
106. Lu, J., Fu, J., Zhong, Y., Yang, Q., Huang, J., Li, J., & Guo, C. (2018). Association between ABCA1 gene polymorphisms and the therapeutic response to donepezil therapy in Han Chinese patients with Alzheimer's disease. *Brain Research Bulletin, 140*, 1–4.
107. Pola, R., Flex, A., Ciaburri, M., Rovella, E., Valiani, A., Reali, G., & Bernabei, R. (2005). Responsiveness to cholinesterase inhibitors in Alzheimer's disease: A possible role for the 192 Q/R polymorphism of the PON-1 gene. *Neuroscience Letters, 382*(3), 338–341.
108. Scacchi, R., Gambina, G., Moretto, G., & Corbo, R. M. (2009). Variability of AChE, BChE, and ChAT genes in the late-onset form of Alzheimer's disease and relationships with response to treatment with Donepezil and Rivastigmine. *American Journal of Medical Genetics Part B: Neuropsychiatric Genetics, 150*(4), 502–507.
109. Lu, J., Fu, J., Zhong, Y., Chen, P., Yang, Q., Zhao, Y., & Guo, C. (2006). The roles of apolipoprotein E3 and CYP2D6 (rs1065852) gene polymorphisms in the predictability of responses to individualized therapy with donepezil in Han Chinese patients with Alzheimer's disease. *Neuroscience Letters, 614*, 43–48.
110. Clarelli, F., Mascia, E., Santangelo, R., Mazzeo, S., Giacalone, G., Galimberti, D., & Martinelli Boneschi, F. (2016). CHRNA7 gene and response to cholinesterase inhibitors in an Italian cohort of Alzheimer's disease patients. *Journal of Alzheimer's Disease, 52*(4), 1203–1208.
111. Sonali, N., Tripathi, M., Sagar, R., Velpandian, T., & Subbiah, V. (2014). Impact of CYP2D6 and CYP3A4 genetic polymorphism on combined cholinesterase inhibitors and memantine treatment in mild to moderate Alzheimer's disease. *Dementia and Geriatric Cognitive Disorders, 37*(1-2), 58–70.

112. Miranda, L. F., Gomes, K. B., Tito, P. A., Silveira, J. N., Pianetti, G. A., Byrro, R., & Caramelli, P. (2017). Clinical response to donepezil in mild and moderate dementia: relationship to drug plasma concentration and CYP2D6 and APOE genetic polymorphisms. *Journal of Alzheimer's Disease, 55*(2), 539–549.
113. Liu, M., Zhang, Y., Huo, Y. R., Liu, S., Liu, S., Wang, J., & Ji, Y. (2014). Influence of the rs1080985 single nucleotide polymorphism of the CYP2D6 gene and APOE polymorphism on the response to donepezil treatment in patients with Alzheimer's disease in China. *Dementia and Geriatric Cognitive Disorders Extra, 4*(3), 450–456.
114. Klimkowicz-Mrowiec, A., Wolkow, P., Sado, M., Dziubek, A., Pera, J., Dziedzic, T., & Slowik, A. (2013). Influence of rs1080985 single nucleotide polymorphism of the CYP2D6 gene on response to treatment with donepezil in patients with Alzheimer's disease. *Neuropsychiatric Disease and Treatment,* 1029–1033.
115. Pilotto, A., Franceschi, M., D'Onofrio, G., Bizzarro, A., Mangialasche, F., Cascavilla, L., & Seripa, D. (2009). Effect of a CYP2D6 polymorphism on the efficacy of donepezil in patients with Alzheimer disease. *Neurology, 73*(10), 761–767.
116. Lu, J., Wan, L., Zhong, Y., Yu, Q., Han, Y., Chen, P., & Guo, C. (2015). Stereoselective metabolism of donepezil and steady-state plasma concentrations of S-donepezil based on CYP2D6 polymorphisms in the therapeutic responses of Han Chinese patients with Alzheimer's disease. *Journal of Pharmacological Sciences, 129*(3), 188–195.
117. Zhong, Y., Zheng, X., Miao, Y., Yan, H., Wang, B., & Wan, L. (2013). Effect of CYP2D6* 10 and APOE polymorphisms on the efficacy of donepezil in patients with Alzheimer's disease. *The American Journal of the Medical Sciences, 345*(3), 222–226.
118. Albani, D., Boneschi, F. M., Biella, G., Giacalone, G., Lupoli, S., Clerici, F., & Forloni, G. (2012). Replication study to confirm the role of CYP2D6 polymorphism rs1080985 on donepezil efficacy in Alzheimer's disease patients. *Journal of Alzheimer's Disease, 30*(4), 745–749.
119. Seripa, D., Bizzarro, A., Pilotto, A., D'onofrio, G., Vecchione, G., Gallo, A. P., & Pilotto, A. (2011). Role of cytochrome P4502D6 functional polymorphisms in the efficacy of donepezil in patients with Alzheimer's disease. *Pharmacogenetics and Genomics, 21*(4), 225–230.
120. Sonali, N., Tripathi, M., Sagar, R., Velpandian, T., & Subbiah, V. (2013). Clinical effectiveness of rivastigmine monotherapy and combination therapy in Alzheimer's patients. *CNS Neuroscience & Therapeutics, 19*(2), 91–97.
121. Blesa, R., Aguilar, M., Casanova, J. P., Boada, M., Martínez, S., Alom, J., & Massó, J. F. M. (2006). Relationship between the efficacy of rivastigmine and apolipoprotein E (ε4) in patients with mild to moderately severe Alzheimer disease. *Alzheimer Disease & Associated Disorders, 20*(4), 248–254.
122. Weng, P. H., Chen, J. H., Chen, T. F., Sun, Y., Wen, L. L., Yip, P. K., & Chen, Y. C. (2013). CHRNA7 polymorphisms and response to cholinesterase inhibitors in Alzheimer's disease. *PLoS One, 8*(12), e84059.
123. Braga, I. L. S., Silva, P. N., Furuya, T. K., Santos, L. C., Pires, B. C., Mazzotti, D. R., & Smith, M. C. (2015). Effect of APOE and CHRNA7 genotypes on the cognitive response to cholinesterase inhibitor treatment at different stages of Alzheimer's disease. *American Journal of Alzheimer's Disease & Other Dementias®, 30*(2), 139–144.

124. Yoon, H., Myung, W., Lim, S. W., Kang, H. S., Kim, S., Won, H. H., & Kim, D. K. (2015). Association of the choline acetyltransferase gene with responsiveness to acetylcholinesterase inhibitors in Alzheimer's disease. *Pharmacopsychiatry, 25*(03), 111–117.
125. Lee, K. U., Lee, J. H., Lee, D. Y., Youn, J. C., Kim, J. L., Moon, S. W., & Jhoo, J. H. (2015). The effect of choline acetyltransferase genotype on donepezil treatment response in patients with Alzheimer's disease. *Clinical Psychopharmacology and Neuroscience, 13*(2), 168.
126. Unnikrishnan, A., Freeman, W. M., Jackson, J., Wren, J. D., Porter, H., & Richardson, A. (2019). The role of DNA methylation in epigenetics of aging. *Pharmacology & Therapeutics, 195*, 172–185.
127. Guo, J. U., Su, Y., Shin, J. H., Shin, J., Li, H., Xie, B., & Song, H. (2014). Distribution, recognition and regulation of non-CpG methylation in the adult mammalian brain. *Nature Neuroscience, 17*(2), 215–222.
120. Li, X., Zhao, Q., Wei, W., Lin, Q., Magnan, C., Emami, M. R., & Bredy, T. W. (2019). The DNA modification N6-methyl-2'-deoxyadenosine (m6dA) drives activity-induced gene expression and is required for fear extinction. *Nature Neuroscience, 22*(4), 534–544.
139. Griñán-Ferré, C., Corpas, R., Puigoriol-Illamola, D., Palomera-Ávalos, V., Sanfeliu, C., & Pallàs, M. (2018). Understanding epigenetics in the neurodegeneration of Alzheimer's disease: SAMP8 mouse model. *Journal of Alzheimer's Disease, 62*(3), 943–963.
130. Morris-Blanco, K. C., Kim, T., Lopez, M. S., Bertogliat, M. J., Chelluboina, B., & Vemuganti, R. (2019). Induction of DNA hydroxymethylation protects the brain after stroke. *Stroke, 50*(9), 2513–2521.
131. Li, E., & Zhang, Y. (2014). DNA methylation in mammals. *Cold Spring Harbor Perspectives in Biology, 6*(5), a019133.
132. Seritrakul, P., & Gross, J. M. (2017). Tet-mediated DNA hydroxymethylation regulates retinal neurogenesis by modulating cell-extrinsic signaling pathways. *PLoS Genetics, 13*(9), e1006987.
133. Xiao, X., Liu, X., & Jiao, B. (2020). Epigenetics: Recent advances and its role in the treatment of Alzheimer's disease. *Frontiers in Neurology, 11*, 538301.
134. Hou, Y., Chen, H., He, Q., Jiang, W., Luo, T., Duan, J., & Wang, H. (2013). Changes in methylation patterns of multiple genes from peripheral blood leucocytes of Alzheimer's disease patients. *Acta Neuropsychiatrica, 25*(2), 66–76.
135. Iwata, A., Nagata, K., Hatsuta, H., Takuma, H., Bundo, M., Iwamoto, K., & Tsuji, S. (2014). Altered CpG methylation in sporadic Alzheimer's disease is associated with APP and MAPT dysregulation. *Human Molecular Genetics, 23*(3), 648–656.
136. West, R. L., Lee, J. M., & Maroun, L. E. (1995). Hypomethylation of the amyloid precursor protein gene in the brain of an Alzheimer's disease patient. *Journal of Molecular Neuroscience, 6*, 141–146.
137. Brohede, J., Rinde, M., Winblad, B., & Graff, C. (2010). A DNA methylation study of the amyloid precursor protein gene in several brain regions from patients with familial Alzheimer disease. *Journal of Neurogenetics, 24*(4), 179–181.
138. Barrachina, M., & Ferrer, I. (2009). DNA methylation of Alzheimer disease and tauopathy-related genes in postmortem brain. *Journal of Neuropathology & Experimental Neurology, 68*(8), 880–891.

139. Siegmund, K. D., Connor, C. M., Campan, M., Long, T. I., Weisenberger, D. J., Biniszkiewicz, D., & Akbarian, S. (2007). DNA methylation in the human cerebral cortex is dynamically regulated throughout the life span and involves differentiated neurons. *PloS One, 2*(9), e895.
140. Monti, N., Cavallaro, R. A., Stoccoro, A., Nicolia, V., Scarpa, S., Kovacs, G. G., & Fuso, A. (2020). CpG and non-CpG Presenilin1 methylation pattern in course of neurodevelopment and neurodegeneration is associated with gene expression in human and murine brain. *Epigenetics, 15*(8), 781–799.
141. Bakulski, K. M., Dolinoy, D. C., Sartor, M. A., Paulson, H. L., Konen, J. R., Lieberman, A. P., & Rozek, L. S. (2012). Genome-wide DNA methylation differences between late-onset Alzheimer's disease and cognitively normal controls in human frontal cortex. *Journal of Alzheimer's Disease, 29*(3), 571–588.
142. Piaceri, I., Raspanti, B., Tedde, A., Bagnoli, S., Sorbi, S., & Nacmias, B. (2015). Epigenetic modifications in Alzheimer's disease: cause or effect? *Journal of Alzheimer's Disease, 43*(4), 1169–1173.
143. Furuya, T. K., da Silva, P. N. O., Payão, S. L. M., Rasmussen, L. T., de Labio, R. W., Bertolucci, P. H. F., & Smith, M. D. A. C. (2012). SORL1 and SIRT1 mRNA expression and promoter methylation levels in aging and Alzheimer's Disease. *Neurochemistry International, 61*(7), 973–975.
144. Foraker, J., Millard, S. P., Leong, L., Thomson, Z., Chen, S., Keene, C. D., & Yu, C. E. (2015). The APOE gene is differentially methylated in Alzheimer's disease. *Journal of Alzheimer's Disease, 48*(3), 745–755.
145. Lunnon, K., Smith, R., Hannon, E., De Jager, P. L., Srivastava, G., Volta, M., & Mill, J. (2014). Methylomic profiling implicates cortical deregulation of ANK1 in Alzheimer's disease. *Nature Neuroscience, 17*(9), 1164–1170.
146. Nagata, T., Kobayashi, N., Ishii, J., Shinagawa, S., Nakayama, R., Shibata, N., & Nakayama, K. (2015). Association between DNA methylation of the BDNF promoter region and clinical presentation in Alzheimer's disease. *Dementia and Geriatric Cognitive Disorders Extra, 5*(1), 64–73.
147. Chang, L., Wang, Y., Ji, H., Dai, D., Xu, X., Jiang, D., & Wang, Q. (2014). Elevation of peripheral BDNF promoter methylation links to the risk of Alzheimer's disease. *PloS One, 9*(11), e110773.
148. An, S., Khanna, K. K., & Wu, J. M. (1994). mRNA levels and methylation patterns of the 2-5A synthetase gene in control and Alzheimer's disease (AD) fibroblasts. *Biochemistry and Molecular Biology International, 33*(5), 835–840.
149. Yu, L., Chibnik, L. B., Srivastava, G. P., Pochet, N., Yang, J., Xu, J., & Bennett, D. A. (2015). Association of Brain DNA methylation in SORL1, ABCA7, HLA-DRB5, SLC24A4, and BIN1 with pathological diagnosis of Alzheimer disease. *JAMA Neurology, 72*(1), 15–24.
150. Carboni, L., Lattanzio, F., Candeletti, S., Porcellini, E., Raschi, E., Licastro, F., & Romualdi, P. (2015). Peripheral leukocyte expression of the potential biomarker proteins Bdnf, Sirt1, and Psen1 is not regulated by promoter methylation in Alzheimer's disease patients. *Neuroscience Letters, 605*, 44–48.
151. Coppedè, F., Tannorella, P., Stoccoro, A., Chico, L., Siciliano, G., Bonuccelli, U., & Migliore, L. (2017). Methylation analysis of DNA repair genes in Alzheimer's disease. *Mechanisms of Ageing and Development, 161*, 105–111.

152. Ma, S. L., Tang, N. L. S., & Lam, L. C. W. (2016). Association of gene expression and methylation of UQCRC1 to the predisposition of Alzheimer's disease in a Chinese population. *Journal of Psychiatric Research, 76*, 143–147.
153. Tannorella, P., Stoccoro, A., Tognoni, G., Petrozzi, L., Salluzzo, M. G., Ragalmuto, A., & Coppedè, F (2015). Methylation analysis of multiple genes in blood DNA of Alzheimer's disease and healthy individuals. *Neuroscience Letters, 600*, 143–147.
154. Silva, P. N., Furuya, T. K., Braga, I. L., Rasmussen, L. T., Labio, R. W., Bertolucci, P. H., & Smith, M. C. (2014). Analysis of HSPA8 and HSPA9 mRNA expression and promoter methylation in the brain and blood of Alzheimer's disease patients. *Journal of Alzheimer's Disease, 38*(1), 165–170.
155. Grosser, C., Neumann, L., Horsthemke, B., Zeschnigk, M., & van de Nes, J. (2014). Methylation analysis of SST and SSTR4 promoters in the neocortex of Alzheimer's disease patients. *Neuroscience Letters, 566*, 241–246.
156. Furuya, T. K., da Silva, P. N. O., Payão, S. L. M., Rasmussen, L. T., de Labio, R. W., Bertolucci, P. H. F., & Smith, M. D. A. C. (2012). SORL1 and SIRT1 mRNA expression and promoter methylation levels in aging and Alzheimer's Disease. *Neurochemistry International, 61*(7), 973–975.
157. Mendioroz, M., Celarain, N., Altuna, M., Sánchez-Ruiz de Gordoa, J., Zelaya, M. V., Roldán, M., & Echávarri, C. (2016). CRTC1 gene is differentially methylated in the human hippocampus in Alzheimer's disease. *Alzheimer's Research & Therapy, 8*, 1–9.
158. Ozaki, Y., Yoshino, Y., Yamazaki, K., Sao, T., Mori, Y., Ochi, S., & Ueno, S. I. (2017). DNA methylation changes at TREM2 intron 1 and TREM2 mRNA expression in patients with Alzheimer's disease. *Journal of Psychiatric Research, 92*, 74–80.
159. Blanco-Luquin, I., Altuna, M., Sánchez-Ruiz de Gordoa, J., Urdánoz-Casado, A., Roldán, M., Cámara, M., & Mendioroz, M. (2018). PLD3 epigenetic changes in the hippocampus of Alzheimer's disease. *Clinical Epigenetics, 10*, 1–10.
160. Hu, W., Lin, X., Zhang, H., & Zhao, N. (2017). ATP binding cassette subfamily A member 2 (ABCA2) expression and methylation are associated with Alzheimer's disease. *Medical Science Monitor: International Medical Journal of Experimental and Clinical Research, 23*, 5851.
161. Mercorio, R., Pergoli, L., Galimberti, D., Favero, C., Carugno, M., Dalla Valle, E., & Pesatori, A. C. (2018). PICALM gene methylation in blood of Alzheimer's disease patients is associated with cognitive decline. *Journal of Alzheimer's Disease, 65*(1), 283–292.
162. Salcedo-Tacuma, D., Melgarejo, J. D., Mahecha, M. F., Ortega-Rojas, J., Arboleda-Bustos, C. E., Pardo-Turriago, R., & Arboleda, H. (2019). Differential methylation levels in CpGs of the BIN1 gene in individuals with Alzheimer disease. *Alzheimer Disease & Associated Disorders, 33*(4), 321–326.
163. Smith, A. R., Smith, R. G., Burrage, J., Troakes, C., Al-Sarraj, S., Kalaria, R. N., & Lunnon, K. (2019). A cross-brain regions study of ANK1 DNA methylation in different neurodegenerative diseases. *Neurobiology of Aging, 74*, 70–76.
164. Xu, C., Liu, G., Ji, H., Chen, W., Dai, D., Chen, Z., & Wang, Q. (2018). Elevated methylation of OPRM1 and OPRL1 genes in Alzheimer's disease. *Molecular Medicine Reports, 18*(5), 4297–4302.

Index

B

C